WORLD FIRE

BOOKS BY STEPHEN J. PYNE

Grove Karl Gilbert
Fire on the Rim
Dutton's Point
Introduction to Wildland Fire

Cycle of Fire
World Fire: The Culture of Fire on Earth
Fire in America: A Cultural History of Wildland and Rural Fire
Burning Bush: A Fire History of Australia
The Ice: A Journey to Antarctica

WORLD FIRE

The Culture of Fire on Earth

Stephen J. Pyne

Henry Holt and Company
New York

Henry Holt and Company, Inc.
Publishers since 1866
115 West 18th Street
New York, New York 10011

Published in Canada by Fitzhenry & Whiteside Ltd.,
195 Allstate Parkway, Markham, Ontario L3R 4T8.

Library of Congress Cataloging-in-Publication Data
Pyne, Stephen J.
World fire: the culture of fire on earth / Stephen J. Pyne—1st ed.
p. cm.
Includes bibliographical references and index.
1. Wildfires. 2. Fires. 3. Fire ecology. 4. Wildfires—Prevention and control.
5. Fires—Prevention and control. I. Title.
SD421.P95 1995 94-33546
304.2—dc20 CIP

ISBN 0-8050-3247-9

Henry Holt books are available for special
promotions and premiums. For details contact:
Director, Special Markets.

First Edition—1995

Designed by Paula R. Szafranski

Printed in the United States of America
All first editions are printed on acid-free paper.∞

1 3 5 7 9 10 8 6 4 2

To Sonja, Lydia, Molly
who have made the hearth fire
into a world fire, too

Contents

Illustrations

Olle Zackrisson, "Vegetation Dynamics and Land Use in the Lower Reaches of the River Umeälven," *Early Norrland* 9, pp. 7–74.

7. The suppression of Swedish fire. Data from Brian Stocks, "The Extent and Impact of Forest Fires in Northern Circumpolar Countries," pp. 197–202, in Joel Levine, ed., *Global Biomass Burning* (MIT Press, 1991), and Gustaf Lundberg, "Om skogselddess förebyggande och bekämpande," *Skogsvårdsforeningens Tidskrift 1915,* p. 114.

8. Silviculture cleans house. Data from Lars Östlund, *Exploitation and Structural Changes in the North Swedish Boreal Forest 1800–1992.* Dissertations in Forest Vegetation Ecology 4. Sveriges Lantbruksuniversitet (Umeå, 1993).

9. Area burned in Greece, 1922–1990. Data from D. S. Kailidis, "Forest Fires in Greece," pp. 27–40, U.N. FAO, ECE, ILL, *Seminar on Forest Fire Prevention, Land Use, and People* (Ministry of Agriculture, Athens, 1992).

10. The competition for biomass. Redrawn from F. Castro Rego, "Fuel Management," in U.N. FAO, ECE, ILL, *Seminar on Forest Fire Prevention, Land Use, and People* (Ministry of Agriculture, Athens, 1992), pp. 209–21.

11. *La Nueva Reconquista.* Data from P. Martinez Hermosilla, "Enfoque Historico de Los Trabajos de Restauracion," *Ecologia,* Fuera de Serie no. 1 (1990), pp. 367–71.

12. Wildfire's terrain. Reproduced with permission from R. Tarrega Garcia-Mares y E. Luis Calabuig, "A Problematica de los Incendios Forestales y Su Incidencia Sobre Los Robledales de *Quercus pyrenaica* en La Provincia de León," *Ecologia,* Fuera de Serie no. 1 (1990), pp. 223–37.

13. Red Skies. Data from *Avialesookhrana.*

14. Revolution and inflation. Data from *Avialesookhrana.*

15. *Nataraja.* Photo by author.

16. Industrializing India. Data from *Report from the Railways Boards, 1940–41* and *Annual Return of Statistics Relating to Forest Administration in British India 1938–39.*

17. Population spiral. Data from Elizabeth P. Flint, "Changes in Land Use in South and Southeast Asia from 1880 to 1980: A Data Base Prepared as Part of a Coordinated Research Program on Carbon Fluxes in the Tropics," unpublished paper presented at "Polluted or Pristine?" conference at East-West Center, 1993.

18. Initial Attack: the Forest Service and fire. Data organized by author.

19. Cooperative fire protection. Data from U.S. Forest Service.

20. Cooperative fire extinction. Data from U.S. Forest Service, *Historical Statistics of the United States,* and *Statistical Abstract of the U.S.* (annual).

21. Field of Fire I. Data from U.S. Forest Service.

22. Field of Fire II. Data from *Historical Statistics of the United States* and *Statistical Abstract of the U.S.* (annual).

R. Smith, *Biofuels, Air Pollution, and Health. A Global Review* (Plenum Press, 1987).

35. Global biomass burning, estimated for 1990. Data from Meinrat O. Andreae, "Biomass Burning: Its History, Use, and Distribution and Its Impact on Environmental Quality and Global Climate," pp. 3–21, in Levine, ed., *Global Biomass Burning*.

36. The global geography of combustion. Reproduced with permission from G. Marland et al., "CO_2 from Fossil Fuel Burning: Global Distribution of Emissions," *Tellus* 37B (1983), pp. 243–58, and R. A. Houghton et al., "The Flux of Carbon from Terrestrial Ecosystems to the Atmosphere in 1980 Due to Changes in Land Use: Geographic Distribution of the Global Flux," *Tellus* 39B (1987), pp. 122–39.

SMOKE REPORT

The New World Order on Fire

It surpasses all wonders that a day goes by wherein the whole world is not consumed in flame.

—Pliny, *Natural History*

Listen to me: I've seen the ashes of past fires, fires of thousands of years old. I know what's going to happen. The Flames will come. I have no doubt about that whatsoever.

—Isaac Asimov and Robert Silverberg, *Nightfall*

He was moving through a new order of creation, of which few men had ever dreamed. Beyond the reaches of sea and land and air and space lay the realms of fire which he alone had been privileged to glimpse. It was too much to expect that he would also understand.

—Arthur C. Clarke, *2001*

We are uniquely fire creatures on a uniquely fire planet. Other planetary bodies in the solar system have elements of combustion—Jupiter has an ignition source in lightning; Mars, traces of free oxygen; Titan, a methane-based fuel. Only Earth, however, has all the essential elements and the means by which to combine them—only Earth has life. Marine life pumped the atmosphere with oxygen, and terrestrial life stocked the continents with carbon fuels.

Fire and life have fused biotically ever since, a relationship aptly symbolized by the ponderosa pine, germinated in fire's ashes, pruned by fire's heat, fed by fire's liberated nutrients, and perhaps consumed by fire's final flame. So the earth—its lands stuffed with organic fuels, its atmosphere saturated with oxygen, its surface pummeled by lightning—is metastable. More disturbing, the planet possesses a creature not merely adapted to fire's presence, or even prone to exploit and encourage fire, but one capable of starting and stopping it. A fire creature came to dominate a fire planet. The history of the planet is unintelligible without the history of fire.

The concept of a world fire has, like almost all fire practices and fire beliefs, a basis in ecological reality. The earth is a fire world, but even more, for humans fire has made possible a special relationship to that world. Fire and humans have coevolved, like the bonded strands of a DNA molecule. The prevalence of humans is largely attributable to their control over fire; and the distribution and characteristics of fire have become profoundly dependent on humans. Fire and humanity have become inseparable and indispensable. Together they have repeatedly remade the earth.

It comes as no surprise then to learn that among human societies the belief in a world-creating and world-ending fire is nearly universal. From the Nordic Ragnarok to the Aztec New Fire, from the Great Fire of the Stoics to the Christian Apocalypse, from the chained Prometheus to the Aboriginal Dreamtime, the pine-pitch torch to the nuclear firestorm—the fate of humans and the fate of fire have joined. But increasingly, as humans have urbanized and devised new technologies to segregate combustion from fire, the vision of a world-ending fire has endured, while the image of a world-creating one has vanished. The original premise was otherwise. It held that fire destroyed one world in order to make another. Behind this vision lay the incontestable authority of nature, so much of which degenerated without periodic burning, and the sense in which anthropogenic fire was the means by which humanity claimed nature, rendered it habitable, refashioned the wild into the domesticated, and asserted a unique identity. The threat of a world overwhelmed with fire was no greater than the threat of a world without fire.

Over the past decade this ancient mythology has assumed a new avatar. Fire imagery has vivified the multiple specters of environmental catastrophe, perhaps on a global scale. Nuclear winter, which proposed that fire-spawned smoke would plunge the earth into the cold and dark of an anthropogenic ice age. Greenhouse summer, its complement, whose heat-trapping gases promise to convert the earth into a crock pot. The Cretaceous-Tertiary boundary event that extinguished countless species—among them the dinosaurs—an episode apparently accompanied by widespread fires that foreshadow the flame-induced mass extinctions of today. Uncontrolled slashing and burning in Amazonia, Indonesia, India—all threatening biodiversity and belching greenhouse gases. Smoke plumes that flood the Amazon Basin, and pastoral burns that pull the Sahara southward. Multimillion-acre wildfires throughout the boreal forest, and endless, annual savanna firings throughout the tropics. The Great Black Dragon fire that incinerated China's Hinggan forest. The East Kalimantan fires that reduced nine million acres of Borneo rainforest to a smoking hole. Australia's Ash Wednesday fires that, within hours, brought an industrialized nation to its knees. Conflagrations through Yellowstone National Park, a fiery black hole that sucked in more than $130 million without effect. Forest-powered holocausts through Oakland and Malibu, California. Oil fires in Kuwait that replaced the traditional imagery of a scorched earth with that of a scorched sky.

The atmosphere, in fact, has been critical to this new mythological metamorphosis. It has leveraged local events into global consequences. Air joins fire anywhere with fire everywhere. Fire's effects are no longer confined to a particular site, or even to traditional off-site transport by wind and water because gases and aerosols enter a globe-girdling atmosphere. In return fire has given environmentalism a vivid iconography of wide popular appeal. Whatever their message of destruction, activists have sought to animate it with fire for television or photojournals. The predictable result is that associating fire with various kinds of devastation has identified fire itself as a devastator. The new world order of fire appears to be an order set on fire.

But fire is an end as well as a means. Accept that neither the earth nor humanity can shed fire and be what they are, and the contemporary geography of fire assumes a different interpretation. Fire results from a special biochemistry—at times a catalyst for anthropogenic practices, at other times the product of human-catalyzed reactions. Too much fire is as likely as too little, and too little can be as damaging as too much. The earth's fire problem is one of maldistribution. There is too much of the wrong kind of fire in the wrong places or at the wrong times, and not enough of the right kind of fire at the right places and times. Probably there is too little fire and too much combustion. Almost certainly, contrary to popular belief, there is a lot less fire on the planet than when Columbus sailed.

In the new world order, or what passes for it, fire is as much a victim as a villain. The unsettling of the planet's ecology has thrown into confusion the role of fire and of the humans who, by evolutionary right and responsibility, ought to manage it. The unwarranted suppression of fire has damaged the earth as much as its promiscuous misuse. The elimination of fire in the world will not save the planet from destruction but only abolish the regeneration that it once promised should follow.

It is the ambition of this book to survey the emerging geography of global fire. It proposes to do so by using fire as both an object of understanding and a device for analysis. The premise is that one can (and must) study history in order to trace the evolution of fire, and equally that one can (and should) exploit fire in order to learn more about history. Much as fire binds humans to the natural world, so fire scholarship spot-welds the humanities to the natural sciences.

The book's design metaphorically exploits the methodology of fire control as a means of controlling its own propagating thoughts. The order of appearance of each section and its character mimics the phases of firefighting.

- "Smoke Report" announces the phenomenon. It is typically smoke, not fire, that observers first sight. The informing fire requires more searching to locate and reach.
- "Size-Up" assesses the question of how to attack the problem, how to match tools with needs. In this case it sketches the theory and practice of fire history.
- "Hotspotting" refers to the triagelike operations that occur at a fire scene in which the most critical parts of a fire are identified and attacked before they become completely uncontrollable. In this context the concept describes a rapid, global sampling of major fire situations.

- "Control" is the process by which a fire is systematically contained, its spread halted, and its perimeter secured. Done properly, the task is time-consuming, full of the adrenaline-flushed ideas and dedicated tedium that make for good scholarship. *World Fire* blocks off only parts of the total perimeter, particularly focusing on America. The section concentrates on the larger history by which the United States, through its Forest Service, established an infrastructure and policy for fire, and on the two problem blazes that have dominated the fire establishment in recent decades.
- "Mop-Up" is the self-descriptive task of actually extinguishing the fire, or enough of its perimeter that it can be considered safe if the lines are then patrolled. A summary essay on the history of humanity and fire is the result.
- "After the Last Smoke" refers to the practice by which firefighters return to the scene of a burn twenty-four hours after they sight the last smoke for a final inspection before declaring the fire officially out and writing up the requisite reports. It asks, here, what the experience has meant and what kind of reports or literature have resulted.

The strategy is ironic in that obsessive fire control has become itself a disordering process. In much of the world it is the control of fire that is out of control. Consider the book an attempted backfire against irony.

The world is a big place, and *World Fire* a small endeavor, a torch thrust into dark woods, the intellectual equivalent of attacking a snag fire. An understanding of fire on earth calls for a greater scholarship. It is my promethean hope to preserve and propagate that understanding through a larger saga that I call *Cycle of Fire*. The concept of a global survey grew fitfully, a kind of Leatherstocking tales that now holds four books: *World Fire, Fire in America, Burning Bush,* and *The Ice*. (The last was written to other purposes, but as a continent smothered in ice, Antarctica proclaims the death of fire and suitably ends the cycle.) To this corpus *World Fire* should stand as an introduction, progress report, reconsideration, promise, and apology.

. . .

This book—the scheme for Cycle of Fire—became possible because of the MacArthur Fellows Program. I hope the administrators of this extraordinary institution can take pleasure in what has resulted, and in the promise of what should soon follow. Additionally, a Fulbright Senior Fellowship sent me to Umeå, Sweden, for three months of research. A Dahlem Konferenzen in Berlin forced me to consolidate my thoughts, and introduced me to the inquiry of atmospheric scientists about the status of fire on the planet. To all I am grateful.

Many other persons have helped during my travels. I can only name them, in inadequate appreciation for their assistance: Johann Goldammer, Ricardo Vélez, Anders Granström, Gunilla Oleskog, Richard Eaton, Cathy Frierson, V. V. Furyaev, Eduard Davydenko, Nicolai Andreev, Jabe Sandberg, Helena Lucarelli, Mike Fosberg, Wally Covington, Jimmy Hickman, Henry Lewis, Gardner Ferry, Mary Zabinski, Alexander Dimitrakopoulos, Boone Kauffman, Brian van Wilgen, Winston Trollope, Phil Cheney, Heloiso Figuero, Bob Mutch, Ronaldo Soares, and Tom Swetnam. To the staff of Fletcher Library—the whole shelf of them, but especially Carol Hammond, Sharon Rhoades, Sondra Brough, Dennis Isbell, and Elizabeth Smith—my gratitude for their indefatigable labors and infinite resourcefulness. To ASU West, the little campus that could, my appreciation. To Bill Strachan, appreciation for his ever-graceful editing. To Gerry McCauley, thanks for his bottomless patience.

Parts of the text have appeared previously in different venues, among them: "Flame and Fortune" as "Monsoon Lightning" in *Antaeus;* "The Summer We Let Wild Fire Loose" in *Natural History;* "Firestick History" in *American Historical Review;* "Consumed by Either Fire or Fire" (as "Keeper of the Flame") in *Fire in the Environment: Its Ecological, Climatic, and Atmospheric Chemical Importance.* All these essays have been revised for *World Fire,* some beyond recognition. To those from whom I needed permission, my thanks.

To my family—Sonja, Lydia, and Molly, who kept the home fires burning while I hotspotted in distant lands—my thanks. Again.

SIZE-UP

Firestick History; or, How to Set the World on Fire

The natives were about, burning, burning, ever burning. One would think they . . . lived on fire instead of water.

—ERNEST GILES, *AUSTRALIA TWICE TRAVERSED* (1889)

The past is a bucket of ashes . . .

—CARL SANDBURG, "FOUR PRELUDES ON PLAYTHINGS IN THE WIND"

Why did the Australian Aborigines not adopt farming, as virtually all the peoples around them did? Anthropologist Rhys Jones decided that they had; or at least that they had come up with a working analogue, a mode of production so elemental, ubiquitous, and misunderstood that observers failed to recognize it for what it was. Aborigines used *fire* to massage the indigenous environment so skillfully that they became, in effect, cultivators of that landscape. "Fire-stick farmers," Jones called them. The term has endured, an apt expression for aboriginal fire practices everywhere. It conveys, too, something of the power of the firestick as an enabling device by which to restructure whole landscapes. Certainly its Aborigines redesigned Australia well before the advent of agriculture or Europeans.

By analogy I propose a variant of scholarship—call it "firestick history"—by which to describe the interaction of humans with their environment. It is a metaphor for a history that explores fire, and for a methodology that allows fire to illuminate and reshape history. Like fire itself, which propagates by converting heat sinks into heat sources, the study of fire becomes both a means and an end. A land once burned is more readily burned in the future. Fire, like humans, remakes the world into forms that better suit it. This kind of interaction is what occurred in Australia, a fire-branded continent. And it is what has happened wherever humans and their firesticks have roamed.

The capture of fire by the genus *Homo* changed forever the history of the planet. Nothing else so empowered hominids, and no other human technology has influenced the planet for so long and so pervasively. A grand dialectic emerged between the fire-proneness of the earth's biota and the fire capacity of humans such that they coevolved, welded by fire to a common destiny. Humans assimilated nature's fire into their biological heritage as a species; equally, almost all biotas have come to accept anthropogenic fire, and not a few demand it. Deny anthropogenic fire and you deny humanity and many of its biotic allies a legitimate place on the planet. Ignore fire history and you dismiss one of the truly defining attributes of *Homo sapiens.*

But what, exactly, is fire history? It properly begins with fire. It includes fires, their history and geography. It absorbs the record of fire practices, the ways in which humans apply and withhold fire. It extends to the evolution of fire regimes, those peculiar patterns that fire etches into a biota, the dynamics of a particular fire ecology. This kind of fire history lies well within the dominion of the natural sciences.

What makes firestick history, however, is the ability to use fire as a means to understand humans better, to describe the character of *Homo sapiens* as a fire creature. Here—by constructing moral universes—firestick history makes its bid as genuine history, history as humanities. For such inquiries the special attributes of fire and its species monopoly by *Homo sapiens* are ideal.

By studying fire—events, practices, regimes, images—one can extract information from the historic record that might otherwise be inaccessible or overlooked. Just as burning often flushes infertile biotas with nutrients and as cooking renders palatable many otherwise inedible foodstuffs, so fire can remake raw materials into humanly usable history. It can drive out of archival scrub the vital character of humanity. One can reinterpret familiar events by fire much as Aborigines employed their firesticks to reshape their surroundings. The geography of fire and the geography of humans are thus coextensive. Track fire history, and you track human history. Around that informing fire humans tell the stories that make up their history, that say who they are.

In effect a pact was struck, the first of humanity's Faustian bargains and the origin of an environmental ethos. Humans gained fire and through fire a privileged access to the world's biota; that biota in turn got a new regimen of fire, one transformed by passage through human society. Everywhere that humans went—and they went everywhere—they carried fire. The hominid flame propagated across the continents like an expanding ring of fire, remaking everything it touched. Within that ring lived humans; beyond it lay the wild. Humans occupied preferentially those sites susceptible to fire and shunned, or tried to restructure, those that were less amenable. They sought out what needed to be burned, and burned it. Their first defense against wildfire was their own controlled burning. A land unburned was a land uncared for.

This relationship was reciprocal, however. If humans became dependent on fire (and in some cases intoxicated by, even addicted to, it), it is no less true that many biotas came to depend on anthropogenic fire practices for their own survival. Organisms adapted not to fire in the abstract or to individual fires but to particular fire regimes, and these were wholly or in good measure the product of anthropogenic practices. A sudden change in that regime—its load, its frequency, its seasonal timing and intensities—could propagate throughout the system. Control over this process was exactly the revolution that human firebrands made possible.

But if fire granted early humans new power, it also conveyed new responsibilities. It was vital that the flame neither fail nor run wild. Domestication thus began with the domestication of fire, and this in turn demanded the domestication of humans. Fire could not thrive unless it was tended, sheltered, fed, nurtured. The most basic social unit consisted of those people who shared a fireside. The hearth was the home.

· · ·

The quest for fire was a quest for power. Virtually all fire-origin myths confirm this fact. In words that in one form or another find echoes everywhere Aeschylus has Prometheus declare that, by giving fire, he founded "all the arts of men." To possess fire is to become human, but fire is almost always denied, only rarely granted by a fire-hoarding potentate, more typically stolen by some culture hero through force or guile. Once acquired, however, fire begins to shift the balance of biotic power in favor of the otherwise meagerly endowed genus *Homo.*

For all humanity's feebleness, fire compensated, and more. It made palatable many foodstuffs otherwise inedible or toxic; with smoke and heat it made possible the preservation of foods that would soon spoil; it promoted a cultivation of indigenous forbs, grasses, tubers, and nut-bearing trees; it stimulated hunting; it hardened wooden tools, made malleable shafts to be rendered into arrows or spears, prepared stones for splitting, kept at bay the night terrors, promoted and defined the solidarity of the group, and made available the evening for storytelling and ceremony. It even allowed humans to reshape whole landscapes as, in effect, humans slowly began to cook the earth. Everywhere humans went, fire went also as guide, laborer, camp follower, and chronicler.

But fire's danger matched its power. If untended, once-domesticated fire could go wild. The extraordinary pervasiveness that made fire so universally useful also threatened humanity's ability to control it. The relationship was truly symbiotic. If humans controlled fire, so also fire controlled humans, forcing the species to live in certain ways, either to seize fire's power or to avoid its wild outbreaks. Fire's power could come to humans only by their assuming responsibility for fire's care. The danger of extinction was ever present. The feral fire lurked always in the shadows.

Even today the control of fire is far from satisfactory. Imagine, for example, the reception that would greet someone who announced that he or she had discovered a process fundamental to the chemistry and biology of the planet, a phenomenon that could grant to humans the power to intervene massively in the biosphere and atmosphere, that could, in effect, allow humans to reform the living world. Some critics would object on the grounds that humans were congenitally incapable of using such power wisely, that they could only harm themselves and the world. Even those who would cheer such an announcement might be quieted by the qualifying fine print, which would stipulate that control would be inevitably incomplete; that even in the built environment lives and property would be lost; that the hazards attendant with the use of this process would be so great that they would require that special crews and expensive equipment be stationed every few city blocks, that buildings possess emergency alarms and exits illuminated by separate power sources, that every residence be outfitted with instruments to detect it, that special insurance schemes be in place—the list goes on and on. Anyone advocating the universal adoption of such a discovery would be denounced as a lunatic, or locked up as a menace to society. The process itself would never make it through federal regulatory agencies. Yet of course this is exactly our relationship with fire. Its power was too great to refuse, and its nature too protean to control completely.

As domesticated fire became indispensable, moreover, its flame had to be kept inextinguishable. Perpetual fire-keeping, or the tending of eternal flames, expressed not only the continuity of human society but also its differentness. The sacred fire of hearth and tribe had to be kept pure. If fire was universal to humans, it was also exclusive to them. Almost certainly no other creature will ever be allowed to possess it. However inadequate we may be as a fire-endowed organism, we humans will never voluntarily surrender our fire monopoly.

With power came choice, and with choice anthropogenic fire entered a moral universe. The human capacity for colossal power through fire lacked an equivalent capacity for control. Humans are genetically disposed to handle fire, but we do not come programmed knowing how to use it. Environmental conditions imposed some limitations on the ability of the land to accept fire, and human societies established still other parameters under which their cultures could absorb fire. But the range of options remained huge, and individual choices neither obvious nor singular. Those choices reflect values, institutions, beliefs, all the stuff of more traditional histories. The capture of fire became a paradigm for all of humanity's interaction with nature.

For fire there was no revealed wisdom, only an existential earth that could accommodate many practices and a silent creator who issued no decalogue to guide proper use. After all fire had not come to humans with stone-engraved commandments or gift-wrapped with an operating manual; it had been seized, and often stolen. Humans were on their own. The fire they grasped they had to maintain. They could nurture their special power into a vestal fire for the earth, or use their torches as the spray cans of environmental vandals. They did both.

Anthropogenic fire defined, as nothing else could, the relationship between humans and the lands they lived on. From medieval Icelanders to the twentieth-century Kwakiutl, people have carried fire around their lands to announce their claim to them. Like Australia's Gidjingali, aboriginal peoples often explain their expansive use of fire as a means of "cleaning up the country," of housekeeping, of exercising their ecological stewardship. Wherever humans and ecosystems have long coexisted, not to burn can be as irresponsible as improper burning. Good citizens use fire well; bad ones, poorly or not at all. In this way fire has become a pyric projection of human life, thought, and character. In its flames the biologic agency of humanity can be judged.

So the two aspects of fire history become one. Take away fire and humanity is quickly reduced to helplessness. But studying fire, particularly in the contemporary world, without reference to humans and the peculiar ways they behave, is quixotic. One might as well study the hypothetical fires of Titan or Ganymede. Fire and humanity are incomprehensible without each other. Consider two diverse examples.

In *A Sand County Almanac* Aldo Leopold calls for "an ecological interpretation of history," and gives, as an example of how natural processes interacted with humans, the story of Kentucky bluegrass. The winning of the transappalachian West began with the clearing of bottomland canebrakes and their replacement by bluegrass, a pioneering species that mimicked and aided the frontiersmen spilling over the Appalachians. Bluegrass was ideal fodder for cattle and made possible the kind of livestock-centered settlements that followed. Leopold then asks, "What if the plant succession inherent in this dark and bloody ground had, under the impact of these forces, given us some worthless sedge, shrub, or weed?" American history would look different.

Donald Worster ably adumbrates this parable by noting that Kentucky bluegrass was itself an exotic, a weed brought from Europe and spread particularly in the droppings of cattle. Bluegrass sprouted first around salt licks, then disseminated through disturbed canebrakes in a biotic synergism between grass and grazer. Part of America was remade biologically to resemble the Europe from which cattle and bluegrass had come. Seizing on Alfred Crosby's concept of "ecological imperialism," Worster demonstrates how the "new field of ecological or environmental history" has rejected "the common assumption that human experience has been exempt from natural constraints, that people are a separate and uniquely special species, that the ecological consequences of our past deeds can be ignored."

But much as Worster adds a codicil to Leopold's example, so fire history might add one to environmental history. If it matters that Kentucky bluegrass is an exotic, it also matters that the mechanisms of disturbance involved fire, that fire was as much a part of this historical synergy as the interaction among pioneers, imported grass, and domesticated herds. Prior to European contact, anthropogenic fire had assisted foraging for nuts, tubers, and firewood; it had sustained slash-and-burn cultivation for maize, squash, and beans; it had participated in warfare and defense; and it had made possible large-scale hunting. Broadcast fire had kept the hunting grounds of Kentucky in browse and grass. The great Barrens—interior seas of prairie—survived because of routine burning by American Indians. Those fire-sustained hunting grounds became the preferred grazing sites for American herders. When those fires were removed, the grass lakes silted in with trees.

Bit by bit the old fire regime disintegrated, along with its practitioners. Intensive grazing removed the fuel necessary to propagate flame; roads and plowed fields interrupted sweeping firefronts; exotic flora and fauna interacted with burned sites to force them into new pathways. A reform in fire use and fire control could not alone have wrought these results, but without a change in fire practices the landscape could not have assumed the shape it did. Of what followed, anthropogenic fire was a necessary although not a sufficient cause.

Still, it remains a useful guide, shared by humans and nature and by all the various cultures who entered the region. Grasp that flickering torch and the environmental and human history of the region are quickly illuminated. The reformation of Kentucky began long before Europeans arrived, and then the process assumed new forms. What happened with canebrakes and bluegrass and the Barrens, moreover, can stand for thousands of fire-mediated transformations in American history. Some sites became more driven by fire, or by fire associated with herder, farmer, and warrior. Others vanished when their sustaining fires departed. Tallgrass prairie crowded by avid weed trees; savannas of longleaf pine, overgrown with understory rough and lacking the sanitizing fires that retarded blue-spot fungus; even-aged forests of jack pine and lodgepole, whose serotinous cones burst open when caught in the flash of crown fires; sequoia groves that sprouted in the ash beds of their fallen progenitors, now crushed by fire-bearing canopies of fir and pine; the grasslands of the Great Basin, distorted into sagebrush, cheat grass, and juniper; the oak parklands of Wisconsin and California, their clear mosaics smeared into tangled throngs of brush and trees—the list of landscapes sensitive to changes in fire marches on.

Go further still, and use fire not only to examine the story but the storyteller. Scan Aldo Leopold's writings and you will discover, in fact, a high background count of fire references. Thus Leopold alludes to the 1871 Peshtigo fire, when the railroads brought industrial-strength slash-and-burn to the North Woods, a cameo of New World forest settlement. Early in his career he writes forcefully about the need for fire control, propagating the official Forest Service critique of "Paiute forestry." Then he reverses himself, his opinion of fire control paralleling his intellectual conversion on predator control. He notes how the invasion of southern Arizona by woody plants coincided with the elimination of free-burning fires, which in turn correlated with the introduction of cattle, the suppression of the Apache, the discovery of workable mines. Fire exclusion was part and parcel of a general reformation that had degraded the region. Elsewhere he relates how landclearing and fire scoured off peat in states bordering the Great Lakes (when America did to its boreal forests what Brazil is now doing to its rainforests). He refers ambivalently to new fire-roads punched into wilderness in the name of forest protection. He notes how overgrazing and exotics from the Asian steppes, such as cheat grass, have inspired a biotic invasion in the intermountain West, one prone to flash fires hostile to any life but its own. Leopold appreciated, in particular, the place of fire in the eternal wars between midwestern grasslands and forests. "Each April, before the new grasses had covered the prairie with unburnable greenery, fires ran at will over the land, sparing only such old oaks as had grown bark too thick to scorch." As a professor of game management, Leopold came to recognize the power of broadcast fire for wildlife. (Big game creatures do not eat mature trees—they feed on sun-drenched browse, new grass, and lush regrowth, all of which rely on fire to rewind their biotic clocks. Remove fire from many nutrient-poor sites and you will propel that ecosystem into a downward spiral, an ecological depression for which a Keynesian infusion of fire may be the best stimulant.) As a young man Leopold fought fires for the Forest Service. He died at age sixty-two while fighting a brush fire that threatened his beloved pine plantings. Trace fire and you trace a good bit of Leopold's life and thought. In the same way, American fire history becomes a torchlight procession through our national experience.

The power of fire to reveal can take more direct forms, too. During the drought of 1992, a wildfire broke out on the Swedish island of Gotland. The fire began at Torsburg, an elevated mesa ringed by an immense wall that dates back to the great migration era. The fire brigade from Kräklingbo soon extinguished the blaze. The next day, however, flames reappeared some seventy or eighty meters away, whether as a reburn or as an autonomous start no one knows. Winds freshened, torching trees became fiery catapults, and the fire raced toward the sea while volunteers thronged helplessly, bulldozers punched fireline after futile fireline, and the authorities readied farmers and villagers for evacuation. Then the winds died down, and the main burn, with more than fifteen hundred hectares contained within its perimeter, ceased its progress. The following day the fire pushed outward along its flanks. Then, stripped of fuels and no longer favored by the weather, it ended.

The fire stunned the local communities, as much as if seaborne raiders had suddenly descended on the island after a thousand-year hiatus. Even so, their response to the crisis was curiously exaggerated. It is no surprise to discover that they were ill prepared for a conflagration; not since the 1950s had Gotland experienced a fire of similar dimensions. But the reaction went far beyond the necessities of firefighting. The damages wrought by the fire paled besides those conducted in the name of public safety. The fire conjured up a threat greater than smoking ash and heat-killed pine. More than any published manifesto or act of the Swedish Riksdag, the response to the fire distills the essence of Swedish environmental thinking.

The destruction began with the firebreaks, cleared swaths intended to check a fire; these were huge, severe, and indifferent to circumstances of site and fire behavior. Mechanized line construction is rough on soils, and where, as in Gotland, the sparse topsoil rests on limestone, bulldozers can quickly scrape off centuries of gritty accumulation and leave white rock like a field of exhumed bones. Much of that surface harbored one of the premier archaeological sites in Sweden, its instructive stones now as mute as a razed castle. Firelines ripped around meadows as well as through forest with callous disregard for the mechanics of fire spread. Worse, after the fire had made its major run, the firelines were widened into enormous wastelands, broad as soccer fields. Bulldozers stacked the debris, chocked with rock and bristling with trees, to both sides in eerie mimicry of the stone walls around Torsburg, the rooted stumps and stripped tree trunks forming a macabre *abatis*. In places the tractors scoured the ground so deeply that the gouges filled with water, forming a moat outside the wall of forest berm.

And then, with the fire out, the devastation continued. Every dead tree along a road or fireline was pushed over out of fear that it might subsequently fall onto pedestrians. In fact the vast majority of the trees died from crown scorch, not basal burning. They could have been logged where they stood, or the road closed until the danger was fully assessed and the hazards remedied. Instead bulldozers crudely felled them, overturning soil and pitting the landscape, the toppled trees banging into still-standing ones and crashing into surface ruins. Along every point of entry to the burn, signs warned of the danger within.

The signs and the actions that accompany them speak of the site's potential hazards to humans. They say nothing about the threats posed by humans to the site. But that is perhaps the larger, symbolic significance of the Gotland fire. The savage firelines and numbing lanes of toppled trees testify to contemporary social values as fully as the armbands and axes excavated from Torsburg testify to Iron Age Europe. Those firelines are exploratory trenches, part of an archaeology of the Swedish psyche.

How could a nation that treats its citizens so humanely, that has claimed special status as a conscience for human rights, behave so badly toward its environment? One answer of course is that the fire was simply beyond the expertise of the Gotland *Räddningtjänst.* Not having experienced crown fires, the local brigades applied the only methods they knew and attacked torching trees as though they were flaming mosses and blueberry thickets. But this fails to explain the obsessive attack that continued after the fire had made its major run. Here, in the absence of published manuals and specific training, the authorities fell back on instinct; they did what subconsciously they knew to be the right thing; regardless of cost, they made the land safe for humans.

At Torsburg, Sweden suffered the vice of its virtues. In remaking the landscape in its own image, Swedish society has taken the same principles that have ordered its human relations and extended them to its relationship with the natural environment. The land—all of it—exists to serve humans either through direct exploitation or through assuring certain amenities that help sustain the quality of Swedish life. Environmental protection thus means protecting humans from the hazards around them, particularly the threat of outside violence. Sweden has moved boldly against acid rain, the greenhouse effect, nuclear fallout—all of which constitute threats to human health and which, not incidentally, have their origins outside Sweden. Environmental abuses, it is reasoned, follow from social abuses. The proper land ethic is to establish a just society and remake nature in its image. What is good for people is good for nature. Have nature serve a righteous society and nature will serve itself. The fantastic sense of security that is the obsession of Swedish society, and the triumph of Swedish politics, must be transferred to the natural world.

There is no room here for anything wild, erratic, or dangerous. There are no values outside those of humans and human needs; no relationships that transcend those of human society; no sense that environmental protection also means protecting nature from humans. Instead nature exists to be used, domesticated, and, once shorn of threats, enjoyed. Sweden's vast forests are in reality minutely manicured tree farms—and the source of more than half of Sweden's export income. Nature appreciation favors birds and flowers, symbols of the garden. The center of Sweden's moral geography resides in the farm, not the wilderness; the *trädgård* of Skåne, not the *vildmark* of Norrland.

Grant these propositions, and the logic of the Gotland fire scene becomes clear. The authorities—the community at large—moved to stop the violence, and then to remake the landscape in ways that would restore wealth and security. But what Swedish environmentalism saw as restoration, a more biocentric philosophy could see as vandalism. The principles that have ordered Swedish society may not apply to wildlands or biological reserves. Today it is estimated that, because of Swedish silviculture and environmental utilitarianism, two hundred or more species face local extinction within the boreal forest, an impressive figure for a biota characterized by great genetic variation within a few species. Among them are beetles equipped with infrared sensors that seek out burned sites.

The moral geography of wild land has different contours than that of domesticated landscapes. The principles that best govern nature's economy may differ from those desirable in human ones. The values that serve human relations may not apply to that endlessly unsettled and vexing relationship between humans and nature. The natural landscape needs dead wood, disease, rot, insects, parasitic fungi, paludified soils, predators, windfall, snowkill, snags, floods, disturbances large and small; it thrives amid a diversity of species, age classes, biotas; it adjusts to fire; often, even in Sweden, it needs fire. It is only in recent decades—curiously coincidental with the period during which Sweden rapidly established its fabled welfare society—that fire has become anathema.

Thus, what happened at Gotland is a kind of assayer's flame that reveals something of the character of contemporary Sweden. It was not the fire but its suppression, the determination to hammer nature back into a form that suited Swedish society, that inflicted the truly serious, enduring damages. Paradoxically, even as it champions global human rights, Sweden may well become an object of international censure on matters of natural rights. Like the fire beacons that once guarded Torsburg, the Gotland fire radiates a message. It tells how firestick history, properly applied, can fertilize fresh insights, like stalks of rye thrusting up through the ash of a felled and fired forest plot.

If anyone doubts the power of fire in nature and society, there is a simple test that extinguishes doubt: remove fire and see what remains. Humanity would plunge immediately into a darkness that would make Hobbesian man almost Olympian in comparison. Light, heat, and power from combustion sources; furnaces, ovens, stoves, and hearths for cooking food, melting metals, and firing ceramics; flame for clearing fields, removing woods, cleaning away debris and pests; fire for driving and attracting game, for flushing pastures, for potash, tar, and fertilizer; fire as a weapon against predator and enemy, and as a shield against the night; fire as inspiration, as a bright magnet for gathering family and tribe, as a symbol of aspiration and thought—all would vanish. What would remain is a large, talking chimpanzee, one reduced to following the spoor of nature's fires, a forager stirring the ashes of an Other's abandoned camps.

Even that would be lost if fire were removed from the planet. So long has the earth experienced fire, and so fully (with humanity's help) has fire penetrated every landscape and niche, that its expulsion is almost unthinkable. Experiments to exclude it have almost universally ended in failure. The earth is built to burn, and things made from the earth share in that combustibility. Even that most synthetic of human environments, the city, has failed to banish fire; and then, such is humanity's dependence on burning, urbanites must substitute some other less volatile form of combustion for fire itself. In the rural environment only the advent of fossil-fuel engines has allowed the removal of open burning from agriculture. In wildlands, where public agencies have attempted to exclude fire, the biota typically degenerates, and eventually burning reasserts itself in the form of wildfire. Even when humans leave the planet to voyage through interplanetary space they do so on plumes of fire.

Still there is one colossal natural experiment in fire exclusion. Since it acquired its ice sheet, Antarctica has not burned. Here is a continent from which nature has banished fire. The outcome is the most inhumane landscape on earth, relentless in its reductionism, horrible in its hostility. Where fire animates, ice reduces. Fire is a source; ice, a sink. That epigram aptly describes the human meaning of both phenomena and the lands they shape.

We are an Ice Age creation but a fire creature. Put a block of ice in one place and light a fire in another and see where people gather. Stare into ice, and you find a mirror that reflects back what you bring to it. Stare into fire, and your projections dance in the flames, reverie takes hold, and the sense of what is clearly Other and Self dissolves into flickering lights and shadows. The perceived fire is not a mechanical reflection but an active alter ego, a pyric double. It cannot be observed apart from its observer.

The observed fire does not work exclusively on the level of societies, cultures, and civilizations. It affects individual persons as well. For a fraction of modern Americans fire continues to shape a way of life. It orders their existence no less than it does the ponderosa forests of the Mogollon Rim or the relict tallgrass prairie of the Midwest. For them fire is a job. For a few it is also something more: an informing principle by which they not only order their lives but illuminate the world around them. Call them pyromantics. Count me among them.

It began for me with fifteen summers on a fire crew at the North Rim of Grand Canyon ("You light 'em, we fight 'em"; "Get there slow and let 'em grow"). I needed to know fire, and I wanted to bring some sense of intellectual rigor to a life of flame and fortune. I did not choose fire history as a means of exploring metaphysical and historiographical questions; those questions were forced on me by the character of fire history. But I learned why more people have ruined their health with a pencil than with a shovel.

Overall, it was easy, even exhilarating, to apply the tools and concepts of historical scholarship to new materials and to reinterpret old materials by firelight. (In his allegory of the cave, Plato thought that knowing the world by fire and fire's shadows was the human condition.) But I discovered that what made fire attractive also made it elusive. A fire history demands equal understanding of fire and humans, each of which is a partial amalgam of the other. It is difficult to keep the two components in sync. Typically narrative bangs and whirls like a washing machine too heavily loaded on one side. Free-burning fire is not a precision instrument. There is an old saying about playing with fire.

The paradoxes and dilemmas of using fire are as much intellectual as practical. The fire cycle traces a flow of information and syllogisms of logic as much as a flow of nutrients and a succession of organisms. Firestick history must answer the charge that it is circular, trivial, even parodic. If means and ends become interchangeable, then the fire cycle can become a circumference without a defining center, and historical scholarship a Möbius strip. The narratives of fire history may come to resemble those M. C. Escher prints in which an ascending staircase leads downward to another ascending staircase that again leads downward and eventually returns the traveler to his point of origin. If fire can be found everywhere, then it becomes banal, no more revelatory than the discovery of air, water, and earth wherever life is found. If fire is unique, then its power to elucidate is limited to particular phenomena. If it is universal, then it explains little about everything or much about nothing. The intellectual chemistry is as delicate as the chain reactions that sustain combustion itself. Without some ironic distance, an inflamed rhetoric can easily end in epic parody. What should speak to the species concludes only in private obsessions.

Storytelling—history—probably originated around an open fire. But there is something very peculiar about such storytelling when fire is itself the object of the story. Is firestick history a subset of other histories? A special genre of scholarship? A gimmick adopted out of personal obsession? I don't know. Gaston Bachelard believed that the human reverie induced by fire made its rational study impossible, and then proved his case by writing *The Psychoanalysis of Fire.* Ishmael interrupts his telling of *Moby-Dick* to warn: "Look not too long in the face of the fire, O man!" There are good reasons why so many myths end their universe with fire, and why a scholarship that seeks an alliance with fire must accept the unpredictable and the dangerous.

I do know that fire is not a generic tool. While it interacts with other human implements and ecological processes, it is not interchangeable with them. It remains stubbornly special. In most human technologies, from the furnace to the field, fire is an often indispensable agent, a critical catalyst. Its ecological consequences cannot be duplicated by other means. Its history cannot be told through indirection. But its peculiar promise for scholars lies in its maddening symbiosis with humans. Fire itself has become an amalgamation of the natural and the cultural, no longer a stable Other, an autonomous Nature that clearly exists apart from human artifice and ambition. Its history may even transcend irony.

Besides, fire is fun. To anyone interested in wordplay, fire history promises an inexhaustible reservoir of puns. To a pyromantic it offers a wonderful opportunity to fight fire with more fire. To scholars it extends a distinctive mode of historical production, the immense excitement of strolling through archives with an intellectual firestick. All that tinder. All that incendiary rhetoric. No wonder fire is eternal.

HOTSPOTTING

Fire Flume (Australia)

> Fires moulding the landscape? It [Australia] is indeed a strange world but who knows which is the obverse of which?
>
> —RHYS JONES

> Drought, dry seasons, and, more than all—the deadliest weapon of the tyrant—the bush-fire, reduces and selects the life of the country.
>
> —W.H.L. RANKEN, *THE DOMINION OF AUSTRALIA* (1874)

Australia is, more than any other, a fire continent, and because of that fact it shows with special clarity the power and limitations of the pact humanity has made with fire.

Australia's fires have influenced the whole temper of its history with a singularity not true elsewhere. Fire evolved beyond a presence into something like an informing principle. The other continents fire touched; Australia, it branded. Fires dapple Australian geography and punctuate Australian history. There are fires of all kinds and for all occasions, but some have an archetypal quality, at once defiant, oracular, and ineffable. In its southeastern quadrant—from Adelaide to Sydney, from the Riverina to Tasmania—Australia has concentrated fire, fuel, wind, and human settlement. Bushfires in this imploding compression chamber acquire an intensity, even a viciousness, that stands alone in the world. When the conditions are right, the region becomes a veritable fire flume as flames rush through the bush like water from a ruptured dam.

The flames cascading through the fire flume assume a demonic character, a trying force that burns away the dross and purifies. This is not a comfortable fire; it emanates from a part of Australia that defies human control and must remain unalterably alien. These are parts of nature beyond the circle of humanity's campfires and its fire drives, powers that humans can help kindle but cannot contain. And while this is universally true, it is not widely admitted. In Australia, however, it is unavoidable. Here the distance between natural fact and cultural illusion widens into a crack that, like a ruptured fuel line, threatens to explode all of spaceship earth.

Australia is a fire continent in the same sense that Antarctica is an ice continent. The two lands even share a bizarre symmetry, cratonic twins that split off from ancient Gondwana and endured a special isolation. Antarctica spiraled southward around the pole, acquired an ice sheet thick enough to deform the planet, and in the process lost almost all vestiges of terrestrial life. Australia rafted into the tropical Pacific, progressively dried and burned with increasing ardor, and reshaped rather than obliterated life, an evolving ark so peculiar that Charles Darwin likened it to a separate Creation.

Each is sui generis. Each had to seek a distinctive identity without direct linkage to other lands. Their natural histories show an irreversible, self-reinforcing commitment to one or several processes over all the others, and that choice comes to dominate its core. The terror of Antarctica derives from its passivity, the reductionist ice that leaches everything else away. By contrast, the Great Australian Emptiness, as Patrick White termed it, never sundered its links to life, and if Australian terror showed a greater aggressiveness, it also allowed, through fire, for a greater degree of human agency.

Since the breakup of Gondwana, fire had progressively insinuated itself into Australian natural history. Relentless leaching had impoverished the soil and shifted the biota toward sclerophylly, favoring the tough, the hoarder, and the scavenger. Then aridity arrived, reinforced that trend, and readied portions of the biota to burn according to the rhythms of dry season and drought. A biota salted with fire-aggressive species, pyrophytes, ratcheted notch by notch toward further fire-proneness. Each event seemed to join them further, like a screw driven into wood. As fires continued, the changing conditions, some of which the fires made, favored still more fire. The country's fires brought to a boil the whole biological billy that was Old Australia.

By the end of the Pleistocene, *Eucalyptus,* originally a minor element of the ancient rainforest but an opportunist of fabulous proportions, was readied for a biological explosion. A fire weed had discovered a fire continent. Once torched, the burning bush resembled a spiral nebula, its fuels and fires like paired arms locked in an accelerating vortex. Eucalypts dominated the scleroforest of Australia, and more than any other process fire drove the dynamics of the eucalypts. A likely source of that detonation was *Homo.*

With fire, Aborigines had access to most of the Australian biota; only the wettest rainforest and most barren stony desert resisted. It is extraordinary, in fact, that on this, the hottest and driest of the vegetated continents, its indigenes—nomads all—habitually walked around with flaming firebrands that dribbled embers everywhere and that required constant rekindling by igniting forest litter and grassy tussocks. Carry a gun and you'll shoot it. Carry a rock and you'll throw it. Carry a firestick and you'll set fire to the landscape around you. Aboriginal burning ensured not only the pervasiveness of Australian fire but its permanence. Their fires were inextinguishable, burning year in and year out, season after season, day and night. The firestick became a flaming lever that, suitably positioned, allowed the Aborigine to move a continent.

It seems implausible, but that is only because our civilization has replaced the universal fire with ever more refined pyrotechnologies, including fossil-fuel combustion, and reshapes the landscape through these tools rather than directly with flame. Fire remains just as indispensable but it is indirect, and fire ecology is adumbrated with the nutrient flows of a global economy and the energy pathways of internal combustion engines. The Aborigines used pure fire.

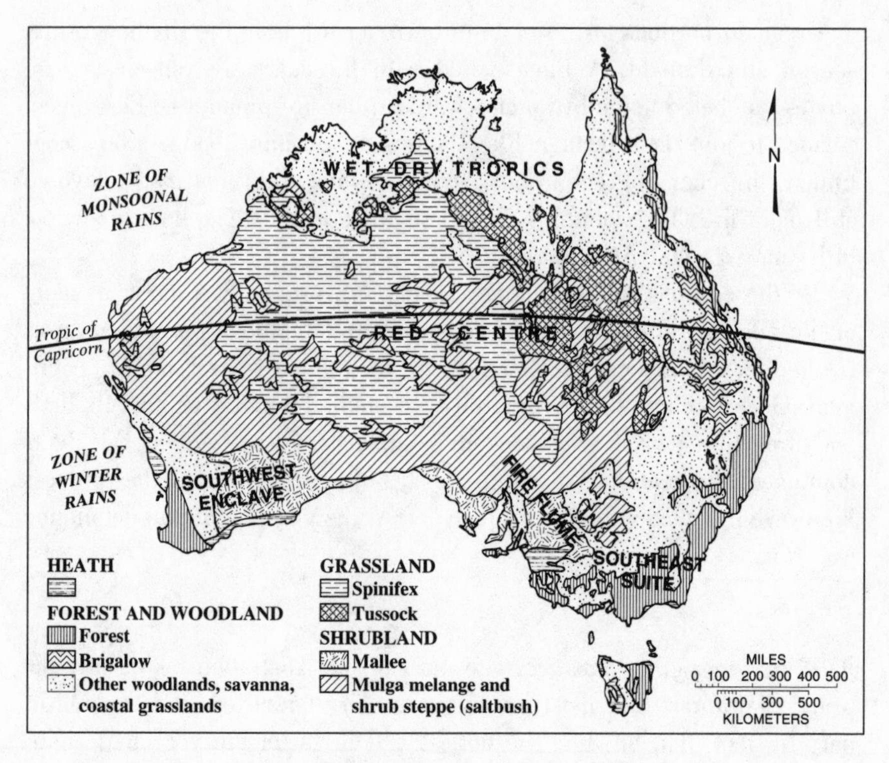

Burning Bush: biotic types of Australia.

The story that anthropologist Henry Lewis tells of Maria Island in the Gulf of Carpentaria is a cameo of the continent. When Commonwealth Scientific and Industrial Research Organization (CSIRO) scientists visited the island, it had been abandoned for almost three decades and overgrown with a forest cover unlike anything they had seen. They were enthralled with what appeared to be a resurgent rainforest. Aborigines with the party, however, saw the scene differently. It was, as one put it, a "bloody mess," overgrown, untended, disgraceful. Within minutes they were setting fires to reclaim the land into a form in which humans could live.

By the time Europeans discovered Australia, fire was universal there. The H.M.S. *Endeavour* under Captain James Cook reported "smokes by day and fires by night." Governor Arthur Phillip of the First Fleet discovered evidence of fire everywhere he traveled, and he explained to Vis-

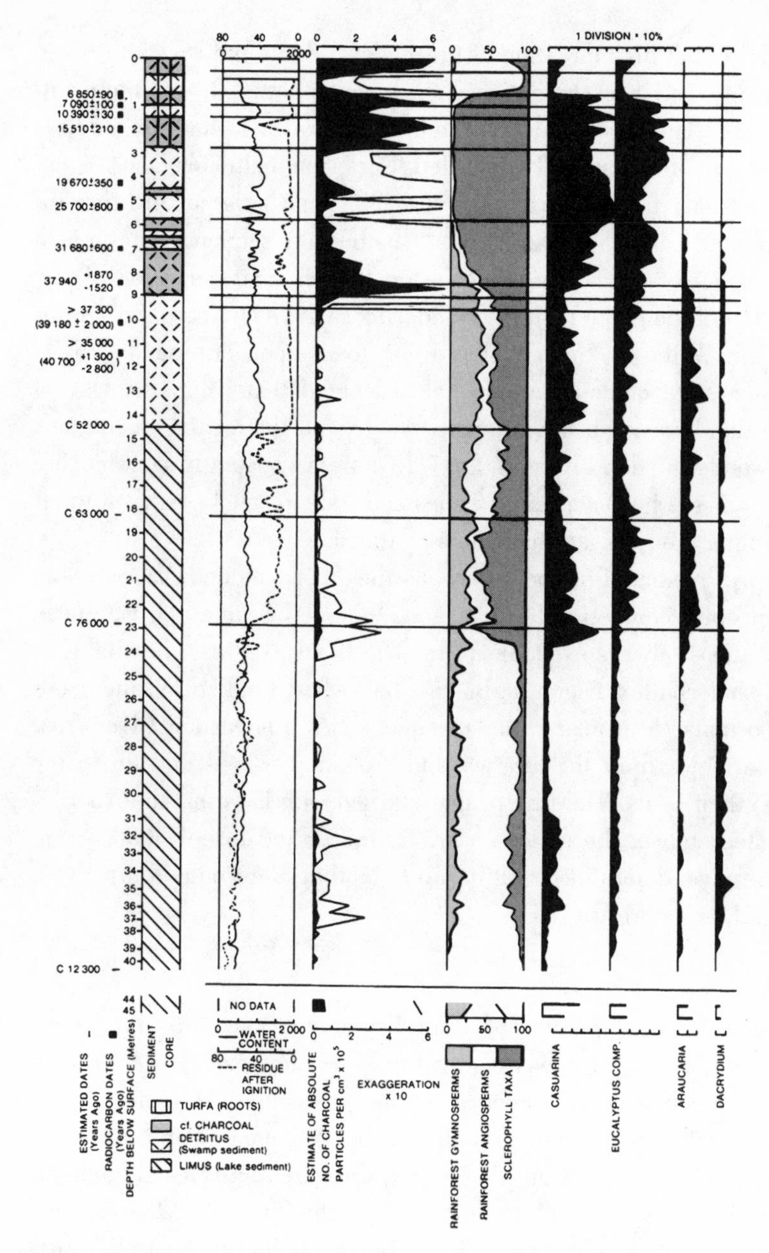

Pollen record from Lynch's Crater, northeast Queensland. Note the spike of charcoal around 38,000 years ago, roughly coincident with the earliest authenticated remains of Aborigines. Humans, firesticks, and a biotic revolution, as recorded in the shift among genera represented, all occur at the same time.

count Sydney that his "intention of turning swine into the woods to breed have been prevented by the natives so frequently setting fire to the country." Captain (later Governor) William Bligh worried that the endless "fires made by the natives" imperiled European cultigens and "every thing that cannot bear a severe scorching." The Aboriginal firestick inspired squatter Edward Curr to doubt "whether any section of the human race has exercised a greater influence on the physical condition of any large portion of the globe than the wandering savages of Australia." Archdeacon John Wollaston thought they were "bent upon putting the whole earth into a state of conflagration" and decided that when the Day of Judgment finally arrived, the Almighty would find the means for the task in "this wonderful depository of fire." Explorers constantly equated fire with Aborigines, whose caches of flames and trails of smoke marked water holes, hunting grounds, and corridors of travel.

Their fires mediated not only between the Aborigine and Australia but between peoples competing for that land. When Aborigine and European met, they almost always exchanged fire. The Black War was, inevitably, a fire war. That conflict began at Botany Bay when English marines confused Aboriginal firesticks with "musquets." At Endeavour River frustrated Aborigines fired the grasses under offshore winds and drove the English to their boats. The Europeans returned with firearms; the Aborigines retreated and set the woods on fire for miles; and Joseph Banks, with ominous resolve, determined that in future landings he would "burn Every thing round us before we begin."

That is not a bad description of what the British colonists did. Like ancestral gods, they remade the world with fire. The European qua Australian became as much a fire-addicted scleromorph as the Aborigine. If he wanted to survive in Australia, so shaped by millennia of anthropogenic fire, he had to burn, and in burning he acquired the same casual habits as the Aborigine. He carried his firesticks in his pocket. Wherever he stopped, he kindled billy fires. The "firestick habit in Australia," explained E.H.F. Swain in 1927, began with the Aborigine. Quickly, however, the white pioneer discovered that "by using the aboriginal method" he could improve his cattle; the new selector found fire "to be a settler's blessing"; and since then "Australia has been burned and grazed, and burned and grazed as it never was burned and grazed before." The rural population, he concluded, is "so enamored of the use of fire that wherever he now goes, and in the periods of highest fire hazard . . . the bushman blithely distributes his matches, and the schoolboy is learning to follow in father's footsteps." To a royal commission investigating the horrific 1939 Black Friday fires, a self-styled bushman noted with a verbal shrug that "the whole of the Australian race have a weakness for burning."

But the issue went beyond the "firestick habit." Colonization massively restructured Australian fire regimes. Eurasian species demanded a Eurasian environment. To get it colonists suppressed some fires, co-opted others to new purposes (for hunting to herding, for example), and exploited their own firepower to destroy an indigenous biota and promote their preferred cultigens. The "intrusion" of cattle alone, Surveyor-General Thomas Mitchell marveled, set in motion "such extensive changes in Australia as never entered into the contemplation of the local authorities." Because those authorities transported exotic flora and fauna (to say nothing of exotic humans)—wheat and weeds, sheep and rabbits—the disturbed fire regimes could not return to their previous state. The removal of Aboriginal fire destabilized ecosystem after ecosystem, and there is good documentation to suggest that fire exclusion by Europeans has led to floral and faunal extinctions. In Tasmania it accompanied the extinction of the Aborigines.

Still, not all fires could be domesticated, eradicated, or redirected to the designated purposes of the Europeans. What began as a biotic revolution soon unleashed a reign of terror. Holocaust fires raged where previously the land had only known, by and large, the simmering flames of chronic underburning. Bushfires even inspired a kind of liturgical calendar of environmental horror, the great fires named for days of the week—Ash Wednesday; Red Tuesday; and Black Sunday, Monday, Thursday, Friday, Saturday. Such fires threatened more than physical improvements. They questioned the very legitimacy of the society that experienced them.

Confronting the conflagrations helped transform Europeans into Australians. Their relationship to the bush was what made the colonists more than Europeans in exile. But defining their identity through the bush meant that they also defined it through bushfire. The bushfire became a set piece of Australian literature and art, an icon of all that was alien, unassimilable, and threatening about this "land of contrarities." In his novel *The Tree of Man,* Patrick White (Australia's Nobel laureate) describes one aftermath. "The fact was, the fighters had become not only exhausted but fascinated by the fire. There were very few who did not succumb to the spell of the fire. They were swayed by it, instead of it by them. . . . Because they looked into the fire, and seen what you do see, they could rearrange their lives. So they felt."

How Australians have responded to bushfire is an intensely practical matter but it is also a metaphor for how they relate to their environment, which is to say, a metaphor for who they are. In traditional geographies Australia stands to the side, not off the map like Antarctica but certainly off center, its concerns marginal to the great themes of human and natural history. But its forced obsession with fire reverses that order in at least one singular regard. As a parable of humanity and fire, Australia stands to the rest of the world as a diamond to coal.

In good measure humans succeeded in Australia through their control of fire. Theirs was, however, a Faustian bargain that further fed Australia's ravenous pyrophilia, its almost biotic addiction to fire. The firestick became a universal tool, and burning a universal solvent. Despite abuse, however, despite a distorting singularity to the relationship that resulted, their common reliance on fire did bond humans to Australia's fire-intoxicated flora, a codependence from which neither could, nor wished to, escape.

But even this was not sufficient. Neither partner enjoyed complete control over the other. Abusive human fire practices and endless tinkering with the fuels that sustained fire rippled through and destabilized whole regimes. Likewise, the peculiar geography of Australia—its fire flume—encouraged explosive bushfires that belonged with hurricanes, earthquakes, and volcanic discharges as natural eruptions against which no human agency could prevail. Such holocausts threatened to wipe out in days, if not hours, a presumed equilibrium between the state of nature and its hominid inhabitants. The land was metastable.

The firestick had its limits, even within a dominion like Australia's so singularly informed by fire. Paradoxically, the firestick had been itself the greatest damper on free-burning fire in that constant burning had checked the buildup of fuels, at least in places frequented by humans. The dam might rupture before the battering of the desert winds, but the lake of fire behind it was kept low. While European colonization disrupted this pattern and magnified the volume of fuels available, the fire flume remained outside human control on the scale needed to dismantle it—within humanity's reach but beyond its grasp. The firestick could illuminate the scene but not ignite it. However symbiotic the alliance between humans and fire, however maddening humanity's almost-but-never-quite-complete control over fire regimes, the prospect for holocausts persists. Fire has an identity beyond its instrumental value to humans. The continent that more than any other has been shaped by fire reveals this fact with dramatic clarity.

Large fires are not restricted to Australia. Plentiful fuels, stubborn drought, strong winds, and abundant ignition—wherever these occur, large fires follow. In the early 1980s they afflicted Canada like a plague, reversing, with shocking suddenness, a trend in which decade by decade the number of burned acres had steadily plummeted. In 1983 fires reduced eleven million acres of Borneo rainforest to a smoking hole, then returned in 1987. In the spring of that year fire roared through twelve to fifteen million acres of boreal forest on both sides of the Amur River. The Chinese hurled everything including the People's Army at the fires, with minimal outcome; confronted with much larger burns, the Russians simply withdrew. A year later and again in 1994 the American West suffered some of the largest outbreaks of fire since the 1930s, this despite the best-financed firefighting organization in the world. Australia has no monopoly on large wildfires.

What it does have is a set of environmental circumstances in which very large, very intense fires can recur with some regularity. The fire flume is a vast region, a subcontinent really, much larger than other mediterranean-climate regions that experience episodic fires driven by desert winds. Unlike Europe, it has no Mediterranean Sea to put much of the sirocco over water or to modulate its effects. Unlike Southern California, the winds are not confined, like so many fire avalanches, to terrain-defined channels. Nothing breaks the onslaught of desert winds drafted ahead of the cold front. It is as though Australia were a great centrifugal pump that periodically discharges its blasts through a southeastern funnel. And once they have passed, the southern burster—a violent shift in winds that accompanies frontal passage—begins its work, completing a deadly one-two punch. The flanks of fires formerly driven southward now break free into multiple new heads and race in eastward-probing tongues of flame.

There is nothing novel in the mechanism. The cold front is a common fulcrum by which to lever small fires into large, but no place else combines the force of desert winds, heavy fuels, and human habitation on this intensity and scale—and with such dismaying regularity—as does Australia. Its geography routinely bunches together the elements of disaster fires, and when drought grips the land, they crowd together like plutonium edging toward critical mass. If fire informs the Australian biota, then the fire flume defines the vital axis of Australian fire.

All the great holocausts of Australian history rush through it, each highlighting some disruptive presence by European colonization. The 1851 Black Thursday fire followed massive grazing, and preceded by days a gold rush. The 1898 Red Tuesday fires that spanned the Bass Strait built on decades of reckless ring-barking and landclearing for marginal farms. The 1926 Black Sunday fires attacked reserved forests and crown lands. But these paled before the Black Friday fires of 1939 that seemed to sweep all the wreckage and violence of settlement into one colossal maelstrom. Its timing, more than its size, made Black Friday a millennial fire, the fire of reference for a generation. Because it was followed so quickly by World War II, this fire, unlike the others, would not be shrugged away. It bonded with the war into a common cry for reform, a vast reconstitution of Australian society and lands that amounted to a new colonization. Eventually the specter of Black Friday inspired an "Australian strategy" of bushfire protection.

How do you cope with fires of this magnitude? There is no one answer. The administrative conundrum is that a single failure, under the worst conditions, will wipe out decades of successful protection. Prevention programs can eliminate most casual ignitions, but there remain many opportunities for accident or arson, and where folk burning is ubiquitous, fire exclusion demands the exclusion of folkways, a reformation quite beyond the power of any fire establishment. Places where endemic folk burning flourishes either lack the meteorological machinery for holocausts (for example, Brazil or Africa) or else fuels are controlled so tightly that little can escape. This is true for much of the Mediterranean, where fire becomes just another implement of gardening, like a spade or a rake. Disciplined fuels, moreover, require a disciplined population.

Another popular approach is to rely on rapid detection and attack such that most fires can be extinguished while they remain small. But suppression is expensive if done properly and ineffective if done poorly, and it can be overwhelmed by saturation starts such as those that follow dry lightning storms. Consider Southern California. Even with dedicated crews and the most sophisticated firefighting technology in the world—a rapid-deployment force larger than the military of some Third World countries—fires still escape initial attack, grow large, and rip through suburbs as privileged as Malibu, as they did in the fall of 1993. Additionally, to be effective in combating large fires the organization must face them routinely. A once-a-generation holocaust may suffice to shape a biome but it will not drill a fire establishment into fighting trim. Without constant exercise you can't maintain a sufficient force in sufficient readiness to meet, at instant call, a crisis of this magnitude. As with earthquake forecasts, it is possible to predict place but not time. Besides, very large fires reflect regional conditions that typically translate into complexes of large fires that overwhelm local forces staffed, at best, for "average worst" conditions. The only long-term solution is to change the fire environment.

This is easy in principle, hard in practice. Drought, rain, the seasons, storm tracks, winds, and mountains—none can be manipulated at will. That leaves the biota itself. Reduce the overall fuel load. Replace flammable flora with less combustible species. Restructure the fuel patterns so that fires start less readily, burn more slowly, and refuse to propagate so wildly. For all this there is ample precedent. Essentially this is how Europe confined its fire problems.

But Australia is not Europe, not even an antipodean Europe. Only so much of it could be farmed. Some domesticated Eurasian plants, notably grasses, could survive if nurtured sufficiently with fertilizers, legumes, and supplemental water. Hardwood trees assumed a role as ornamentals. North American pines thrived on plantations. But, unlike Britain, Australia could not be remade into a garden. Australians could not weed fire out of their bush.

The limitations were only partially, if initially, environmental, however. Those fires also burn within a cultural landscape. They reflect values, beliefs, knowledge, desires, misinformation, ambition, fears—all the ideas and emotions that motivate human societies. Increasingly fire has responded to economic and cultural factors as much as to lightning storms and desert winds. It is physically possible to clear and plow more bush for alien cultigens, but only at a prohibitive cost. More than an availability of rain and soil, a global economy dictated the viability of wheat, wool, and wood. Similarly, Australian society discovered cultural values in its native bush that argued against converting its pyrophytes into something else. It wanted—needed—that indigenous biota.

This reflected a profound reexamination of what it meant to be Australian. All in all, European Australia had not been well served by its intellectuals, who confronted bushfires with the intellectual equivalent of burlap bags and a box of matches. The billy fire, not the hearth fire, was the symbol of settlement; the intellectual joined the shearer, the digger, the selector and squatter as a kind of transient, a literate swagman on walkabout through the bush, a sojourner through the Outback of Empire. That bizarre bush existed to be transfigured into something else that belonged somewhere else. That it could not be remade into Britain was a mockery.

So it was also with bushfires or the circumstances that rendered the Australian fire special. Intellectuals—ever suspicious of fire anyway—were not disposed to inquire what the fire flume might say about life in Australia, or in what ways its lurid flames might illuminate larger questions about humanity. If Australians were to cope with its eruptions, that had to change. And finally, in the postwar years, it did. When Australians celebrated their bicentenary, they actually ringed the continent with bonfires that would, so Geoffrey Blainey hoped, "honour that most powerful, majestic and frightening force in our history—the force of fire."

. . .

The Australian strategy that emerged in the postwar years solved these concerns by appealing to wholesale burning for fuel reduction. Strip firing could protect cultivated fields; broadcast burning, expanses of native bush dedicated to forestry, grazing, or recreation. The manifold adaptations of the scleroforest to fire made burning a benign treatment. Routine burning was something volunteer bushfire brigades could do without major capital outlays, it could embrace large areas, and it evolved out of ancient fire practices. The postwar generation gave the folk practice rigor, science, and a moral fervor, a conviction that only hazard-reduction burning on a massive scale could prevent a repetition of 1939. In 1965 foresters invented an ingenious system by which incendiary capsules could be dropped from light aircraft. Now Australians had the means to reinstate the wholesale, routine burning of the continent. To advocates the device quickly became an icon of an emergent Australian nationalism, distinctive from that of either Britain or America. Britain ignored fire, Australia accepted it. North Americans sought to suppress fires with water bombers; Australians, with aerial burning.

But those cultural values soon conflicted with others. Australians strove to redefine their identity by redefining their relationship to their unique bush. That meant, first, that they had to preserve the bush as close to its native condition as possible, which in practice meant to its pre-European state, so far as that could be understood. With breathtaking sweep Australia began to convert much of its public lands to nature preserves and parks and, in a somewhat different sense, to Aboriginal reserves. This took the torch out of the hands of those groups, notably foresters and bushfire brigades, who had devised and who executed the Australian strategy. It also meant a redefinition of which practices best suit the newly reserved bush.

Here the issue of fire practices boiled over into controversy. Some conservation groups, particularly those nurtured on British models or appealing to Britain's intelligentsia, did not want fire. They saw no distinction between the prescribed burning advocated by the Australian strategy and rural traditions of burning off. If not evil, bushfire was at least an embarrassment. They wanted none of it. The proper model of a preserve was a garden.

Others accepted the evidence for fire—that fire was endemic to Australia—but distrusted the philosophy and techniques by which hazard-reduction burning was practiced. Instead they advocated a more biocentric fire, burning adjusted to the needs of ecosystems, not just to reduce aggregations of fuel. They wanted a mixture of fires, some mild, some intense, all varied according to the seasons. Routine burning, they feared, would homogenize the biodiversity they sought to encourage. They wanted fire applied or withheld according to the requirements of particular locations, not released wholesale from the skies like chemical rain. They wanted to restrict, not expand, aerial ignition. After all, the precious reserves that housed a flora and fauna found nowhere else had been established for more reasons than to protect them from fire.

Behind these arguments were larger differences in the fundamental questions about what it meant to be an Australian and on what basis Australian society could claim legitimacy, issues as ineffable as the bush itself. The debate exposed the Australian strategy as logically true but not necessarily true in a pragmatic sense. In theory, hazard-reduction burning could retard large fires, lessen damages, and multiply the effectiveness of bushfire brigades; in practice, it could never be applied on the scale necessary to prevent another Black Friday holocaust. But the controversies also exposed the inadequacy of Australia's cultural resources, which had demonstrated roughly the intellectual power of its bushfire brigades. The confusion over fire practices, their ends and means, only reflects a much deeper confusion about what Australia should be and how it should be understood. But then the Aborigine had forty thousand years or more in which to work out, through fire, a modus vivendi with Australia; the European, a little more than two centuries.

Regardless, the upshot is that no consensus exists about how to control the large fire. The bush will not be converted wholesale into something else nor will its fuels be reduced on the scale demanded. There is every reason to expect that large fires will continue to splash through the fire flume—as, in fact, they did in the 1983 Ash Wednesday conflagration, which actually exceeded the norms laid down by Black Friday and took even more lives, and as they did so vehemently in the suburbs of Sydney in the summer of 1994. The effort to preserve the bush will also preserve fire.

. . .

The traditional Australian response has been fatalistic. If a fire is not large, then it is not truly dangerous and does not demand special attention. If it is large, then nothing can be done. The great fires burn all places alike, randomly, malevolently; only the lucky survive. In fact a great deal can be done in particular locations. It is possible to shield individual dwellings, farms, and reserved sites from fire, not completely but within reasonable bounds. People can build bush houses that can endure fire, and in which they themselves can survive a conflagration. What they cannot do, under the existing circumstances—the dynamics of the fire flume, the demands of a fermenting Australian nationalism—is abolish large fires at their source. Conflagrations will continue. The millennial fire may in fact transcend decades or even centuries of fire history and rewrite the biotic archives of a region overnight.

In the end, it is less their actions and words that matter than the fact that Australians do debate, decide, and act. If an Australian is someone who engages in a special way the Australian bush, then he or she must also engage bushfire. It is the character of the interaction that matters, because this Australians can control, rather than the actual outcome, over which they have mixed influence. To eliminate the large fire would require a kind of environmental holocaust, the wholesale destruction of the bush and, with it, any prospect for an autonomous Australian identity.

Clearly, the fire flume is a distinctively Australian condition. But it poses a universal dilemma for fire management—not only that conundrum by which the protection of large wildlands creates the conditions for large wildland fires, but that more perplexing relationship that exists between humans and fire. Elsewhere in the industrialized world, peoples have dodged the issue. They have exorcised large wildfires by eliminating their wildlands. Or they have diluted the problem by mixing the elemental fire into other, less volatile compounds. Or they have disguised, deferred, relocated, and watered down the issues by insisting that only a few more dollars, a new fleet of airtankers, better patrols against arsonists, a bevy of greenbelts, or a refined prescription punched into a computer would yield final control. A philosophy of natural prescribed fire argued that predicting fire behavior was itself a sufficient "control" over wilderness fires. Yellowstone National Park elected to abdicate any moral responsibility at all, scrambling out of reality's quagmires for the higher ground of ideology. The fire establishment sent out the political message that, granted the right technology, large fires are controllable. Australia's fire flume, however, vaporizes such dewy fantasies, and recent developments in Australian nature conservation make it even less likely that large fires can be avoided. The smart money says that Australians will continue to face conflagrations. They do so as stand-ins for all of us.

Probably they will muddle through. Australians would like to join most of the rest of the world in tabling the matter. Their historic fatalism has bred a sardonic temper: "giving it a go" too often means "making do." But the fires keep coming, and if the Aussies choose, they have the opportunity to pass through those flames. They need to look once again, as Patrick White enjoined them, into the fire. Then they need to tell the rest of us what they have seen.

Veld Fire (South Africa)

This is where fire and humanity first joined, and that fact makes Africa the same as everywhere, only different.

There are, naturally, fires of all kinds and plenty of them. Nearly every manifestation of fire on earth is here present; every biome burns, save the perennially wet and the perennially dry. The firing extends across virtually every climatic zone, from the tropical to the mediterranean. Africa amasses every fire practice, every means by which humans exploit fire to extract materials from their environment, to reshape their surroundings, to understand their world. The magnitude of African fire—the aggregate of its biomass burned annually—exceeds that of every other continent. Viewed from space, sub-Saharan Africa is, during its dry seasons, a galaxy of lights, the grandest display of free-burning fire on earth. The burning of the African veld has global consequences.

All this makes African fire big, though not unique. What justifies the distinctiveness of Africa is the antiquity of anthropogenic fire, for this is, literally, the hearth of hominids. The African biota has not evolved with fire per se but with anthropogenic fire, and fire has here—more than elsewhere, and for a far longer time—mediated between humans and the natural world. That relationship may date as far back as 1.6 million years ago. Almost always the fossil evidence for *Homo* displays signs of fire or exists because burning has preserved it. As often as not the presence of fire at a site—hearths, charred bones, fire-cracked stones—is sufficient to attribute the associated remains to ancestral humans. Fire joined an enlarged brain and specialized tools as a defining attribute of *Homo erectus* and his successors.

Around the African hearth, fire took on a different quality than elsewhere. As they moved out of Africa, marauding humans carried fire along with their lithic tool kits; they used it to drive game, to shape habitats, to recalibrate the odds between hunter and prey. Their cooking of large game began by literally cooking the earth. The arrival of technologically advanced hominids, like tidal waves set off by the African quake, appears to have coincided with massive extinctions among the larger megafauna that populated the Pleistocene at its peak. In anomalous Africa, however, fire-wielding hominids and a menagerie of megafauna managed to escape extermination, perhaps because they had coevolved in elaborate symbiosis, a mutual ritual of adaptations that danced around a common fire. The wondrous wildlife of Africa thrives because of the fabled African veld; the veld flourishes because, according to various schedules, it burns; and for tens of thousands of years, perhaps tens of hundreds of millennia, humans have wielded the torch and directed how exactly that veld should burn.

It is a power at once subtle, brutal, almost limitless in its ecological range—a power that humanity bears uniquely, and for which it must assume a unique responsibility. Anthropogenic fire was not only a biotic necessity, and the contriving torch a unique ecological process; veld burning assumed the status of a moral act, the origin of a land ethic so ancient it should have been encoded in our genetic memory. With fire, humans could restructure the African biota in ways that no other creature could. The interactions between humanity and nature were no longer contained within the prescriptions that related carnivore to prey, scavenger to carrion, herbivore to plant, tree to grass. Fire touched everything. Like some ecological virus, humans had seized a critical process close to the nuclear core and through it redirected whole biotas to their own purposes. Lands depended on anthropogenic fire as termites depended on the bacteria in their guts or algae needed fungi to make lichens. Change the character of fire and you change the character of that environment. Eliminate fire—fail to stoke the biotic boiler—and you subject that machinery to a heat death, a slow cooling into extinction.

Of all this southern Africa is a concentrate. Imagine Africa as a great flask, its broad top collecting the flora, fauna, people, and practices of the Old World—filled, as it were, with a primordial sap from the Tree of Life. Set that flask to boil. Watch as a hard residue collects at the bottom, the distilled essence of the African biota. Southern Africa is that crystalline residue, and the veld fire the means of distillation.

The classic African veld is a savanna. Grasses shape veld biology and define veld burning. Grass carries the flames, grass sustains the marvelous menagerie of herbivores and carnivores, grass makes the veld cognate with the steppes, *campos,* prairies, *cerrados,* and sweeping savannas of the other continents. But clinging to the coastal fringe of the Cape, like barnacles crusting the deep hull of the African ark, spans another veld, a shrubland known as *fynbos* (literally "fine bush"). This biome belongs to a global fraternity of scleromorphic scrub that infests mediterranean-climate lands in Europe, Australia, Chile, and California, all similar, yet all distinctive, like coastal forts erected by some colonial power in different lands. Fynbos, like the others, is a fire-climax community.

The biotic density of the fynbos is fantastic. The fynbos biome brings together floral elements from stressed, nutrient-poor landscapes scattered throughout southern and eastern Africa, usually sandstones, often mountains. But the boiling down of the African biota has here yielded a distillation so pure that it constitutes a unique domain, the Cape Floral Kingdom, one of only six such kingdoms on earth. At least 8,500 plant species crowd into its condensed dominion, of which 5,800 are endemic. On Cape Peninsula alone, roughly the size of the Isle of Wight, there are 2,000 species, a floral bouquet that exceeds the bounty of all the British Isles. The Hottentots-Holland Mountains, some believe, exhibit the richest plant biodiversity in the world. Even within the fynbos the endemism of particu-

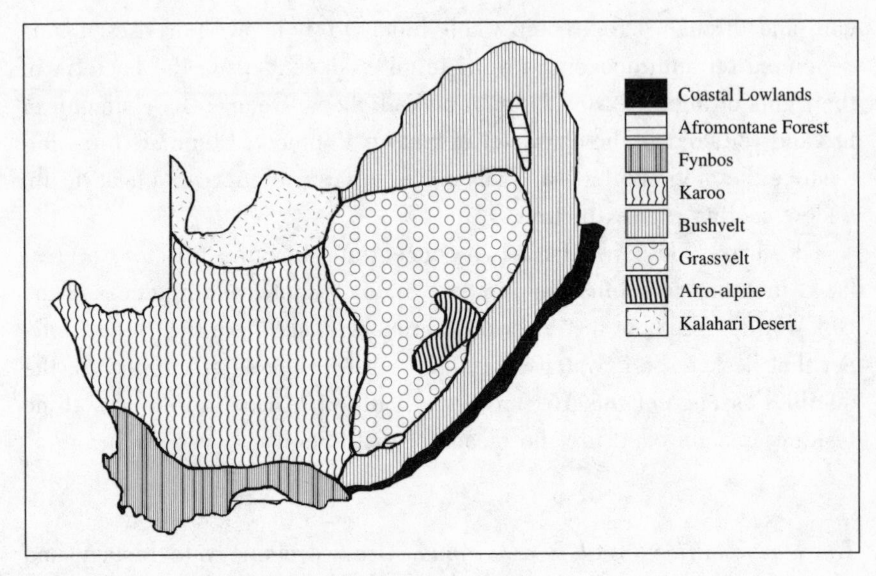

Flaming veld: the biomes of South Africa.

lar species to particular sites is phenomenal, with some restricted to a small valley, mountain slope, or moist spring. The whole biome is miniaturized, not only in its structure but also its functions, like a computer shrunk to a microchip.

What drives this system—what pushes scarce nutrients as though they were electrons through this biotic circuitry and stimulates the germination of a tough, wary flora—is fire. The fynbos burns. Routinely, necessarily, often intensely, fire scours the scrub, cloaking the Cape region in smoke and outlining its mountains in fiery silhouette. Fire comes as routinely as winter rains, droughts, and frosts. Without fire the fynbos senesces and ultimately disappears.

The adaptations to fire among its constituents are dazzling. Fynbos geophytes often flower after flame; the fire lily, for example, blossoms only after burning, and then within twenty-four hours after a fire. If burned, many geophytes with edible corms will fission such that one bulb can be harvested and another remain to sustain the crop, like the yeast in sourdough starter. Many fynbos plants reproduce by seed, and require fire to stimulate seed release or germination. The seeds of *Leukodendron*, for example, remain in serotinous cones until flame melts the waxy seal and releases them. The inexhaustible Proteas of the fynbos typically reproduce from seeds stored in protected buds or in the ground. Some seeds have a coating attractive to ants, which carry the seeds to their underground nests, eat the veneer, and leave the kernel, where it lies dormant until a fire stimulates its germination. Fire is as fundamental to the machinery of this biome as spark plugs to an automobile.

Fynbos management thus begins with fire management. It is vital to recognize, however, that fynbos is adapted not to "fire" in the abstract but to particular regimens of fire—to fire in certain seasons, with certain intensities, with frequencies that vary by year and decade. Randomly firing the plugs won't drive an engine; the sparks must be timed, and the timing will vary with the engine speed and the flow of fuel into its combustion chambers. In fynbos the flow of fuel is measured by biomass and regulated by organic pumps that follow the life cycle of the plants that make it up. The profusion of plants argues for a profusion of burning regimes.

The usual prescription is to burn the veld every four to twenty-five years, with the particular regime set by local conditions. More frequent fire may be impossible or destructive, either because fuels have not rebuilt to the point that they can carry flame or because incessant burning, or burning out of season, can allow grasses to invade and convert scrub to savanna. Conversely, less frequent fire—fire exclusion, say, on the order of thirty to forty years—is sufficient to allow forests to reclaim the landscape from their fire-shielded refugia in rockfall or wet *kloof.*

No spark, no power. No fire, no fynbos. Clearly forests can grow in much of the geographic dominion claimed by fynbos. They don't, in large measure, because the fynbos burns too frequently and intensely for the indigenous forests to leave their fire-safe sanctuaries. But once converted to grasses, in those places where the nutrients are sufficient, or to forests, where rainfall is adequate, new fire regimes are established that make reclamation by fynbos difficult.

Prior to human occupation, lightning and, in selected areas, rockfalls accounted for ignition. They remain potent sources even today. Jonkershoek Valley, for example, has seen its entire north-facing slope burned out over a five-year period from lightning storms. The Easter 1991 fire bust, begun with a dry-lightning storm on Good Friday evening, kindled

10,000 hectares of fynbos fire over a four-day period. Ignition by rockfall, in which cascading stones throw sparks, is unique but persistent to the region. The 1969 earthquake shook loose flints that kindled whole hillsides. The fynbos is too ancient, too compressed to have emerged only within the era encompassed by fire-wielding hominids.

For eons, however, *Homo sapiens* has controlled the flame. Known occupation by hominids dates back 125,000 years. At least 2,000 years ago Khoi pastoralists entered the fynbos shrubland with sheep, a few goats, and cattle. The full range of their burning is not known, but they clearly used fire for foraging, for hunting, for the stimulation of selected browse in an annual cycling through the landscape, burning different "patches" of the fynbos as each would carry fire. At a minimum, they did not prevent the veld, however ignited, from burning more or less freely. Because fynbos proper makes poor feed for livestock, pastoralists could not successfully convert fynbos fuel into fodder, which meant that herding could not, by itself, abolish fire. Early European settlement involved an almost seamless transfer of that torch. Originally Dutch colonists relied on native herds for meat; even the far-ranging voortrekkers used indigenes as herders, and this necessarily meant a continuation of indigenous fire practices. Even in the nineteenth century overseers had instructions to burn the veld annually "to obtain better pasturage for the stock." From Khoi-Khoi pastoralists they learned to move flocks to the mountains in the summer and to the Karoo in the winter, firing the mountain fynbos a month or so before their travel. By whatever means European settlement disturbed the Cape Floral Kingdom, fire exclusion was not one of them.

But intellectuals hated and distrusted fire. Their vision of proper husbandry had been nurtured in the Low Countries, where sedentary farmers integrated livestock with crops, converted wasteland heath into fields, and intensively, obsessively gardened whole landscapes. They saw fynbos only as an African analogue of Jutland's heaths, England's moors, Holland's *heide,* all eventually subdued by a relentless reconstitution of soils, flora, and fauna. Suspicion intensified with the arrival of European forestry, which had extended the principles and moral order of intensive agronomy into woodlands. Regulated fields meant a regulated society.

Free-burning fire threatened all of this. To the minds of critics it vaporized the organic humus, which alone made land fertile; it dried soils and promoted drought; it discouraged the adoption of winter pasturage and fodder crops such as lucerne by offering an alternative, low-labor, low-capital means of fertilization; it emboldened a nomadism that easily merged into uncontrolled wanderlust or, if capped, rebellion. A 1687 edict by the Dutch East India Company punished veld burners with scourging, two-time offenders with death by hanging. The threat to crops and property alone made veld burning indefensible. A European-style polity was impossible without European-style agronomy. Fire threatened the vision of a subdued veld as it did the veld-thatched houses of a subdued peasantry.

By the early twentieth century, the campaign against veld burning reached a crescendo. Rarely did critics distinguish between traditional veld burning and burning combined with other practices or promulgated outside traditional seasons. No one asked how it was that the fynbos had survived millennia of fire, or in what ways the very diversity valued by botanical collectors might have coevolved with fire as a necessary forcing mechanism. Typically schooled in German botany, biologists collected, named, classified, registered, and nurtured species; they grouped aggregations of plants into ideal "types"; books cataloged Proteas, Ericas, Aloes—the exuberant flora of a unique floral kingdom. But no one really investigated the dynamics of fynbos ecology. Then drought and cumulative farming abuses threatened Cape agriculture. The Drought Commission of 1923 condemned "unequivocally" the practice of veld burning. Nature conservationists joined agronomic theorists to demand that veld fires end. When naturalists succeeded in establishing reserves for rare fynbos endemics such as the marsh rose *(Orothamnus)*, they immediately blockaded the sites with wide polygons of firebreaks. The protection of the reserves first demanded protection from veld fires.

What happened was no less dramatic for being, in hindsight, entirely predictable. Within a few years, *Orothamnus* became increasingly rare, and by the time concern escalated into alarm, the plant had nearly vanished from the reserve established to nurture it in perpetuity. Not found elsewhere, *Orothamnus* might well have become extinct. But individual plants survived in the enveloping firebreaks, which were cleared and burned routinely. Then, after a wildfire broke through the barriers, the marsh rose erupted in profusion. The inextinguishability of fynbos fire had saved the fynbos reserve. Grudgingly its human stewards recognized that among their principal duties was to ensure that the veld was properly burned.

That charge has become vastly more complex, however. Alien plants such as *Hakea, Pinus,* Australian *Acacia,* and *Eucalyptus* have infested many sites and often outcompete the indigenous species after a burn. Thus not burning risks eliminating the fynbos; but so does burning under a regimen that favors the invasive aliens. The indigenous ants, too, are under siege from Argentine exotics who consume the enticing seed coats but do not store the seed itself underground and thus break the reproductive chain. Humans harvest Proteas for flowers and seeds, preventing reseeding after a burn. And of course the legacy of misused fire and fire exclusion has, in many places, so altered their ecology that recovery is impossible and a new future must be evolved. However problematic the appropriate regimen, it was not fire but fire exclusion that nearly triggered the biotic meltdown of the Cape Kingdom.

Much vaster and more typical is the grassy veld of Africa. If the fynbos is crystalline, like the matrix of a fine-grained igneous rock, the grassveld is plastic and more tolerant, a half-lithified conglomerate. The capacity of grass to combine with other flora means that the veld spans many environments. This malleability is reflected in the word itself, cognate with the English *field.* Thus *veld* shows an almost infinite capacity to combine with other terms—*grassveld, bushveld, thornveld, highveld, lowveld,* ad infinitum. It is a vast dominion over which rainfall is variable, soils mixed, and the biota adaptable. Summer rains and winter fires replace the winter rains and summer fires of the Cape's mediterranean climate. But throughout, the critical element is grass. To reach it fire must compete, and interact, with Africa's glorious megafauna.

Veld fires influence almost every aspect of this environment. The geography of residual woodlands, outliers of the great Afromontane forest, closely parallels the geography of berg winds and burning. Sheltered sites retain forests; exposed lands surrender them to grassveld. If unburned for several decades, the indigenous forests will reclaim adjacent lands, and can so alter the understory and microclimate that they will repel veld fires that advance upon them. In the bushveld the timing and intensity of fires largely dictates the relative distribution of scrub and grass. In general, frequent hot fires favor grasses; infrequent cool fires promote woody shrubs and trees. Even within the grassveld, the regimen of burning shapes the relative proportion of a species and its overall vigor. Sourveld requires routine, even annual burning to stave off bush encroachment and to prevent rapid decadence among its grasses; without fire it quickly becomes unpalatable and is shunned by all except termites. In contrast, sweetveld can retain its fodder value without frequent fire, but burning promotes that most valued of all pasture grasses, *Themeda triandra;* and fire fights off the pressures that would otherwise propel the system toward bush, forest, or weed such as the fern *Gleichenia,* a plant easily extirpated by burning but otherwise deadly to any other flora. Exclude fire and this green cancer will metastasize over whole regions. Banish fire and the indigenous grasses will decay or discover themselves weeded out of existence by choking bush. Remove fire and the African veld becomes quickly uninhabitable to Africa's fabled fauna.

The fauna interact everywhere with veld and fire. There are species of grasshoppers whose reproductive rhythms are timed by veld burns. The lesser black-winged plover nests only on burned ground. The bald ibis feeds on half-cooked tortoises and other carrion caught by fires; the ibis's distribution correlates exactly with the distribution of lightning fires, and it vanishes where fire is excluded. Nearly all grazers and browsers, excepting oddities such as the giraffe, feed on short new growth, exactly the outcome of properly timed veld burns. In sourveld, unburned grasses soon become a faunal desert, inedible to nearly all decomposers, until eventually it is either overrun by asphyxiating bush or incinerated by wildfires that feed on its bloated fuels. To restore vitality the system must wait for storms or elephants, a kind of biotic bulldozer, to trash the woods, thus allowing grasses and veld fires to reinvade and restart its cold engines. And so it goes, over and again, not simply repeating itself endlessly like some ecological mandala, but recombining annual variation with idiographic changes induced by climate, cyclone, drought, pestilence, locust plagues, and a stream of exotic organisms, the tremendous diversity of plants and animals shaping an equal diversity of veld burns, linked not by any rigid chain of causality but through a vast, often atonal choreography.

Imagine, in such an environment, the emergence of an opportunistic species that could, improbably, control the pattern of veld burning. The power of such a creature would be enormous. Yet this is exactly what early humans accomplished, with consequences as profound for hominids as for the veld. Used carefully, veld burning could enhance, even stabilize, this environment; abused—by overuse or underuse—it could quickly transfigure the landscape in perhaps irreversible ways. That humans probably acquired fire in Africa, that they first possessed it only in association with a primitive technology of stone and bone, that they coevolved as unique fire creatures among the Pleistocene menagerie of the veld, has apparently granted a special character to African fire. Anthropogenic fire has been active for so long and so persistently that all the elements of this biota have mutually adjusted to it, as fire has to them, without terrific dislocation. In fact, humans became not merely tolerated but essential. They served as stewards of the veld fire on which so much depends.

The shock came when *Homo* returned to Africa with newly acquired biological booty in the form of domesticated animals, then plants. Livestock competed with other herbivores and had to be shielded from resident carnivores. Herds required a subtle shift in the patterns of grazing, and this meant that the veld was burned in somewhat different ways. Two thousand years ago pastoralism had penetrated to the Cape of Good Hope. The more fundamental transformation, however, occurred with the appearance of a full-blown economy based on mixed farming in which slash-and-burn agriculture, assisted by iron tools, combined with livestock herding to restructure the gross geography of woods and savannas such that the proportion of grassland to woodland increased. This occurred with the steady migration of Bantu-speaking peoples from West Africa to the east and south.

The process was still in motion, still curving around the southern arc from east to west, when European explorers reached the Cape. For roughly the last two hundred years European-derived colonists advanced from the

Cape eastward with an environmental economy remarkably similar to that of the Bantus. The political cataclysm of their encounter profoundly influenced the environment because it disrupted the evolved equilibrium of the veld. Entire tribes came and went, throngs of wildlife were shot, grasses ferociously overgrazed, forests leveled, weed species introduced and even cultivated before running wild, fires set outside traditional controls or, worse, not set at all. What had been a choreographed ritual disintegrated into a melee.

Again, intellectuals schooled in European agronomy misread and distrusted veld burning. They argued that, as in England and Holland, farming and herding should proceed without fire. They even installed agricultural models for the management of game reserves and parks. Wildlife, they culled; the veld, they divided into paddocks; aridity and drought, they countered with stockponds and tanks. They sought to limit overgrazing by limiting overburning. After the Soil Conservation Act of 1946, veld burning in nature reserves required permission from the Department of Agriculture. A new conservator to Kruger National Park even announced that the time had come to wean the park from veld burning once and for all.

This celebrated experiment defied all practical experience. As late as 1940 Colonel Stevenson-Hamilton, conservator at Kruger for almost fifty years, had declared categorically that "in a sanctuary for wild life" it was "essential to burn old long grass, but this must be done methodically." Anyone intimate with the veld knew as much. Even poachers baited their snares with small patches of burned veld to which hungry ungulates gravitated, eager for the succulent young growth. Hunters, game, veld, and fire were interlinked by a ritual that dated back to *Homo erectus*.

But the new order knew better. In 1948, arguing from pastoral models, the park board decreed a new policy that sought to shift burning from autumn to spring, to restrict such fires to no more than once every five years, and to eliminate accidental fire. The outcome was calamitous. Animal populations plummeted, and large numbers, starving, left the park for burned lands outside it, where hunters slaughtered them in droves. Bush encroachment reached epidemic proportions, establishing itself as nearly ineradicable. Fuel buildup led to catastrophic wildfires that swept half a million hectares at one time and killed animals in alarming numbers. By 1954 the park board recognized that their fire policy was insupportable. Kruger Park reinstated triennial burning and launched a research program to discover the best regime to suit the park's purposes. Traditional veld burning had supported traditional veld uses; as new purposes evolved for reserves, managers had to develop new patterns of veld burning. By 1990 that meant a more "natural" order established along more "ecological" principles in which, ideally, whatever veld could be burnt would be.

Denying fire to the veld was not merely a practical but a moral failure. The acquisition of fire by humans, a unique power, had brought with it a unique imperative to see that the fire upon which the veld and its creatures depended was suitably applied. The Kruger experiment demonstrated conclusively how fundamental that duty was. It illustrated how, in southern Africa, fire control meant control over the patterns of burning, not control over whether the veld burned or not. It had to burn.

In such an environment human practices become, necessarily, fire fables. This holds even for such celebrated examples as that of the honeyguide bird *(Indicator indicator).* The symbiosis between this bird and its human collaborators has long stood as testimony to the phenomenal interdependence between African wildlife and African humanity. The bird feeds on honeycomb, which it can ingest thanks to bacteria it carries in its intestines. But it requires help to disperse bees and expose the hive. For this it relies on the ratal (honey badger) and humans—and recent evidence suggests an overwhelming preference for humans. It locates a hive, then tracks down a human, calls him to follow, and directs him to the discovered hive. (A variation allows also for the man to call the bird.) People expose the hive, take the honey, and leave the comb to *Indicator indicator.* There the story normally ends.

But like most veld stories this one includes a fire subtext as well. The means by which humans penetrate the hive is by smoking out the bees. Once completed, the burning grasses or brand is then tossed to the ground. In the dry season the veld takes fire. Honey collecting is, in fact, documented as one of the principal means of veld burning in wooded regions. The veld fire, in turn, keeps the woodland open, stimulates the flowering of those plants the bees require for pollination, and thus sustains the fantastic collusion among the pollen-seeking bees, the human-seeking birds, and the fire-brandishing humans. On a grand scale, that is what evolved throughout the veld, and it is what makes African fire unique and instructive.

The earliest reports of European explorers to southern Africa allude to fires. Rounding Africa, Bartholomeu Dias sighted the *Capo de Fume,* Cape of Smoke; Vasco da Gama, farther east, named the headlands *Punta das Queimadas,* Point of Fires. They associated the fires, correctly, with humans, but they did not appreciate the complex, venerable symbiosis those fires advertised.

Here humans did not readily dominate the biota. They had to compete with other carnivores, browsers, scavengers. Their special niche was to keep the flame for all creatures; their unique role, to ensure that the veld was properly burned. If humans failed, then the system faltered. Lightning could eventually impose a new regime, but not before the biomes changed, the system ran down, wildfires destroyed once flourishing veld, and woods choked out the vast megafauna of the savanna. For tens of thousands of years the veld had adjusted to the regimen of anthropogenic fire, not to that of lightning. Fire exclusion would be disastrous, an act not only of ecological madness but also of moral vandalism, an abdication of one of the defining traits of humanity. What the vestal fire was to ancient civilizations, anthropogenic fire was to Africa.

Outside Africa fire had a coarser character, an invasive quality. Elsewhere hunters employed fire drives to force prey into traps, spears, or arrows; in Africa, fire had a gentler touch, used more to coax than to compel. Although African hunters did use fires to herd elephants into favored kill sites and to flush out bucks, partridges, and hares, in general ungulate herds were not driven by flaming fronts so much as they were enticed by the lure of the succulent growth that thrived on the selectively burned veld, each patch in turn, over the long seasons, granting to target species the fodder they most craved. Until the advent of high-powered rifles and a global market for horns, tusks, and skins, humans did not rule so much as they counseled the veld.

European contact broke these supple links. The wave of megafaunal extinctions that swept continent after continent as humans entered from their African hearth now returned in a vicious rebound to Africa. The slaughter was unbelievable. Strand after strand of an ancient web that defined the veld unraveled—human societies upended in turmoil, animals slain by the tens of thousands, the veld ravaged by mistimed fires or not burned at all. Savanna and woodland were a shambles. Some of the wreckage was deliberate; much was a byproduct of ecological invasion—the fire, famine, death, and pestilence of an African apocalypse.

A similar uncertainty afflicts attempts at reform, which now take two directions. One is nature conservation, in which land is set aside to preserve or restore the biodiversity of pre-European (and even preagricultural) times. The other is the integration of the farmed and grazed veld into a global economy such that the critical nutrient cycles span other environments, even other continents, beyond the range of tree roots, bird-dropped grass seed, or roaming elephants. Money can partly substitute for fire to power this nutrient flow and furnish the pieces otherwise lost in the chaos of colonization. Both strategies propose a new paradigm of veld fire.

There is a scene in Isak Dinesen's *Out of Africa*—geographically set somewhat north, but ecologically and psychologically congruent—in which her Kikuyu servant, Kamante, rushes into her bedroom and shouts for her to awaken.

> "Msabu," he said again. "I think you had better get up. I think that God is coming." When I heard this, I did get up. . . . He gravely led me into the dining-room which looked West, towards the hills. From the door-windows I now saw a strange phenomenon. There was a big grass-fire going on, out in the hills, and the grass was burning all the way from the hill-top to the plain. . . . It did indeed look as if some gigantic figure was moving and coming towards us.

No European ever mistook the burning veld for the face of God. Instead, following the Portuguese, they planted their stone *padrãoes,* laid claims, and began the incredible task of integrating Africa into a new global order of which neither they nor anyone else had much control. The old bonds that had fused humans and veld broke apart, and now as the hybrid society of southern Africa tries to redefine itself—as it attempts to restructure not only the relationship among its diverse peoples but between all of them, as humans, and the natural world—it must reclaim veld burning. It must rediscover the proper proportions of fire, redefine the right prescriptions to guide the special stewardship of fire that humans alone can perform.

To that difficult task the European padrãoes stand as sad, stark symbols. They proclaimed the coming of a European agronomy that distrusted fire, and a global maritime economy that would relocate species and redefine biotic relationships without regard to the ancient links forged by the region's fire ecology. European farmers had, over centuries, domesticated a flora and fauna for which fire was unnecessary or hostile, and of course free-burning fire was anathema to any economy based on fixed landownership. Wherever they took that hothouse biota, they had to shield it from fire, much as they had to protect domesticated sheep from local carnivores and hybrid wheat from indigenous graziers.

The imagery of those installed pillars is instructive. No African artist painted a scene, as Piero de Cosimo did in Italy at the time the Portuguese first toured the Cape, in which a forest fire segregates humans from the rest of creation. More typical is San rock art, with its celebration of hunter and beast. Those images have survived in part because they were baked onto the stone by veld fires, not unlike the way fire had preserved the interplay between hunters and hunted on the landscape. The padrãoes, however, proclaimed a new order, guided by a different logic in which fire had to be removed rather than integrated. What fire had fused had, through the abolition of fire, to be rent. The inscriptions on the padrãoes were transferred to the land in the form of fuelbreaks, slashed with similar geometrical rigor across mountain and veld, announcing an imperial claim as surely as the text on the weathering stones.

Queimada Para Limpeza (Brazil)

They are always there, somewhere—pilot flames of Brazilian life.

They burn in gutters, on street corners, in yards. They can be found, smoking from the earth like fumaroles, in dusty *bairros,* in the vacant lots of cities, on the outskirts of metropoli. They burn whatever is at hand—garbage, grass, shrubs, trees, agricultural chaff, industrial scraps. When the dry season comes, they seize new fuel and spread outward and blossom along roadsides, in ditches, on *fazendas* and in forests. Farmers burn their fields. Ranchers burn their pastures. Loggers burn their slashings. As the rainy season approaches, the fires become a frenzy. From Amazonas to Paraná, from Rondônia to Bahia; in rainforest clearings, in savanna and scrubland, in coastal *mata,* in biological reserve and industrial forest; on agricultural sites, large and small, for subsistence crops or for market—everywhere that rural Brazil thrives it does so with the help of the *queimada para limpeza,* "the cleaning fire." It is the generic Brazilian fire, and it describes, as well as anything can, the generic relationship of Brazilians to their land.

Because of it, however, Brazil has also lost some control over the power to define that relationship. Probably no other country has, over the past decade, experienced so much scrutiny regarding its fire practices. The reasons are several. Brazil is vast, claiming almost half the landmass of South America. It has a cornucopia of fire, everywhere visible and nowhere well understood. It is a nation of uneven industrialization and vigorous internal migrations, which is to say a nation undergoing change, the kind of rapid social change that typically translates into abrupt eruptions and extinctions of fire. And it is a country that is experiencing this transformation within the context of a global environmentalism that has targeted Brazil's land-use policies—and its fire practices—for special opprobrium. Burning Brazil has come to symbolize humanity's disequilibrium with the earth.

The historical geography of Brazilian fire begins with the distribution of wet and dry lands, wet and dry seasons, wet and dry epochs. Contemporary Brazil embraces a double-cored environmental ellipse. One core is fixed in the chronically wet, around which spreads classic rainforest; the other is in the chronically dry, an enormous extent of land that ranges between the rainforests of Amazonia and those of the Atlantic coast—an expanse that embraces the scrubby *caatinga* of the northeast, the savannalike *cerrado* of the central plateaus, the grassy *campos* of the south.

The extremes know fire only infrequently. The rainforest of western Amazonia is perennially wet. Outside the core, forests are a jumble of primary and secondary types, the outcome of chronic but low-intensity disturbances, variable rainfall, some seasonality—and fire. At the other pole is the caatinga, a chronically dry region further subject to massive droughts from time to time. The opportunity for routine burning is slight, restricted to outbreaks of precipitation or disturbances that place normally unavailable fuel into a position for burning.

Each of the two dominions is in fact a composite of moisture and aridity. The wetness of the rainforest varies by season, by year, and by

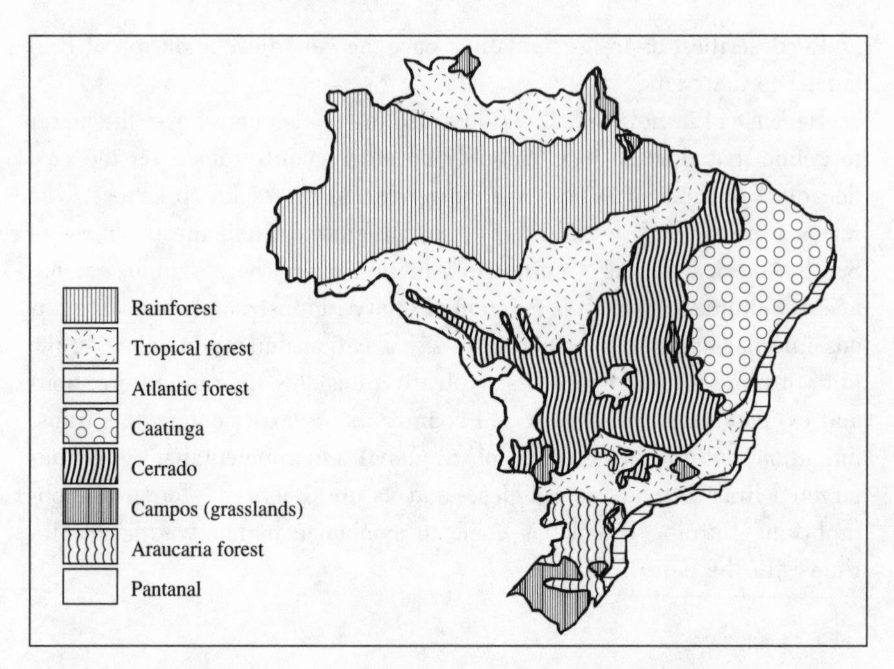

Burning Brazil. Note the extent of the *cerrado* and *campos* vegetation.

epoch. While temperature is virtually constant day to day, rainfall and humidity fall off around August. Great droughts, such as those of 1987 and 1988 which withheld rain from Manaus for seventy-three days, blast the region once or twice a century, about as often as hurricanes smash into New England. That aridity makes a crack into which fire can be driven. Similarly most of the dry dominion is at least seasonally wet, which is to say, stocked with fuels. Brazil's fire season is thus defined differently in each dominion—for the wet, by the period of dryness; for the dry, by the period of moisture. The perfect composite is the massive mixed grassland, the cerrado, that spreads across much of the interior, a complex intermediary between forest and prairie, variable in its seasonal rhythms of rain and drought, the natural epicenter for Brazilian fire.

Other, lesser biomes fringe these. Along the coast the Atlantic forest clothes the rugged Serra do Mar, its native flora ruthlessly reduced to small enclaves. There seems to be little basis for natural fire here, although anthropogenic fire has been prominent since the advent of agriculture. To the south, almost temperate in climate, spreads the forest-steppe landscape of the Brazilian pine *(Araucaria).* Again, only remnants of the original biome remain, but it appears that regular fire was vital to the reproductive success of the species. Bordering Bolivia and Paraguay, the Pantanal is a colossal swamp, seasonally flooded and drained. Its fire dynamics, if any, are not understood, and it is possible that routine flooding may substitute for routine fire. Scattered throughout the perimeter are grasslands, some dry, some wet, and many extensive. The southernmost grade into the pampas.

The present geography of fire, however, reflects a past history of wet and dry epochs, not merely of seasons. Today's biotic types are far from immutable. Some claimed much larger dominions in the past; some were crowded into minor *refugia;* some were mixed in proportions and patterns much different from those they exhibit today. Lands that are wet today were dry, and lands that are today immune to combustion once burned readily. Brazilian fire regimes represent an ancient evolution in which wet and dry extended their cycle of fire over the course of prehistoric or geologic time.

It is believed, for example, that at the onset of the Holocene the Amazonian rainforest existed only in dispersed enclaves, that the climate changes that ended the last glaciation allowed the forest to march over landscapes otherwise opened to cerrado. Almost everywhere in Greater Amazonia the soil is charged with charcoal, frequently baked into lenses like the burned biotic shards of a sacked city. Only when the climate shifted to a chronically wet regime did a *revanchist* rainforest crowd out the flaming cerrado. It is also a fact of tremendous but ambiguous importance that all of the existing landscapes have assumed their contemporary shape in the presence of humans.

The prehistory of the Brazilian environment is understood poorly, and like its biotic history is apparently destined for imminent revision. Recent research suggests that even the rainforest is the product not of unbroken quiescence but of chronic disturbance by meandering streams and slash-and-burn cultivation, each event perhaps small in area but large in its cumulative influence when projected over thousands of years. Archaeological discoveries suggest that the early-twentieth-century depiction of humans in the Amazon Basin—sparse in numbers, marginal in resources—needs rethinking. Excavations now substantiate the report by the earliest explorers that populations along the major rivers were dense and founded on agriculture, including maize. It may be that war, slaving, and diseases instigated a population collapse, not unlike what occurred in Australia and North America, such that much of the forests may be the result of human removal in the decades following the Conquest.

There is plenty of anecdotal evidence from the historic record to suggest that this interpretation deserves consideration. So does historical analogy to better-documented sites in North America and Australasia. And so does anthropological research. Accounts of early explorers describe fire hunting in campos, broadcast burning to convert slashed-and-burned sites into grassy hunting grounds, fire used to harass enemies, to cover raids, to open paths through tall grass and bamboo, and to announce entry into friendly territory. Commentators such as the nineteenth-century Danish botanists Eugene Warming and Peter Lund debated whether, in the face of overwhelming burning, those fires had been sufficient to create, not merely sustain, the extensive campos and cerrados of Brazil. Recent ethnographic studies of cerrado tribes such as the Kayapo demonstrate how these people reshaped the landscape into patches of forest, planted shrub, and grass, and confirms that they did so with liberal doses of fire, often through annual burning. Of 120 plant species around a typical encampment, 118 are used—surely no accident. Where slash-and-burn agriculture was practiced, it injected regular disturbances and exotic species into the scene, percolating the biota. Studies have suggested that, in southern Pará, perhaps 40 percent of the fabled botanical biodiversity is attributable to anthropogenic disturbance, an impact not possible without fire. The Xingu have explained simply that "fire is life."

What has been lacking, however, is hard data of the sort abundant in the northern hemisphere—other than buried charcoal it may not exist. But here also analogies argue for greater complexity. Consider, for example, the fascinating study of charcoal from sediment cores off the Pacific coast of Mesoamerica, probably a paradigm for most of the Americas. The greatest abundance appeared in the decades prior to the Conquest, begun in

1522. Throughout the region it is estimated that the indigenous population collapsed to 5 to 10 percent of its precontact levels; the forest reclaimed swiddened sites; introduced cattle overgrazed savannas and allowed further reforestation; charcoal flux plummeted until 1770 to 1860, the period of record lows, before population increases revived burning for forest clearing. Even today charcoal levels have not returned to preconquest levels, although this also reflects changes in land use for which other technologies have replaced fire.

It is more likely than not that what occurred elsewhere in Latin America also occurred in Brazil. Humans could not repel the postglacial climatic changes completely, particularly where those changes broke down the seasonality that made burning possible; but they could selectively resist those pressures, as they were manifest in the biota; they could modify the revanchist rainforest in parts; they could burn Brazil's grasslands, like grasslands everywhere, as fully as climate and technologies permitted. If humans could not overthrow the emerging biota, neither could it expel humans. What Europeans found was less a state of nature than a landscape under constant negotiation among climate, biota, and pre-Columbian peoples.

That encounter began with a few Spanish conquistadores who crossed the interior, but the Portuguese who settled the coast made the enduring impact. Brazil's rivers allowed the Portuguese to project their influence far beyond the coastal fringe. Diseases; slaving expeditions by *bandeirantes;* attempts to organize the indigenes into missions; wars; gold and diamond rushes; other forays into the *sertão,* as the backcountry was known—all unsettled the demographics of Brazil, and removed for many biomes the contriving hand of humans.

Contact also exposed Brazil to Europe's biological allies, not only disease but also cultigens and livestock. Cattle spread up the São Francisco River with effects that were undoubtedly profound but little recorded. In some cases herders merely displaced hunters, and livestock wild deer and jaguar. Elsewhere herders imposed a new order. The French botanist Netto remarked of Minas Gerais (1866) that the herders were so convinced of fire's value that they "enforced the burning over of as much land as they could." The character of agriculture also changed. Export commodities such as sugar (later coffee) eventually replaced manioc, and large plantations supplanted the smaller fields of shifting cultivators. When Indians died off or fled, African slaves were imported to work the fields. Exotic plants and animals mixed Brazil's biota as much as immigration, slavery, and intermarriage mixed its races. Settlements intensified and spread along the Atlantic and into select sites of the interior.

The land was remade, in some places by direct conversion, in others through the disruption or extermination of the prior regime. But everywhere that process involved fire—fire to convert forest and fallow to field, fire to hunt, fire to forage, fire to strip cover for prospecting, fire to wage war against enemies, fire to boil down cane into sugar, fire to flush pastures, fire to clean every niche and nook.

For the biotic medley that Brazil became, fire was at once a weed, a wedge, and a weapon. Brazil lacks the kind of vicious dry winds that unexpectedly whip calm burns into conflagrations, so only occasionally, as they did in Paraná in 1963, do large fires erupt. Instead fires were small and ubiquitous. The exurbance of the Brazilian biota—the avidity with which it seizes open sites, the shock between it and European cultigens—commended fire as a means of conversion and all but commanded its use to check a biotic counterrevolution. The commitment to a rural existence committed Brazil to queimadas.

Collectively these practices made a composite fire, the queimada para limpeza. Like the Brazilian biota, like its peoples, Brazil's fire was a hybrid. The French naturalist Auguste St. Hilaire studied it during his travels in the 1820s, detailing how it served the rural countryside. He concluded that it originated from the fire practices of the Brazilian Indian, then assumed its modern shape by interbreeding with compatible fires from Portugal and Africa. Certainly it adapted to the roles it was given and the places it occupied. It was everywhere different—and everywhere the same.

Its extraordinary universality has rendered the queimada both benign and malignant. It jolts old pasture out of dormancy, readying it for new growth. It purges fields of debris and the vermin that feed upon it, preparing them for planting. It removes woody waste, whether felled or regrown. Above all it is, as its name claims, a cleaning fire. What Americans do with brooms, rakes, shears, and garbage cans, Brazilians do with fire. Fire season is spring-cleaning on a continental scale. The fires dust, polish, sweep; they scrub, tile by tile, the vast mosaic that is rural Brazil. Without fires the land would be uninhabitable.

Queimadas vary considerably by time and place. Early in the dry season, burns are mild and spotty. Some creep along the surface, felling without consuming long stalks of grass; some catch a freshening breeze and fan into a rush of flame. Night fires burn more quiescently than daytime fires. Dense fuels flare more than sparse, dry fuels more than wet, hillsides more than valleys. Each burn recalibrates the remaining fuels and thus defines the prospects for subsequent fires. As the dry season ripens, the landscape percolates with smoke—now here, now there, some larger, some smaller, some flaming, some sour with the smell of dying fire—until finally, as the rains approach, the fires boil away any residual fuels.

That is the bright side of burning. The dark side has to do with more fundamental questions about appropriate land use and the reflex use of fire for almost any and every purpose. The problem of fire control thus resembles gun control. Fires, as it were, don't kill forests; people kill forests. Rural life dictates the use of fire, not fire the manner of rural existence. But the means do influence the ends. Without fire landclearing would quickly cease. Without fire overgrazing and soil abuse would slow. Without fire the conversion of wild land to rural land and the destructive intensification of rural economies would be difficult. Likely they would be impossible.

The problem is less fire per se than the capacity of fire to interact with other human practices. An infusion of random fire, suddenly dumped on an unwilling landscape, can be the means of a hostile takeover, a biotic buyout by unsecured capital leveraged by junk burns. Once wedged into a site, fire and grazing, for example, can split open a biota. Besides, broadcast fire is a folk art, tempered by trial-and-error prescriptions developed over the course of centuries. Too often, if wrenched outside the traditions that guide its use, burning becomes mindless and promiscuous; the same act repeats itself in too many contexts. Garbage, piled and burned. Roadside debris, piled and burned. Agricultural waste, piled and burned. Forests, piled and burned. Revealingly, the annual burning reaches a climax on August 24, St. Bartholomew's Day, the day the massacre of the Huguenots commenced.

Still, the residue of rural Brazil has to go somewhere. If it is not burned, then it must be removed and disposed of by other means. That can require a restructuring of rural life as pervasive as its reliance on fire. The United States offers at best a compromised counterexample. Surveys of rural America a century ago—say, C. S. Sargent's treatise on forests for the 1880 census, or Franklin Hough's *Report on Forestry*—map a geography of burning that is often indistinguishable from that in contemporary Brazil. In fact, because they regularly coincided with outbreaks of extreme weather, agricultural fires in the United States often escalated into conflagrations far more damaging than Brazil's queimadas. The reservation of public lands and the transformation of America to an industrial state, combined with aggressive fire-control programs, gradually swept open burning from the landscape at large and herded it, along with Indians and bison, onto special reservations.

But combustion has not been eliminated, and the consequences of its abrupt suppression have often been damaging. Older residents of metropolitan Phoenix, for example, can recall how, in the spring, they burned off their dead lawns. New growth quickly shot upward, and a timely burn prevented an unwanted fire set by accident or vandals. Now fire codes prohibit open burning as a hazard; air quality boards condemn its smoke as a pollutant. Accordingly, suburbanites first mow the old grass, then rent a power rake to gouge up the dead debris and scarify the ground. They stash the residue in scores of plastic bags. Sanitation crews collect the bags and deposit them in clogged landfills, where they slowly decompose into methane. The practice makes the queimada look environmentally as well as economically benevolent by comparison. The issue is less the queimada in and of itself than the context within which it coexists.

The requisite change in context came when Brazil, responding to global prods, determined to colonize its immense interior. The international market for commodities such as rubber was one stimulant; geopolitics was another. The surge to remote frontiers in the postwar era—a reflex that sent Australians to their tropical north, Americans to Alaska, Canadians to the Arctic archipelago, Soviets into Siberia, Indonesians to Borneo—sent Brazilians into the planalto and Amazonia. The government declared its intention to secure the lightly populated interior. To symbolize this resolve it relocated the capital from Rio de Janeiro to Brasília, a new city. Then, during a period of military rule, it began to construct a network of roads around and through its Amazonian borders. It was hoped that this would strengthen its frontiers, relieve population pressures (particularly in the calamity-prone northeast), and begin to convert natural resources into national wealth.

What happened was something else. Colonists followed the roads, but not in the numbers required, or from the regions expected. Much of the land was unsuitable for sedentary agriculture and had to be abandoned. Rocketing inflation encouraged land conversion into large ranches. Miners and loggers conflicted with Indian tribes, previously immune from direct contact. As Brazil converted back to civilian rule, the full magnitude of the experiment became apparent, and Brazil's geopolitical impetus to open up Amazonia inspired a global outcry to close it down. Deforestation on this scale, it was charged, ruined the planet's biodiversity, devastated the cultural diversity of Brazil's native peoples, and threatened the climate of both Amazonia and perhaps the earth.

Fire illuminated the ruling images of protest. Burning made the forest slashing vivid. Flames highlighted the conversion of forest to pasture in ways that pictures of Brahma cattle chewing their cud could not. Smoke clouds that obscured Amazonia paradoxically made the region visible to those who shared the planetary atmosphere. With exquisite timing, in 1987 Alberto Setzer, a scientist at Brazil's national institute for space research (INPE), developed a methodology to transform infrared data from the American NOAA satellite into a daily map of active fires. Each day at 3:00 P.M. the satellite's infrared bands recorded the hottest spots on pixels laid out in a 1-square-kilometer grid, thereby sampling, not inventorying, the fire load in Brazil. When the system allowed precise geographic locations of the fires to be sent to responsible agencies within an hour of satellite overpass, the mechanism existed to take at least symbolic action against illicit burning.

The effect was galvanic. It brought to the world proof that Amazonia was burning on a vast scale, contradicting official assurances that fires were few and unimportant and confirming, it appeared, warnings that the earth had crossed the threshold of a greenhouse effect. When a major drought struck Amazonia in 1987 and 1988, the fires became a global scandal. Satellite monitoring of the burns gave protesters the graphic imagery they required. Space shuttle astronauts took photos of smoke plumes the size of Europe. Wildfires in America, culminating in the Yellowstone conflagrations, gave the American public a powerful if compromised point of comparison. The fires linked Brazil with Europe and North America by reducing the planet's biodiversity to ash, poisoning the common atmosphere, making vivid otherwise abstruse points of ecology and philosophy, and establishing pathways of information. They did for environmentalism what MTV did for popular music. Brazil's fires became the pilot flames of apocalypse.

What the program meant for Brazil was more ambiguous. The practical effectiveness of satellite surveillance depended, of course, on the presence of an infrastructure, which did not exist, to manage fire. What the algorithms record is also ambiguous. Inside the INPE facility at Cachoeira

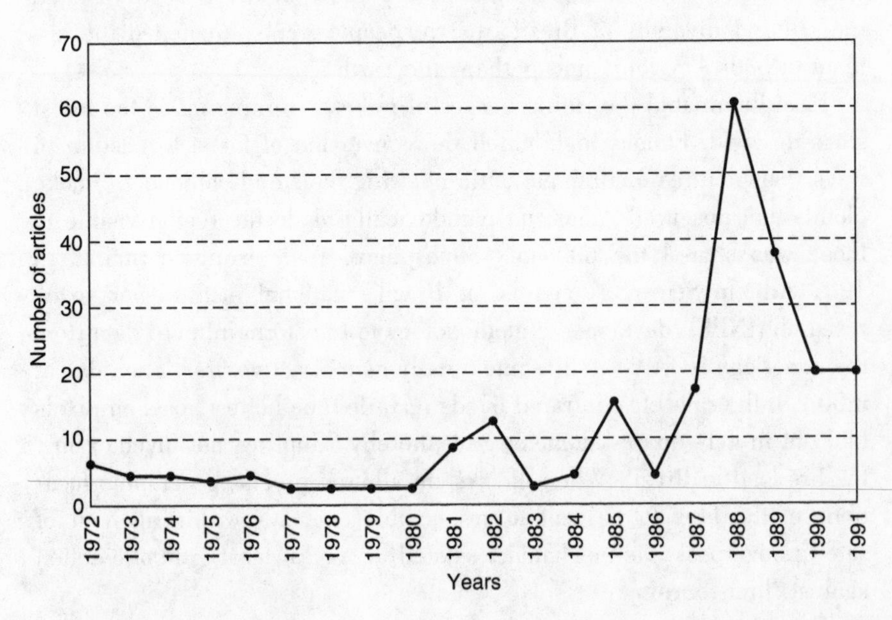

Profile of a world fire: Brazilian media articles on Brazilian burning, both forest fires and queimadas.

Paulista, a computer screen lights up with bright dots, each registering a fire sighted by remote sensors in earth orbit; outside, smoke from a dozen unrecorded queimadas floods the valley. The power of the satellite detection lay not in its record of individual fires so much as its identification of a fire problem, or what was perceived as one.

Both pushed and pulled, Brazil moved to take action to reduce the fires. Its capacity to manage fire on a level beyond folkways, however, was minimal. Only the largest urban centers had a modern firefighting force, and only industrial forest plantations were equipped to control fire on anything resembling wildlands. Elsewhere fire management meant the proper use of queimadas—burning fuelbreaks around protected sites, burning field and stubble, burning slash prior to planting, burning off the miscellaneous pockets of fuels that were otherwise unconsumed by beast or flame in towns, fazendas, and roadsides. Unlike former British or French colonies, Brazil had no tradition of forestry. It had, correspondingly, no resident intellectuals or bureaucrats hostile to fire but neither did it have any institutional apparatus to deal with it. Everything had to be created.

And everything had to be done amid crises that compounded as rapidly as Brazil's foreign debt. The slow conversion to civilian rule, galloping inflation, international furor over the reckless development of Amazonia, all converged to argue for change and to make change difficult. But by 1989, in response to world opinion, Brazil announced *Nossa Naturaleza,* "Our Nature," a program of environmental reform. It convened a national commission that reviewed the situation and recommended, among other measures, that Brazilian states regulate routine burning through a permit system, that government ministries reorganize, in particular that the Instituto Brasileiro do Meio Ambiente e dos Recursos Naturais Renovaveis (IBAMA) assume control over parks, forests, and fire in general through a program known as Sistema Nacional de Prevencão e Combate aos Incendios Florestais (PREVFOGO).

By decree President Sarney confirmed that proposal as official policy. This resulted in some confusion because the Brazilian constitution guaranteed responsibility for fire protection to legally constituted fire services, all of which are urban. (The national fire academy in Brasília subsequently expanded to include wildland tactics.) Its proposal to regulate rural burning was also unenforceable since there was no infrastructure by which to issue and oversee permits or network of rural fire brigades through which to regulate burning. The generic character of the queimada allowed critics to lump together indiscriminately every kind of burning and to demand a single solution. There was none. Meanwhile the state of São Paulo organized a fire-control effort (Operation Mata-fogo), and the World Bank contributed $8.5 million toward firefighting in the Legal Amazon, an area comprising roughly one third of Brazil, and discussions began between the governments of Brazil and the United States about a joint initiative regarding forest management. The initiative would of course begin with fire.

The suppression program started with an emergency plan that joined firefighters with the *policia federal* in what resembled a paramilitary operation. Helicopters and engines swarmed onto sites that were burning illegally, where they were sometimes met with bullets, near-fatal encounters that gave new meaning to "firefight." (Citing concerns that such operations would hurt its image, the Brazilian military declined to participate.) Burned acreage plummeted, although a wet season had more to do with the reduction than high-tech fire control. In 1991, following preliminary surveys, the U.S. Forest Service joined IBAMA in a program of fire research and technology transfer. Exchanges culminated the next year in a major symposium, 1st Seminario Nacional Sobre Incendios Florestais e Queimadas, the first time Brazilians had convened to discuss the full spectrum of fire in their national life. A workshop followed to plot out a strategy for national fire management and further exchanges. Two months later the U.N.-sponsored Earth Summit came to Rio, confirming the special place Brazil had assumed in the geopolitics of environmentalism, and the special place fire had claimed in Brazil.

In all this there was something surreal. Americans explained how fire management derived from land management, and how fire control was incomplete without fire use. Having long ago committed to suppression, thoughtful Americans were struggling to reestablish some system of deliberate burning. Brazilians explained how fire use could no longer substitute for fire control. Immersed in queimadas, they longed for some measure of systematic regulation, if not for what that meant in the field then at least for what it brought in redeemed national pride. What they heard differed from what they believed and what they understood differed, because of international scrutiny, from what they could do.

Media attention promised to transform proposals and programs into a *telenovela.* Each nation was photogenic for the wrong reasons—America for its air tankers, Brazil for its slash fires. A journalist from São Paulo lectured the symposium's fire specialists that Brazilians could not accept a complicated message, that it could accept only the strict abolition of fire. (Why Brazilians could discriminate between a river in a flood and a field watered by sprinklers but not between a wildfire and a queimada was never explained. Why fire-rich Brazil would want to imitate a country like fire-poor Britain and pursue a program of biotic impoverishment was not elaborated.) An indiscriminate censure of fire was pushing officials into an indiscriminate attack against fire.

Brazil's landscape of fire was more complicated. It included fire-assisted land conversion in Amazonia; the air-blackening burning of sugar cane in São Paulo; wildfires in industrial plantations of pine and eucalypt; underburning for regeneration of *Araucaria;* patch burning for the maintenance of cerrado; drought-driven fires in the organic soils of nature preserves; slash-and-burn cultivation in secondary forest and caatinga; pastoral burning in tropical grasses; fires along the urban fringe of wooden slums; agricultural cleaning fires before or after harvest; generic queimadas almost everywhere. All could be subsumed under a common rubric of fire management, but not all required the same response or could be regulated through the same institutions or according to the same prescriptions.

Fire management is a means, not an end. In some instances, effective fire control would mean de facto control over land practices. But it is more likely that fire control will be ineffective until land use is regulated. This is almost certainly the case with fire in Greater Amazonia. In other instances, attempted fire exclusion would probably destroy lands, most quickly those in the cerrado or campos set up as nature reserves. To concentrate on charismatic—that is, photogenic—sites, moreover, ignores the pressing need for some system of rural fire protection. The problem is to determine the proper regimen of fire and then to create institutions that can enforce it. What Brazil requires is appropriate policy, something that can protect the legitimate uses of fire as fully as it can attack the reckless ones. What it needs for implementation is an appropriate technology, something closer to Australia's bushfire brigades than to America's airtankers and hotshot crews.

The alternative to selective action is not failed fire management but irony. Shutting down forest conversion in Amazonia only diverts the process to other, equally rich biomes in Goiás or Mato Grosso. Eliminating patchy anthropogenic fires in nature reserves only sets up the prospect for large, high-intensity fires ignited by lightning, which happened to the Parc Nacional das Emas, where several years of fire exclusion ended with a lightning strike that seared 85 percent of the park in one fiery flash. Replacing slash-and-burn agriculture with industrial forests in Pará has led to charcoal production on a monumental scale, a worsening of problem emissions. And so it goes. Land management cannot succeed without fire management, but fire management is faulty and disruptive, a torch waved blindly, without a basis in land management.

It is no easy task to create a fire management system for half a continent. By the 1990s Brazil was progressing toward an integrated fire program in ways that would have been unthinkable a decade before. An infrastructure was slowly emerging, and through scientific enterprises such as the International Geosphere-Biosphere Program, Brazilian fires joined those on other continents, part of a global assessment of anthropogenic impacts on the planet. Amid massive political, economic, and social turmoil Brazil was struggling to accomplish in years what had taken the United States decades and the European powers several centuries.

It is in Brasília that the disparities between ambition and reality clash most fiercely. In the dry season the surrounding lands burn. The campos burn, the cerrado burns, the fazendas burn. Farms, ranches, parks, empty fields, even grassy yards—all burn. Fires creep into suburbs, sweep through the Parque Nacional de Brasília, lick the tarmac at the international airport. While ministries debate policy and presidents issue proclamations, smoke rises in every direction. A ring of quiescent fire surrounds the capital, shielded in part by the artificial lakes that curl around it like a moat. At the Praça dos Tres Poderes, the government has erected a monument to Tancredo Neves, celebrating the restoration of democracy and freedom, its centerpiece an eternal flame. In the distance, across the Lago Paranoda, queimadas proclaim their own monument to Brazil's true eternal flame.

But there is more. Brasília's designers plunked the brazen capital down on the *planalto central,* close to the arid focus of Brazil's dry environmental ellipse. Fires are abundant here because they have always been abundant in this landscape. In some form, at some frequency, they belong here. Remove fire willy-nilly and you will transform the scene as fully as injecting fire willy-nilly into Amazonias. Snuffing out queimadas would solve Brazil's fire problem in the same way that banning automobiles would solve America's air pollution problem.

Not enough is known about Brazilian fire ecology and history to determine an ideal formula for fire management, but there is sufficient data to declare that fire exclusion is not an answer. Too often fire, not the land use practices that kindled it, has become the object of international outcry and political pressure, and Brazil's success with the latter is judged by its ephemeral successes with the former. Yet rural Brazil, Brazil's indigenous tribes, and Brazil's nature reserves could not function without fire.

Fires will endure in Brazil not only because they can exist there but because they need to do so. This is as true for Brazil's emergent industrial economy as for its enduring rural society, a condition that Brasília, too, magnifies. Near its center rises an Eiffel Tower–like structure, the *torre televisor.* It signifies, as it was intended to broadcast, Brazil's claim to status as a great cultural and economic power. It aspired to epitomize the industrialization of Brazil, the opening of the Brazilian sertão. At the base of the tower vendors sell souvenirs made from the shrubs, trees, and grasses of the surrounding cerrado. The most spectacular products, often exported, are the flowers—orchids, especially—that blossom only after the season's burns.

Svedjebruk (Sweden)

Sweden's difficulty with wildland fire, a problem that grows more serious annually, is that it doesn't have much.

The triumph of Swedish forestry was its adoption of the silvicultural ideals of central Europe and projection of them into the boreal forest, its transformation of *vildmark* into *trädgård,* wildland into garden, and its placement of forest wealth in the service of a social democratic state. With its forests, Sweden industrialized, and from its forests has come the surplus capital necessary to finance the fabled socialism that has made Sweden synonymous with affluence, peace, and security. Along the way the order of Swedish nature took on the order of that society. The country became a vast tree farm.

There seemed to be no place, or need, for fire. A disciplined and stable population, a relatively benign fire climate, a commitment to intensive silviculture—with them, Sweden had apparently divided and conquered its fire problem. It segregated fuels from ignitions, and then reduced each to infinitesimal quantities. Lightning fire became insignificant, starved of its traditional fuels; ancient fire practices were forgotten, no more relevant than the folklore of spruce harrow and *gaffelplog;* wildfire shrank to trivial proportions, as though Swedish nature, too, had withdrawn from confrontation and violence. By the mid-1980s Sweden no longer even kept statistics for wildland fires. The forest volume prepared for the National Atlas declared in 1990 that the "damage which is of the greatest significance" in Sweden is caused by "storm, fungi, moose, and insects"; only "previously" had fire been important.

Elsewhere—both in Europe and in Europe's colonies—it could be argued that European ideals had been perverted by global capitalism, wars, totalitarian politics, imperial delusions, and recalcitrant soils and rebellious floras. But Sweden showed what this ideal might look like if such distortions were smoothed away. Here, along Europe's northern frontier, a stable society could fashion a sustainable landscape. Here nature could convert to something like Swedish socialism: planned, equitable, secure. Here European ideals of stewardship, the governance both of peoples and of lands, could receive a fair test. That test, somewhat unexpectedly, has included fire.

The suppression of Swedish fire is, in fact, a recent phenomenon. The landscapes of modern Sweden have endured fire, even craved fire, since the land emerged from under glacial ice and inland sea, like a whale rising from the ocean, over the course of the Holocene. The oldest of forests then colonized the newest of lands. The boreal forest is ancient, far more so than the tropical forest, and the species that immigrated to Scandinavia in the wake of the last glaciation brought with them long eons of fire-tempered adaptations. They would need them to survive.

The geography of Swedish fire played out across a complex milieu of wet and dry—some of that polarity located in the atmosphere, and some on the earth. Sweden ranges across a powerful gradient of latitudes, its southern lands sharing in the classic European deciduous forest, its northern *fjäll* thrusting above the timberline and beyond the Arctic Circle. Climatically, the Scandinavian peninsula forms a mobile frontier between two great weather regimes, one maritime and the other continental. The first brought moisture and relative warmth from a North Atlantic modulated by the long reach of the Gulf Stream; the second pushed the drier air of the Eurasian landmass westward. In various seasons one or the other prevailed. But this relative dominance also played out over decades as well as years—and sometimes for centuries, or even millennia. The two regimes ebbed and flowed like climatic tides, washing over an emergent land that recorded those changes in an ecological palimpsest.

The geography of wet and dry was no less embedded in an earth fragmented into land and sea, hill and lake, forest and mire. The landscape inscribed a vast mosaic of fire-prone and fire-immune sites. But these are relative designations. They could, and did, change with the seasons or with secular bouts of drought and deluge. What was incombustible one year could burn the next; what burned one summer might smother ignition for the subsequent century. A mire could repel fire during wet times or when flushed with spring melt. In dry seasons it could hold a smoldering burn for weeks, or months, a slow match by which to rekindle surrounding shrub and woods as conditions permitted. A spruce forest could establish itself during wet times and burn during dry. Heath could flourish in dry centuries, swept by periodic fire, then be reclaimed by trees as wet weather returned and fires faded.

Swedish geography thus favors fires, but in a broken, constantly checkered pattern. The fragmented fuels prevented the kind of conflagrations typical in the boreal forests of Canada or Siberia; even crown fires tended to be local; a major fire year would see many intense but relatively small burns rather than a few gargantuan blazes. The climate, too, worked to temper the prospects for wildland fire. Large fires raged probably once a century in the north, perhaps more often in the milder, more exuberant south. Overall it is estimated that about 1 percent of the boreal forest burned annually in prehistoric times.

Such figures alone do not adequately convey the importance of wildland fires, which must be measured relative to the biological work demanded of them. In a land where the time for growth is short, and where rates of decomposition are even slower, fire is a critical catalyst, an accelerant. Fire decomposes rapidly what fungi and insects might require decades or centuries to do. To forests in which the dominant trees can live for a millennium, a major fire once a century can have an ecological impact equivalent to decadal fires in heath or to annual burning in moist prairie. A single season's burning of peat may erase several centuries of accumulation; it may instantly eradicate spruce, except in fire-spared refugia; properly recurring, it can preserve pine and ling indefinitely. By altering the local environment—by exposing the land to more sunlight, by enhancing evaporation, by varying pH, by stripping away insulating humus, by restructuring the relationships among species—a fire can counteract some of the impacts of a deteriorating climate. Conversely the absence of fire allows those processes to advance unchecked. The biota marches inexorably toward spruce and mire.

Certainly the potential for natural fires exists. Where the two climatic regimes clash, fires follow. Thunderstorms ignite fires annually, more abundantly in the south where storm fronts pass than in the north, where moisture must seep into a kilnlike dome of high pressure and coagulate into local thunderstorms, like dew condensing on a cold glass. Still, large fire busts can result, as they did in 1914 and 1933. The evolutionary pressures that encoded themselves into the genetic structure of the boreal biota persist. But they have, in practical terms, become almost trivial in comparison with the colossal manipulation of the environment by humans. The fire history of Sweden has become overwhelmingly the story of anthropogenic fire.

The mixed geography of Swedish fire thus coevolved with humans who, in different places at various rates and with assorted means, profoundly restructured its ensemble of wet and dry. From the beginning there were humans present, outfitted with the initial technologies of the Neolithic revolution, tramping along the fast-marching frontiers of retreating ice and falling sea, as opportunistic as the birch, alder, and pine that seized the disturbed landscape. Unlike them, humans could, through fire, prolong the disturbances. Over the centuries their successors converted forests to fields, drained mires, harvested peat, grazed meadow and marsh, built reservoirs, introduced exotic flora and fauna, constructed roads, logged woodlands, transformed forests to heath and then replanted heath to forests. Every niche experienced some kind of change. Within limits anthropogenic fire could halt spruce invasion, reverse mire formation, inhibit paludification, counteract lake acidification, and stimulate a host of other challenges to what humans considered to be a degrading climate. Only by such means could they, in the end, colonize the new lands.

Without fire that land was virtually uninhabitable. The torch shielded humans from ice, cold, moisture, and the endless winter darkness. People lived in tents or houses to preserve the nurturing fire. Fire heated house and sauna; smoke drove away pestilential insects; flame warded off wolf and witch. Torches helped hunt for bear and assisted fishing in the evening. Fire promoted foraging for berries and herbs, and (for Finns) the harvesting of mushrooms. Pastoral burning penetrated heath, forest, mire, and meadow; through their livestock, humans projected their influence and their fires far beyond the limited range of settlements; and the institution of summer grazing at remote cabins, a kind of transhumance, extended their range of influence even further. Slash-and-burn agriculture reworked forest and peatland. Anthropogenic fires made possible the essential technologies that integrated the region into national and international markets—fire mining; charcoal for smelting and ironworks; tar and turpentine for naval stores; potash and saltpeter for a primitive chemical industry.

Their fires defined relationships not only between humans and the land but between humans. Lapps, Finns, and Swedes all at one time or another used a scorched earth to define the limits of their claims and to prevent encroachment by miners, cattle, or reindeer. Armies fought with fire sleds, fire ships, fire arrows, and fire missiles flung from catapults. Blackmail often took the form of a "fire-tax," with money extorted to forestall arson. The response to plague was prophylactic fire, not only to infected houses but to the countryside. Fire towers, bonfires, or special trees torched on cue announced vital news or warned of new Russian invasions along the Gulf of Bothnia. The fiery cross called villagers to meeting. Ritual fire attended the major ceremonies of Swedish life, from the candle crowns of St. Lucia Day to the great bonfires of Beltane and *Midsommarsdagen.*

People penetrated everywhere, and they forced their fires to compete with those of lightning. The evolved landscapes of Sweden were those that fire, or those that fire-assisted agriculture and industry, helped bring into being. In the south, fire ranged routinely in extensive heaths, the stubble of cultivated fields, forage-producing bogs, and swaled forests; in the north, among grazed forests and mires; and almost everywhere in the many manifestations of fire-based cultivation. New fire regimes proliferated not solely through new ignition patterns but by the interaction of fire with other processes such as grazing and draining. Some provinces of Norrland, for example, once had a third or more of their landscape under mires, sites that when drained became exposed to new kinds of burning.

The range of burning was vast in sum, but intricate in practice. A new mosaic replaced the old one, partly because the emergent anthropogenic landscapes built upon and retained some elements of the old, partly because landownership and use by farmer and village favored innumerable subdivisions. Still, some burning could be extensive, or become large through fires that escaped into surrounding heath or forest. Most were more delicate, like etchings on Orrefors glass. A slope was burned here, a berry patch there; here a farmer fired the slash from a new swidden, there a fallow field, still elsewhere his pastures—an expanse of *Calluna* heath; the blankets of grass, moss, and *ljung* beneath birch and alder forests; the grass-moss-ljung cover beneath pine. Settlers even burned along the slopes of small lakes so that a slight nutrient flush would descend into the slack waters and stimulate sedges useful as fodder or thatch. Almost every component of the biota could be refashioned with some kind of fire, or became exposed to new fire sources that resulted in novel fire regimes. Anthropogenic fire worked and reworked the landscape like lumps of bog iron forged and hammered and fired again into swords of Swedish steel.

But perhaps the most vital, and most characteristic, manifestation of folk fire culture was the practice of slash-and-burn cultivation known as *svedjebruk.* This was the most fundamental expression of environmental burning because it first broke the land to agriculture. *Svedjor,* the sites once felled and fired, supplied the tiles from which a radically different landscape mosaic could evolve, the pieces with which Swedish agriculture could reshape the environment to arable field, meadow, fallow, forest pasture, and shifting cultivation. It made possible the recycling of landscapes.

So prominent was the practice that a number of European observers believed that *sveden* was the origin of "Sweden," that svedjebruk, which had so shaped the countryside, had given one of its names to the country as a whole. Certainly innumerable sites recorded its presence in place-names, towns called Svedja or sites known as Brännland. Certainly, too, *svedje* made Sweden possible.

If svedjebruk remade the land, it is also true that *svedjening* assumed many of its characteristics from the lands it touched. In Scandinavia it assumed two general forms, one typical of Swedes, the other of Finns. Swedish svedjebruk served grazing as much as cereal cultivation; Finnish svedje, primarily the cultivation of rye and turnips. Both obeyed a similar rhythm of cutting, firing, planting, refiring, grazing, and fallow, followed by the repetition of the sequence that gives both expressions their close resemblance. Each, in turn, adapted its specific expressions to the landscapes, fuels, and peoples with which it interacted. One form characterized old-growth forest; others, various varieties of forest fallow; still others, mires, peatlands, and stony hills, any site with agricultural potential. The varieties were legion.

A *svedjed* landscape was complex and intricate, rife with biotic niches. In any one year settlers had an abundance of svedjor in various stages of development—some felled; some freshly burned; some put to rye, oats, barley, or turnips; some grazed; some reburned; some abandoned to fallow. There was room for most of the species that had existed before *svedjefolk* appeared, and room for the new species that followed and served them. But svedjebruk also preserved habitats, a diversity of landscapes and at least some of the processes, fire included, that had made them. If anything the diversity increased at any one site because svedjening speeded up the rates of change. What had burned in a century now burned in decades. Sweden became an environmental kaleidoscope that humans constantly turned with ax and fire.

The torch was in fact indispensable. Only by burning could the land be refashioned into usable forms; only with fire could vital secondary processes such as grazing by cattle be effective. In the absence of adequate animal manure, the mixed agriculture of central Europe was impossible, a condition not overcome in marginal regions until the industrial revolution made chemical fertilizers widely available. Until then burning assumed that function—and others. And although anthropogenic fire could only modulate, not overcome, climate, it helped define climatically "good"

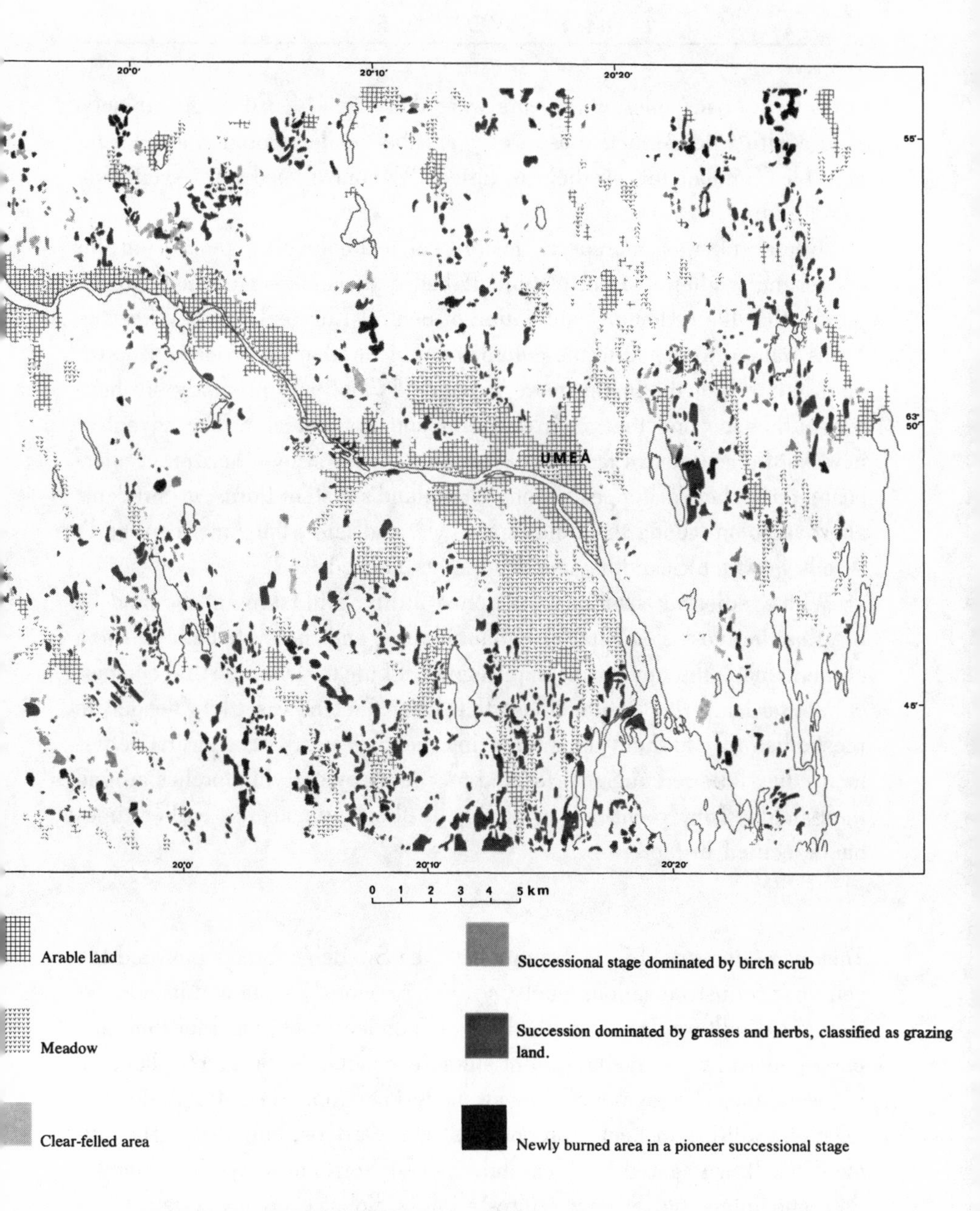

The geography of controlled burning: svedjebruk and recently burned pastoral lands around Umeå, northern Sweden, in the mid-nineteenth century. Already agricultural burning is in decline, especially pastoral burning, but the mosaic is clearly evident. Not included are mires regularly burned for fodder, or sites less than two hectares in size.

years from "bad" ones. Bad years were those in which drought replaced svedje with wildfire, and those wet years that made vigorous burning impossible. Without fire svedjebruk failed. Without a good burn svedjefolk faced famine.

Svedjebruk was, moreover, the critical technology for pioneering, for converting wildland to cultivation. It had a dynamic—a capacity to expand—that the sedentary cultivation of central Europe lacked. When the Vasas wanted to expedite the colonization of Sweden's interior and northlands, they brought in the more dynamic Finns, who promptly slashed-and-burned beyond the grasp of bailiff and tax collector. The advent of new Nordic settlers, in fact, was everywhere marked by a horizon of charcoal—on Iceland's steppes, among Greenland's distant fjords, at Sweden's abortive colony along the Delaware River, and later, the American North Woods to which Scandinavian immigrants flocked.

With svedjebruk symbolism matched utility. The taking of the land by fire was the most elemental of human claims, and the systematic burning of land among the most basic expressions of human dominion and ecological stewardship. It is no accident that when Icelandic settlers needed to reestablish title to their lands following the sixty-year chaos of first settlement, they followed ancient Nordic precedents by carrying torches around their fields. They reenacted symbolically the landnam that earlier times had executed in fact.

Then came forestry. To a substantial degree Sweden's forests powered the country's industrialization, much as it had previously sustained its conversion to agriculture. Ironworks had long depended on charcoal for fuel, and been limited by the availability of suitable woodlands. Sweden's chemical industry, too, had extracted its raw materials from trees. But industrial logging, swelling to hectic proportions in the 1870s, added a further dimension. It aggregated fresh capital, opened Norrland to serious exploitation, and integrated Sweden's forests into a global economy. First as saw timber, then as pulp, industry found new values in the forest, and the forest's wealth, in turn, helped transform Swedish society; between 50 and 70 percent of Sweden's export income derived from forest products. The logger's ax, however, struck Sweden's forests like Thor's hammer. It shattered the old mosaic, leaving behind a broken landscape like a wall of smashed and mottled plaster.

The geography of fire was, once again, restructured in fundamental ways. In southern Sweden, forestry accompanied a slow process of agricultural abandonment, a shift from farm, grazed forest, and heath to planted woodland. Forests became literally tree farms, for which fire had no more value than in fields of growing barley. In central Sweden, forestry redefined the relationships among competing economies, suppressing traditional agricultural burning in an effort to promote industrial logging; the new economics demanded that trees go to lumber, or at least pulp, rather than to ash, fallow, tar, and charcoal. The economic calculus of industrial Sweden redefined svedjebruk from an asset to a liability. It saw svedjebruk not as transforming useless wildland into farms but as perpetuating an antiquated agriculture that wasted wood and fiber. In Norrland, logging competed with fires, deliberate and wild, for wood. Old practices encountered new fuels, and new fire sources fingered into previously inaccessible kindling. Large fires swept much of Norrland in 1851, 1868, 1878, and 1888, some of which burned down local villages and even towns such as Umeå.

But these new eruptions, however individually spectacular, played out against a larger implosion in which the amount of burned lands experienced a dramatic collapse. Industrial settlement drained fire from the forest the way it drained water out of mires. Then, beginning in 1903, restrictive forest legislation brought all wooded lands of all manner of ownership into active production, fusing forestry with farming. A minuscule 2 percent of Sweden's wildlands remained outside the forest economy, mostly in lands unsuitable for silviculture. Industrial forestry, which redefined Swedish land use perhaps more completely than any practice since the Neolithic revolution, did not want, and did not believe it needed, fire to assist its exploitation. Free-burning fire became anathema.

The expulsion of fire proceeded faster in the southern provinces than in Norrland. Mechanization made possible the intensive utilization demanded by the new political economy and predicated by silviculture, and it opened roads everywhere, which improved access for rapid attack on wildfire. But ideology also merged with economics. Academic training and the pervasive influence of German exemplars confirmed the belief that forestry meant the application of agronomic principles to trees, that Germanic traditions of "clean" forests—woodlands stripped of dead wood, snags, and understory—were a measure of good management.

It is interesting that the same word, *hushållning,* applied equally to housekeeping and to land use, including forestry. Dead wood no more belonged in the forest than refuse in a kitchen or oily rags in a basement corner. Even the legendary Linnaeus had to bow to the tyranny. When he wrote that svedjebruk had value in the harsh lands of Småland, Baron Hårlman, an advocate of manure-based agriculture, demanded he remove the offending passages from his published Skåne journal. Linnaeus was forced to agree. A century later forestry was prepared to repeat that official censorship. Sweden's formidable gradient of latitudes would become a gradient of cultivation and of combustion. The agriculture of Skåne would become the silviculture of Norrland.

Still, compromises existed. Too much of the traditional economy, nature's and society's both, had depended on controlled fire to eliminate it overnight. The labor force that practiced forestry had a rural heritage that included burning. The most spectacular, and obvious, adaptation involved the reworking of slash-and-burn agriculture into slash-and-burn forestry, the reformulation of svedjebruk into *hyggesbränning.* Even in the nineteenth century observers saw readily that the best Scots pine grew on sites that had earlier burned. They believed, however, that they could tame the wild elements of fire, break it to the halter of civilization; that they could sow pine in place of rye. With remarkable fidelity to ancient mores, they modified the techniques of svedjebruk to serve the purposes of forestry. Hyggesbränning could assist silviculture as svedjebruk had agriculture; it could pioneer new lands and bring into production sites otherwise unusable. Those revisions, not incidentally, helped preserve the plentiful species and something of the ancient landscapes that generations of svedjefolk had fashioned.

Intellectual enthusiasm for hyggesbränning peaked during the 1940s, then paused, reached its greatest geographic extent in the 1950s, then rapidly imploded into insignificance. In 1958, some 17,400 hectares in state forests experienced hyggesbränning; in 1988, only 131 hectares. The reasons to support it had long been obvious—the flush of self-regenerated forest that sprouted after a suitable fire. The causes of its collapse are

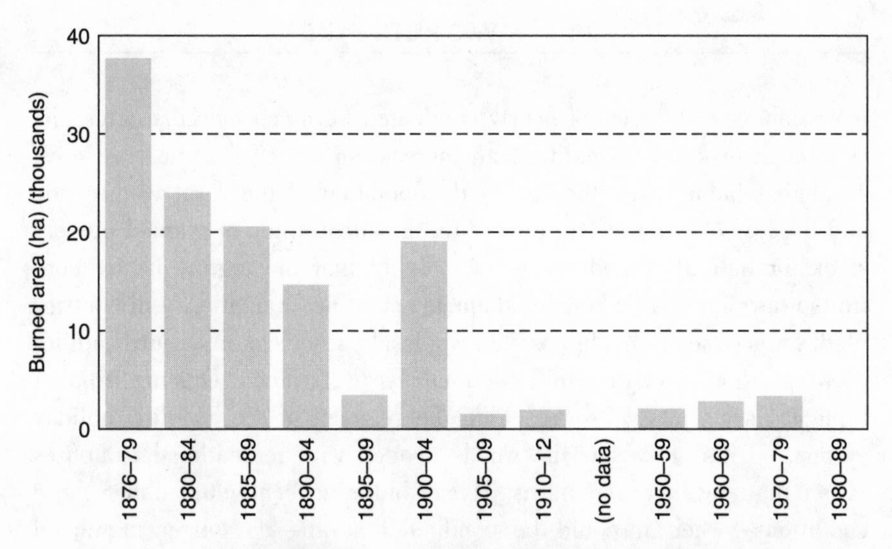

The suppression of Swedish fire: a century of wildfire diminution. The graph does not, however, incorporate controlled burning, which made up some of the deficit until the 1950s when it, too, succumbed to intensive silviculture.

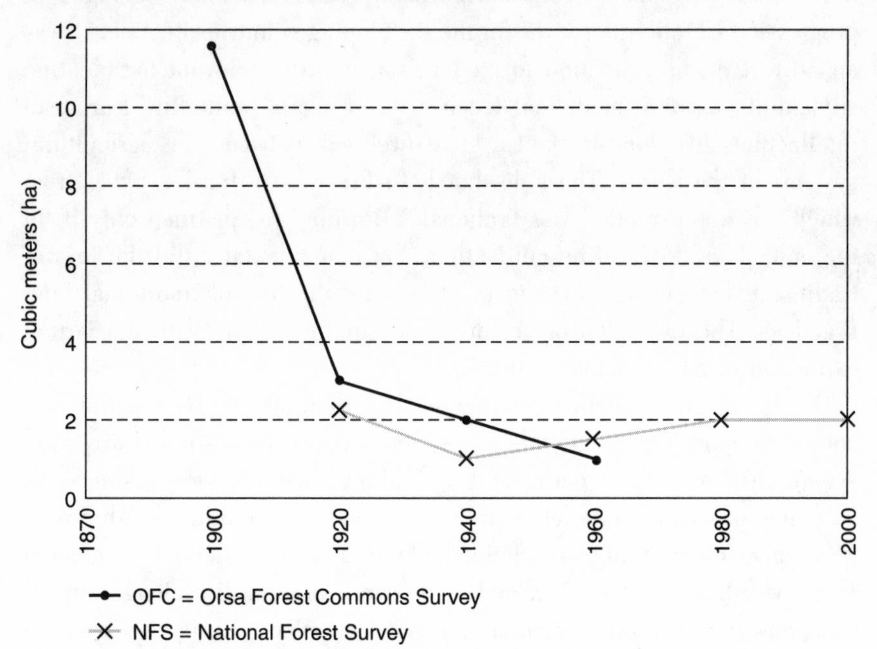

Silviculture cleans house: the reduction of dead wood in Swedish forests as measured by the volume of standing dead trees. This particular site is the Orsa Forest Commons in Jämtland.

more complex. A series of poorly conducted burns on marginal sites, the retirement of key personalities, an increase in the size of clearcut sites, escalating labor costs, the steady depopulation of the rural countryside and the loss of cultural lore about fire, the substitution of planted nursery stock for natural reseeding, and the belief that the arguments for controlled burning still lacked an adequate scientific foundation—all contributed. Large tractors made possible wholesale scarification, a surrogate for plowing, thus strengthening the analogy to farming. Organized labor brought more forestry workers into the regimen of the Swedish holiday system, so that July saw the woods flooded with recreational campfires rather than regeneration burns. Hyggesbränning conducted under these conditions—when it could be conducted at all—no longer mimicked svedjebruk, and no longer yielded its good results.

Perhaps more decisively, the postwar green revolution in agriculture came to Swedish forestry. Hyggesbränning reeked of the archaic, a folk art rather than a science. Slash-and-burn cultivation no more belonged in progressive Europe than it did in the developing countries of Asia. It was an artifact of the transition to modernism, as little relevant to scientific silviculture as were medicinal leeches in an age of penicillin. European intellectuals had long hacked a great firebreak between the agricultural systems of the world. The side that held fire was "primitive"; that from which fire was excluded was "rational." Burning was justified only if, in the end, it led to fire-free cultivation. Not coincidentally did the loss of traditional landscapes go hand in glove with the loss of traditional land practices. The exclusion of fire occurred simultaneously with the virtual extinction of Swedish folk culture.

Forestry agreed, both in principle and in practice. Hyggesbränning could no more serve as the basis for a permanent silviculture than svedjebruk could for a permanent agriculture. Instead, foresters accelerated the transplantation of European agronomic techniques—draining, fertilizing, clearcutting, scarifying, planting with genetically selected stock, weeding and thinning, and of course excluding fire along with insects and growth-retarding fungi. Society demanded security; security required stability; and sedentary agriculture, or its forest cognate, sedentary silviculture, best served a society so ordered.

The dominance of social values was absolute. Nature existed to serve society. Harvested sites rich in slash, as was true with svedjor, were best burned in July, but this put fire in competition with *semester,* the Swedish vacation month. Nature's seasons would have to adjust to society's calendar. Holidays took precedence over hyggesbränning.

The reconstitution of the Swedish environment was pervasive, from the shores of the Baltic to the alpine tundra of the uplands, on private land and public, on what had once been field and forest, marsh and meadow. In a sense, the era replayed the saga of *Storskiftesstadgan*—the great consolidation and redistribution of peasant landholdings, Sweden's version of Britain's enclosure movement, that had from the mid-eighteenth century to the middle nineteenth restructured the Swedish landscape and the society that created it. This time forestry took the place of the "new" agriculture. But while the same pieces were often refitted, they were now all shades of forest green, all parts of a vast plantation of trees, the mosaic of a monoculture. Every parcel of forest land was numbered, cataloged, and assessed, statistical fodder for the mighty maw of commercial forestry.

By 1990 only small fragments of former landscapes remained. Nature preservation was remote from political economics and social conscience, and preserves—little more than vestiges—belonged roughly on a par with the rune stones and log buildings carted off to open-air provincial museums. The largest preserves resided on high fjäll or offshore archipelagos, well beyond the grasp of forestry. Of twenty national parks, twelve were less than ten square miles in size, and seven less than one square mile. The great heathlands of Halland had vanished beneath forests; the preserved fragment at Mästocka, where Nils Holgersson had once watched untrammeled heath roll beyond the horizon, had the dimensions of a municipal playground and survived only because of routine, if reluctant, burning. The popular literature on nature dealt largely with flowers, the symbol of the garden. In Västerbotten, a nature preserve showed off a splinter of the "primeval" forest *(urskog),* most of which dated from the 1831 fires, but found it necessary to identify and give the Swedish words for fire-scarred stumps, standing snags, and rotting logs, all forms of dead wood exotic to Swedish forests. The fires that had once inspired those forests were no less remote from contemporary consciousness.

Thus, when an accidental fire broke out in slash on state-owned forests at Åtmyrliden in 1982, researchers from the Lantbruksuniversitet at Umeå persuaded the authorities to leave a portion untouched so that they might, as an experiment, discover just what kind of species would return according to what kind of processes. The site's fire history included, besides an indeterminate history of lesser burns, a majestic chronicle of major conflagrations dated to 1504, 1555, 1591, 1641, 1785, and 1831, all of sufficient dimensions to leave identifiable scars and to impress themselves on the age structure of the forest.

But what had once shaped Norrland's landscape was now itself shaped by the further evolution of that landscape. Only through special protection of the sort usually given to Bronze Age graves and prehistoric petroglyphs could the indigenous postfire biota survive. Fire ecologists had to fossick around the ruins of past burns not unlike archaeologists excavating the mounds at Old Uppsala or sifting through the shards of a burned and buried Iron Age village.

The Swedish environment assumed the properties of Swedish society. Sweden's "middle way" had no place for wildfire, the environmental equivalent of riots and wars. But that could be extinguished by good planning and forestry practices—wildfire leached from the scene much as ditches drained mires; lightning fire plucked out like rotting stumps. What was less clear was the status of controlled burning. Traditionally, European forestry had sought to remove it as primitive, extravagant, or unreliable. Swedes accepted those precepts, but Swedish society went further. It may be no accident that the expulsion of fire accompanied Sweden's postwar boom and the political triumph of the Social Democrats. Nature could best serve society by resembling it. At this point, however, social visions and ecological complexity clashed.

In the United States the concept of environmental protection has acquired a double meaning. It refers to the need to protect people from environmental hazards, both those of natural origin and those such as pollutants put there by humans. But it refers equally to the protection of the natural environment from humans. It grants special status to nature reserves, endangered species, wild rivers; it encompasses the preservation of natural processes, even disturbances, as well as the natural scenes and species; it grants to nature, within limits, certain rights, analogous to human rights; it even allows legal "standing" in courts of law; it extends concepts evolved out of American liberalism beyond the realm of humans. In practice, the history of American environmentalism is one of constant compromise and conflict, but the principle exists, as it does not in Sweden.

Swedish environmentalism is almost wholly preoccupied with the protection of humans from the hazards around them. For Swedish society security is the ultimate value, and protection from outside violence, the obsession—and triumph—of Swedish politics. During nearly two hundred years of neutrality, Sweden has evolved and prospered while all around it Europe lay in periodic ruin. The one, it appeared, was a precondition for the other. Under the impress of social democratic politics that sense of security expanded to include the uncertainties and travails of economic existence and of human health. And they have, perhaps inevitably, expanded further to include the natural environment. The environment must be cleansed not only of toxins, carcinogens, food contaminants, and unsafe roads, but of such natural catastrophes as floods, bark beetles, allelopathic shrubs, and fires. Where risk persists it must be distributed equitably among the various elements of Swedish society; no group should bear an unfair burden of the hazard. Social equity has been achieved, however, by shifting many of its costs onto the environment.

Sweden's forests epitomize this transfer, and forestry may stand as an exemplar for how Swedish society has sought to reshape the landscape into something like its own image. Swedish silviculture projected social values onto the polity of nature. It is not obvious, however, that the same principles can govern both dominions, and perhaps nothing shows better than fire the unexpected outcomes. What Swedish forestry saw as a hazard and condemned as waste may be a necessity, as essential as rain. The removal of fire has altered the pH of soil and lake, upset the geography of sunlight and shade, changed the insulating properties of humus, and interrupted the dynamics of interspecies competition. Unburned, the boreal landscape can suffer a kind of nutrient anemia as decomposition lags ever further behind production. What was intended as a system of steady-state recycling can slow into a spiral of degradation. What the Swedish state proposed as an environmental extension of the planned society, of economic democracy, has become a formula for biotic homogenization.

Removing fire has had the effect of ecologically leveling Sweden's biomes. For eons the boreal forest had diversified in large degree under the impress of recurring disturbance, often by fires both great and small. Now that discrimination has vanished. Pine, spruce, alder, birch, mosses, flowers, lichens, shrubs—all had sought their own niches in the crannies and cracks of a Swedish geography disturbed by wind, flood, ice, moose, drought, and various regimens of fire. Now, as Swedish stewardship continues to remove those pressures, as it further erases differences between wet and dry, burned and unburned, biodiversity fades into a classless smudge. Some two hundred species are believed threatened by Sweden's brand of boreal forestry, including insects equipped with infrared seeking organs that search out freshly burned sites. Similar losses threaten whole biomes. *Hed, moss,* and *myr, åker* and *fält, skogen*—heath, moor, bog, fields, forest—all have converged toward the slurred norms of a commercial woodland. Banishing disturbances such as fire has abolished much of the diversity and vitality that was their paradoxical legacy.

But if forestry was the vehicle for reformation, it was itself the expression of a deeper philosophy. It had become powerful because it merged with a political order, and it had become pervasive because it sustained a broader social ethos. Beyond its economic value, beyond the essential wealth it put at the disposal of the Swedish state, intensive silviculture imposed an order on nature that echoed, both physically and metaphorically, the order of Swedish society; that proclaimed the universality of social democratic values; that subordinated all relationships to those that governed and sustained human society. Swedish idealism is a philosophy of human relations, an inquiry into the principles that govern behavior among peoples, not between people and nature. Nature consists only of natural resources. Nature's purpose is to serve society.

Those assumptions are now being challenged. The first critic is the Swedish environment. Biodiversity is one aspect; fuels, another. The ancient order imposed by agriculture, later replaced or supplemented by silviculture, continues to erode, a trend that European economic integration will likely accelerate. Yet the principal means by which Sweden has absorbed that surplus farmland—forestry—is becoming less intensive, perhaps harvesting only 70 percent of the annual growth. The world market will probably drive this figure even lower. Meanwhile other lands are being committed to nature reserves but without any special mandate for management other than a doctrine of laissez-faire ecology in which whatever happens in the absence of active human intervention is accepted as "natural" and desirable.

All this is building up surplus biomass, which is to say, fuels. What humans don't claim, nature will, much of it by fire. Decade by decade the number of ignitions incrementally increases; lightning continues its restless foraging for fuel; and it is only a matter of time before spark and kindling meet. Under these conditions the return of wildfire is inevitable. But the fuel mosaic is, to date, widely scattered, and boreal burning is notoriously lumpy in time. Fires flare spasmodically, not in direct proportion to available fuel loads. The magnitude of the landscape changes may lie hidden for decades.

The second critic, an intellectual revolution, is more immediate and compelling; and it is to this that Sweden must first respond. Ecological ethics are proposing new values, some outside the scope of human society, that grant other creatures certain rights. Sweden cannot point to its environment with the same pride and clear conscience that it can to its health care, unemployment programs, and education system. It is not enough to sanction its exploitation on the grounds that the state has divided more or less equally the risks and rewards, any more than a Viking raid up the Seine can be justified because the crew shared the fighting and the spoils. Otherwise, Swedish forestry is no better than biotic plunder, the ecological equivalent of the Danegeld. But reform of environmental relations will force other reforms, both social and intellectual. Swedish society cannot change Swedish nature without itself changing.

Already this is happening. Recreational interest in the outdoors has grown exponentially; universities have created professorships in ecology; provincial governments continue to establish nature reserves, however individually small; environmental literature sprouts like mushrooms, arguing for a "living" land and rivers; and foresters have even expressed a cautious curiosity about hyggesbränning. As yet Sweden's philosophy of nature conservation remains primitive, either administering sites for their recreational values to humans or just withdrawing from active manipulation altogether. The former principle assures that the natural landscape directly serves human society; the latter, that such sites will not entail any direct costs to that society. Apparently it is gesture enough that the land be withdrawn from commercial production.

It is doubtful that such a policy can succeed, however, either as an environmental ethic or as a prescription for practical management. A robust environmentalism demands intervention as active as that by which Swedes maintain their society. Sweden would have to subsidize programs that would not contribute directly to the material wants and security of human society. Its safety net would have to expand to include species as well as economic classes. It would have to accept fire as an environmental presence, use fire as an instrument of environmental preservation, and fund fire management, not simply fire protection, as part of a full range of environmental services. The restoration of prescribed burning—the reincarnation of svedjebruk in more modern forms—is a likely compromise, a middle way of ecological engagement. Controlled combustion can do for forest management what controlled consumption has done for the Swedish economy. The land itself seems to beg for it.

In the early 1980s an accidental fire broke out in the old province of Dalarna, scene of the great colonization by Finnish svedjefolk. The fire was small and quickly contained, the dead timber rushed to mill, the site resown to commercial trees. In the ashes, however, something else fought its way to the surface, like relic stones heaved upward by frost. A plot of rye—ancient rye, long-dormant *svedjeråg*—waved in the summer sun, both poignant and insistent. The scene speaks equally to Sweden's human and natural history; of the obligation that contemporary society has to reconcile them; and of the anthropogenic fires that have long bound them together.

Greek Fire
(Greece)

When Zeus, leading the Olympians, waged war against the Titans, the clash climaxed with an apocalyptic firefight over Crete. According to Hesiod,

> *he [Zeus] advanced with steady pace amid flashes of lightning*
> *and from his stout hand let fly thunderbolts*
> *that crashed and spewed forth a stream of sacred flames.*
> *The life-giving earth burned and resounded all over*
> *and the vast forest groaned, consumed by fire . . .*
> *Wondrous conflagration spread through Chaos . . .*

It was lightning—and the power to start fire—that gave Zeus supremacy. So it is with furious irony that the surviving Titan, Prometheus, gave that supreme weapon to humans. "All the arts of man," which Prometheus claimed came with fire, included warfare.

The association of war and fire came readily to the ancient Greeks. Homer used the simile to describe an enraged Achilles:

> On went Achilles: as a devouring conflagration rages through the valleys of a parched mountain height, and the thick forest blazes, while the wind rolls the flames to all sides in riotous confusion, so he stormed over the field like a fury, driving all before him, and killing until the earth was a river of blood.

Lucretius believed that the "weapons of ancient times were hands and nails and teeth. / Then axes hewn from the trees of the forest. / Flame and fire as soon as men knew them." The Greek tactician Aeneas, who assembled the first European treatise on war (c. 350 B.C.), listed sulfur, pitch, pine, wood, incense, and tar as basic incendiary weapons. Herodotus described how the steppe-dwelling Scythians protected themselves by a scorched-earth policy. Thucydides recounted how, during the Peloponnesian War, the Spartans developed a flame thrower as a siege engine, and how they encircled a Platean army with an enormous fire until a freak rainstorm dowsed the threatening flames. Invading armies would fell and fire forests before withdrawing.

But the litany of fire warfare didn't end with the passage of ancient civilizations, and it was never far from the Hellenic world. Wars and insurrections were fought with fire no less than sword. During Byzantine times a flammable liquid, probably a compound involving pitch and naphtha—the notorious "Greek fire"—became the scourge of besieging armies and wooden naval vessels. (Greeks thus burned their forests secondhand.) Crusaders suffered from attacks by fire, not only those poured from battlements at Acre or set by sappers or against siege engines, but broadcast burns as well. Saladin, for example, fired the scrub ringing the hill held by Guy of Lusignan and the True Cross, and when Guy succumbed, so did Jerusalem. In the Balkans the Turks routinely slashed and burned forests to strip away the protective cover needed by raiders and guerrilla fighters. The German Wehrmacht did the same against Greek partisans in World War II. Cypriot separatists fought the British with "forest sabotage groups" armed with torches; and during the Turkish invasion of Cyprus in 1974, aerial incendiaries burned more than 30 percent of the island's productive forests. Apart from outright war, almost any form of social unrest, from political protest to economic sabotage to insurrection, has quickly translated into fire. Citizens vote with the torch.

The history of Greece is not only a history of fire—virtually all lands tell a fire history—but a history of hostile fire, of fire used violently, of a malevolent and vindictive burning. The unhappy history of this oft-conquered land is written with flame, for fire weaponry was not restricted to assaults on humans. Greek fire was poured on the landscape as eagerly as it was dumped from besieged city walls. Much as invader after invader has left his imprint on Greek life and institutions, so also have his fires shaped the Greek landscape into a kind of biotic war zone.

The Mediterranean Basin is the type case for a category of land, or rather of climate, characterized by long, hot summers and short, wet winters. Typically the land is mountainous. The formula is ideal for scleromorphs—tough, hard-coated plants, characteristically pyrophytic, resistant to drought, rich in oils; for complex ensembles of vegetation, mosaics as finely wrought as a Swiss watch; and for fire—small, large, benign, violent. The biota as a whole seems to distill itself into the fabulous shrubs that mediate between high forest and grassland. The region also shows a concentrated fire that can direct the ecosystem in one of several directions. Like a ball balanced on a hill, the system can fall one way or another, each with its own fire regime.

These features focus with dramatic clarity in Greece. The towering mountains and plunging basins block out a tectonic zone where Africa, Asia, and Europe crash into one other, a zone of geologic tension. Southward, the Mediterranean Sea floods the deeper depressions, and the mountains break into a chain of peaks like a necklace hanging from the Balkans. The summits rise from the sea like triremes. Similarly the region compresses the biota of three continents, and it masses together their human history. Not surprisingly the regional history is one of exceptional turmoil, preserved with melancholy splendor in the ruins of prior civilizations that litter the landscape. But that social violence has also expressed itself as an equally vicious environmental violence—the land, too, lies in ruin.

The landscape of Greece is as much an artifact as its temples, cities, and roads, and like them it rises out of the rubble of the past. Just as city rose upon city, as peoples incorporated old stones into new designs, as, for example, Christians reconstructed churches with the reconsecrated stones of pagan temples and theaters, so the cycles of conquest have accompanied equal cycles of environmental degradation and renewal, each new landscape built out of the biotic rubble of its predecessor, the residual flora, fauna, and landforms. A few sites were spared; sacred groves remote from settlement might survive, much as select monuments of antiquity might escape predation more or less intact.

Yet the dominant pattern was not to preserve but to rebuild, often with the same fatal design flaws. Each revolution did not return the system to its exact starting position, but extracted a cost, a loss of vigor, an environmental entropy. The overall trajectory for both people and land has thus been one of fatigue, a drain of vital nutrients and soils, a slow starvation that has left, at its worst, an emaciated landscape stripped to hollowed skin and bone. The blackened timbers buried at the sites of ancient cities have their counterpart in the charred wreckage of a plundered ecosystem.

The Mediterranean biota is adapted to fire, even to violent fire, but it cannot accommodate with equal facility all fires or fires that accompany other, destructive practices. If the burning proceeds too relentlessly, if it repeats without sufficient pause for renewal, if it joins with poor plowing, reckless logging, or indifferent husbandry, if it accompanies free-ranging pastoralism, especially by goats, then fire can lead to decay instead of renewal. Where humans integrate fire within a pattern of stable stewardship, fire can massage and reinvigorate the biota; it can mediate between the social world and the natural; its regime can preserve a capacity to rebuild. But where humans use fire as a weapon, burning can destroy. It becomes hostile, tearing down without also rebuilding. It is a sad burden of Greek history that here, where three great biotas and civilizations meet, the contact has been violent, and the fires damaging.

The classic concept of biological succession saw progression, the upward evolution of one community to another. But the Mediterranean tradition has reversed that prospect. The best one can hope for is a resurrection of ancient glory; the reality is a record of decay and diminution. The degradation of the regional flora follows a declension as seemingly rigorous as that of a Greek noun. High forest gives way to maquis, a dense shrubland. Maquis degrades into garigue, a low shrub. Garigue withers into phrygana, a dwarfish diminutive of poor palatability. Phrygana succeeds to batha or outright rock. The sequence can ratchet upward, but only given sufficient time and purpose.

For much of Greek history, however, certainly for the past millennium, this land has been oppressed, kept in subjugation through force and a bondage of predatory abuse in which it, like its people caught in a cycle of taxation and debt, saw their community spiral downward into a condition of permanent poverty from which, with massive soil loss, even the hope of recovery appears dim. When Greece achieved independence from the Ottoman Empire in 1830, it is estimated that 48 percent of its lands were forested. A century and a half later, despite a considerable expansion of national territory, only 10 percent remains in forest. Throughout, fire endured—climate and fuels alone demanded its persistence. But over the centuries it changed character. Free-burning fire became more often hostile fire, both a weapon of oppression and protest, each burn thrusting the land into a larger spiral of self-destruction. The transfer of forests to state control by King Otto, for example, set off an orgy of incendiarism.

This—perhaps the true Greek tragedy—is indelibly associated with pastoralism. Livestock has been integral to European economies since the Neolithic revolution. At some times and places, however, pastoralism has come to dominate society and the economy overall. When this happens, herding ceases to synchronize with the other elements that make up the annual liturgy of agriculture; it abandons any semblance to husbandry and surrenders to free-ranging flocks, the environmental equivalent of *banditti* roaming the hills; its requisite burning escapes the confines of field and woodlot and ranges widely, often destructively, not infrequently as a means of protest and intimidation. This promiscuous grazing encourages promiscuous burning. Through broadcast fire, herders can promote browse, terrorize uncooperative farmers, and protest the existing order. Broken free from a mixed economy in which livestock and farming are symbiotically joined—flocks feeding on and fertilizing harvested stubble, for example—herding becomes defiant, lawless, reckless. In 1900 the

Greek Orthodox Church even threatened with excommunication anyone convicted of incendiarism. But in Greece it is not simply pastoralism that has prevailed; it is the goat.

A German forester put it directly when he explained early in the twentieth century that the Greeks could eat their goats or their goats would eat the forests. The Greeks chose their goats. There are today as many sheep and goats as people, roughly ten million. Some goats are tethered, rudely domesticated and incorporated into subsistence farms, but most are not. They range widely and densely, feeding on fire-stimulated browse, devouring new forest regeneration, a four-footed plague that blights the biota, plundering settled agriculture like nomads preying on cities. Protests by sedentary agriculturalists or foresters are answered with more fire. Increasingly, too, the herds, and the fires that accompany them, range across the mountains without supervision, reverting to a semiferal state. The biota degrades. As herds move up slopes, soil washes down.

There is little doubt that this kind of herding damages the land. Foresters, in fact, made the Cypriot goat a symbol of environmental havoc, and fought bitterly to extirpate the free-ranging goat—and his allied goatherder—from demarcated forests. Where they succeeded, the forest returned. On Lesbos, too, the elimination of the goat on half the island has resulted in a dramatic recuperation. Bulgaria and Yugoslavia purged themselves of their goat herds in the 1950s and witnessed a dramatic resurgence of their forest.

Then there is the counterexample offered by the other Mediterranean peninsulas, Italy and Iberia. In Italy, the herds remained better integrated into a village economy, and their place constrained by local cycles of transhumance. In Iberia, pastoralism converted to sheep and organized transhumance on a national scale by chartering a special guild, La Mesta. Whatever its damages, social and environmental—and they were huge—the Mesta at least proposed a pattern of grazing that offered the prospects for control, that installed a *system* of legal privilege and state-sanctioned routes; the violence it inflicted was part of an organized order, a pastoral analogue to the *Reconquista.* In Greece, transhumance dissolved into innumerable guerrilla bands, the environmental equivalent of a social order based on banditry and vendetta. Greek herding had an antisocial bias, an outlaw economy, a part of whose violence was fire applied to the land not in support of a native flora and fauna but for the insatiable goat. The countryside gradually crumbled, like a fortress long and often besieged that finally collapses into rubble.

But the Greeks will not repudiate their goats. Clearly the goat thrives in a symbolic world for which the garigue-clad hills are a poorly materialized copy. The goat saturates early Greek ceremony and society—Pan, the satyrs, the Dionysian festivals that served as both orgy and origin of Greek theater. (*Tragedy* comes from *tragoidia,* which derives from *tragos,* "goat.") The goat romped on the borders of organized society, marginal to the civic order of polis and empire; it taunted reason, and the political economy based upon it that platted the valleys of Attica and Argos and colonized around the eastern Mediterranean; it became the model for the cloven-footed devil of Christian demonology, yet endured; it survived Rome, Christianity, Byzantium, Venice, the Turks. Throughout all the centuries that followed, through the chronicles of oppression and predatory taxation and military conquest, the goat retained its archetypal powers, a symbol of bacchanalia, marginality, and freedom. Greek peasants identified with the goat—lawless, lustful, vindictive, free-ranging. The freedom they could not achieve psychologically or politically they transferred to the goat; the oppression they experienced socially they transferred to the land. Greece bought the goats that its Balkan neighbors disposed of following World War II and added to its swollen herds.

The newest threats to Greece are more subtle, and more pervasive. Industrialization has come to Europe's Third World with profound consequences. A small number of cities have swelled to gargantuan size, new city-states; the ancient rural economy has decayed, and like refugees rural populations have fled from its wreckage; summer homes recolonize coastlines and tourism replaces traditional industries; political unrest targets the rituals of democracy rather than autocracy. The evolving geography of fire reflects all of these developments and more.

The transformation began after World War II and the civil war that followed. Militarily and economically, Greece became integrated with a new Europe. It began to modernize. By the 1960s a slow stream of change had become a white-water rapid. The rural economy stagnated, then began to disintegrate; only massive subsidies and a generational inertia that kept a significant fraction of the rural population, primarily the elderly, on their ancestral lands prevented a total collapse. The markets for fuelwood, resin, olives, wheat, goat cheese, and wool disappeared or faced stiff global competition for which agricultural subsidies were mandatory. Traditional land practices became obsolete or redundant. Land was abandoned. Fields fluffed with weeds. A resurgent biota reclaimed orchards. Forest understories ripened with shrubs, grasses, and regeneration, no longer tended by farmers, fed to livestock, harvested for ovens and handcrafts, or burned in careful plots or piles. Fuel oil and electricity replaced fuelwood. The rural exodus paralleled an increase in rural fuels; a drying climate since the 1960s aggravated hazards by making still more fuels available; these fuels, in turn, have fed a rise in fires. Throughout the Mediterranean the intensity of biomass use by humans and livestock correlates inversely with the prevalence of wildfire. The less biomass that is used, the greater the area burned. The figures for Greece are virtually identical to those for Portugal, Albania, and southern France.

Simultaneously the great urban centers, Athens and Thessalonica, have spawned a rural reclamation in the form of recreational usage and summer homes. Exurban villagers are replacing rural farmers and seasonal tourists, the transhumant herders of the past. Exurbanites have crowded along the coasts, however, leaving the hinterland's downward spiral of decay more or less unimpeded. They have not introduced surrogate land-use practices to compensate for lost traditional practices, so important historically in absorbing surplus biomass. They do not live off the land, do not assist in the controlled burning that has become increasingly mandatory to diminish the progressively unruly fuels, do not assist in the watchfulness that was a part of village life and the firefighting in which villagers traditionally engaged. They are transients who leave to others—or to no one—the responsibility for land stewardship. In all this the exurban outreach has done nothing to compensate for the fuel-and-fire spiral provoked by the rural exodus.

Rather, exurban expansion and tourism add to the hoary arsenal of Greek fire. Because Greece lacks a cadastral survey, landownership more closely follows Ottoman precedents in which site occupancy and improvement qualify as title. To burn and build thus conveys de facto ownership.

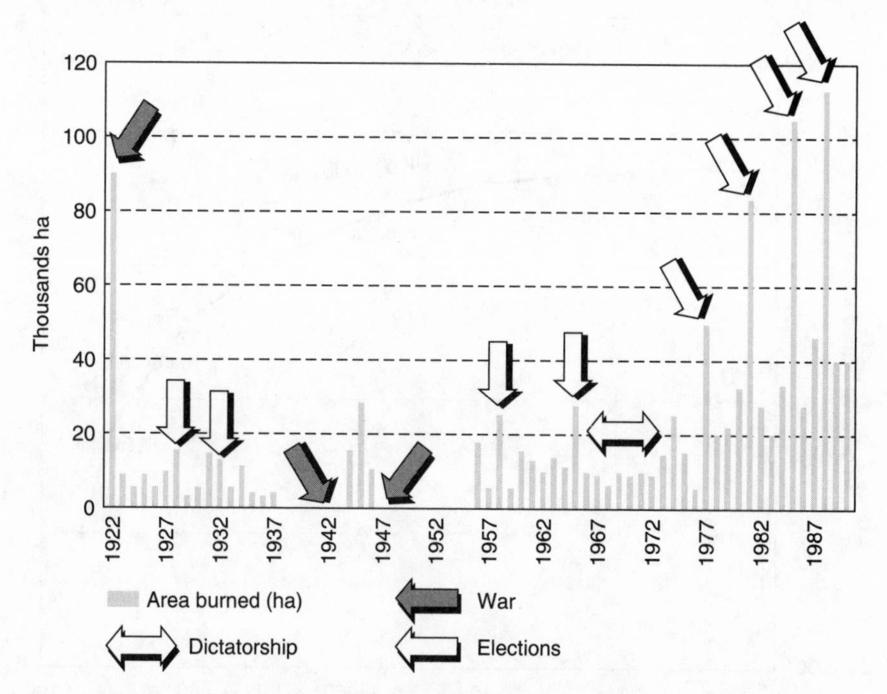

Area burned in Greece, 1922–90. Note the overall increase in fires in recent decades with the unraveling of the rural landscape, and the strong correlation between fires and periods of social unrest.

The incentive for arson is irresistible, and houses sprout like fire weeds after well-placed wildfires. Technically the state could protest; but in practice, through corruption, indifference, and legal uncertainty, it does not. Forest law no longer applies once the forest has been burned away. In fact, the state can barely protect its own gazetted forests, which also lack adequate boundaries and surveys. Instead, at considerable—and rising—expense the state forestry service fields an army of firefighters and a fleet of aircraft to combat the escalating incendiarism. During 1977–87 fire protection claimed 10–15 percent of the entire budget of the Greek Forest Service. In truth, the investment in infrastructure and personnel costs would make a true calculation of the protection budget much higher still.

The result resembles nothing so much as a replay of the chaotic violence typical of regional history. Greek fire has returned. Since 1977,

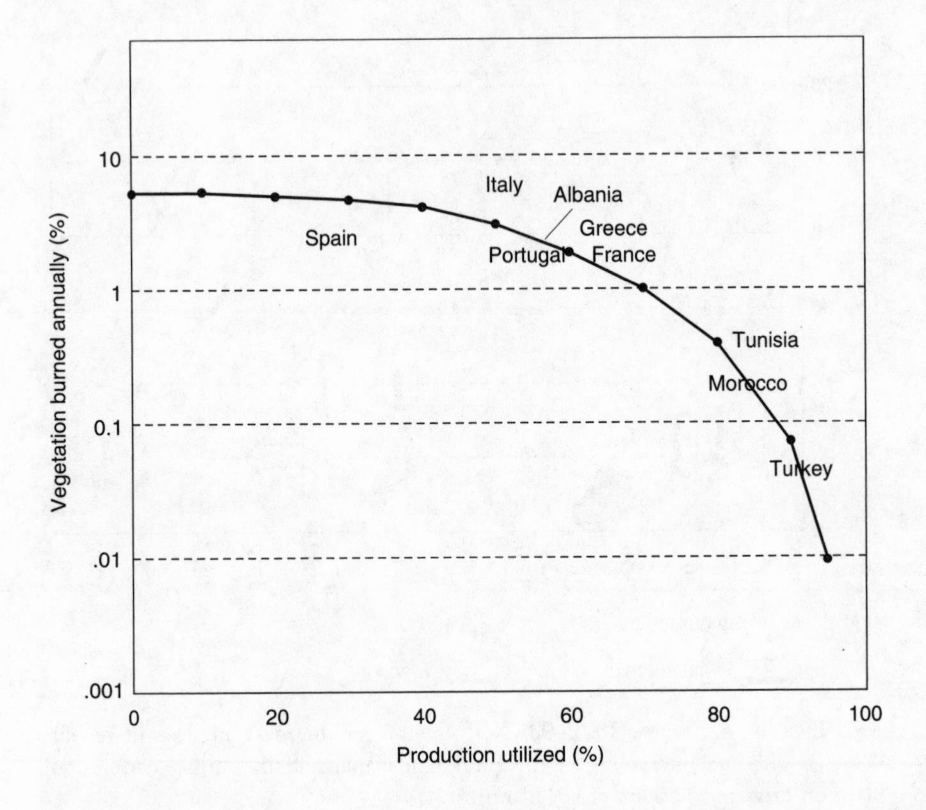

The competition for biomass, a Mediterranean example (1980–85). The greater the utilization of biomass the less the area burned.

when a nun died in a wildfire on Mount Parnassus, fire-related fatalities and injuries have increased alarmingly—and this in a land where traditional building materials such as stone and tile prevent major structural conflagrations. But the social dimensions are broader still. During the 1980s the average burned area per year more than tripled. The worst fires of the century flared in 1981, 1985, and 1988, all of which, not coincidentally, corresponded to political unrest associated with elections. Traditions of political protest by fire have, in effect, been rekindled from new sparks and new tinder.

The breakdown reached a new order of magnitude in 1993 when, compounding four years of drought, Greece passed legislation that legalized houses built on burned forest land (thus sanctioning arson), experienced a stream of refugees from Albania along trails blazed by fire, and suffered through a national election. An arson fire on the island of Ikaria killed eleven fleeing farmers. Not only did the usual suspects, the Mediterranean littoral and Peloponnese, burn, so did the more impervious northern forests. The 2,417 fires (and 47,000 burned hectares) set new records for havoc. Conflagrations even swept the national parks on Mount Olympus.

Overall Greece is squeezed in an industrial vise. Ruralites are reluctant to abandon traditional fire practices, even though they find it increasingly difficult to control their burning or to halt wildfires, while exurbanites are unable or unwilling to substitute new practices. The country is caught between a resurgent wildland flush with fuels and an expanding urban rim; the friction between them has ignited fire in their widening frontier, well beyond traditional methods of control. As one cause recedes, another advances. Even if economics, for example, drove pastoral fires into near extinction, arson fires would likely fill the void. In fact, the dynamics of fire starts is not well understood; a whopping 34 percent of Greek fires have unknown origins. Worse, not only has the total area burned increased but the average size of fires as well. Consider Corinthia. During the 1950s the average size of reported fires was 2 hectares. For the 1980s it was 37.4 hectares. It is unlikely that fire suppression alone can contain the menace. On the contrary, despite escalating budgets, it is failing.

What is needed is a new order, social and environmental—the two cannot be segregated. Greeks will control wildfire only if they can control the shifting fuels and shifting ignitions that have unsettled their land. It is not enough to eliminate traditional sources of burning; it is equally necessary to substitute for traditional sources of control. If all rural usage were abolished, fires would actually worsen because of surging fuels and an absence of firefighters. But it is unlikely that exurbanites can be persuaded to replicate rural practices, or that state agencies, or even the military, could squelch such a simmering rebellion in the countryside. Besides, controlling ignitions does not by itself contain the area burned. Shutting down fires only shovels in more fuel.

Perhaps the major check against wholesale collapse is an elaborate policy of agricultural subsidies—the intention of which is less economic than political—to prevent the social chaos that would follow from the total disintegration of a traditional rural economy exposed, unprotected, to world markets. In this regard Greece is little different from Fiji or Tanzania. The choices involve more than simple economics, of course; they have to do with what kind of society Europe desires, and how much of its identity as an agricultural civilization it is willing to surrender or redefine. Greece faces many of the same choices as Sweden and France, and it expresses many of the same contradictions. Even as markets force the abandonment of aging olive orchards, subsidies promote new groves; both movements bring fire. So also with pastoralism. There are funds to promote firefighting on mountain pastures, yet there are other subsidies to maintain the goat population. Still, the political pressure on all of the European Community to downsize these subsidies is tremendous, and inevitably they must shrink.

As they fall so must fire protection. A program of aggressive suppression can substitute only in select sites; even here firefighting does not put out fires so much as it puts them off. Beyond state forests the power to project armed fire control is limited. What is required is a kind of land management that replicates, in suitably new forms, traditional practices of husbandry and the harvesting of wildland products, that provides an alternative to a social economy dictated by goats. What is required, at a minimum, is some form of controlled grazing and controlled burning. What Greece needs is to rebuild its landscape out of traditional materials but according to new plans.

What Greece is to the Mediterranean, Crete is to Greece. It masses into one island the geology, history, and biogeography of the eastern Mediterranean; it distills the dynamics of contemporary change; and it harbors the most intractable fires. Here the cycles of civilization and conquest, of repression, collapse, and subterfuge, of recovery, erosion, and decay, trace back over four thousand years. The Minoan ruins at Knossos have their environmental analogues in the terraced hillsides, abandoned and reclaimed and washed, century by century, down to the sea. Cretans, it is said, customarily wear black because in this land of vendetta one is always in mourning. But the land, too, is dressed in black. Here Zeus warred with the Titans and savagely burned the earth. Here Greek fire is most stubborn and defiant and senseless.

The land is lush with contradictions and cross-purposes. At one site peasants tend olive orchards in the traditional way, on terraced slopes, the trees carefully pruned and the loose debris burned, the surface kept clean to help the harvest. At another site, however, whole hillsides blossom with abandoned and overgrown orchards, primed with the equivalent of biotic gunpowder and ready to explode with a spark. Elsewhere arson has purged a hill of vegetation, and with the assistance of agricultural subsidies new terraces are bulldozed to support more olive trees.

But the trend is away from traditional agriculture. The increasing population cannot afford to have family farms divided and subdivided again and again among sons; the intensive cultivation and husbandry demanded by traditional agriculture requires an investment in labor that is no longer possible; the rural economy lives largely off tourism and the welfare provided by subsidies. Farming favors specialty crops, such as vegetables, for export to the urban markets on the mainland. Gardens replace fields, and greenhouses, gardens. The macroeconomy relies on an exchange of people—the emigration of Cretans, the importation of tourists. Foothill villages suffer a steady attrition, grated on one side by the expansion of urban developments along the coast and ground on the other by continued herding in the mountains.

The environment, too, shows fatigue. Geologic processes move slowly compared to human history; soils are more easily lost than made. Land is abandoned not only because of new economics but also because of ancient abuse. The pressures have been relentless. From the earliest days of the Neolithic revolution, humans exploited every niche that they could. Every valley. Every slope. Every mountain hilltop. Every woodpile, pasture, and spring. What humans did not harvest, cut, or consume themselves their beasts, weeds, and bees reshaped, their fires and flocks attacked and absorbed. The Mediterranean flora possesses extraordinary resilience, yet even this long-suffering biota has its limits. During those times of unrest when humans could not recycle and restore some of what they had extracted, the system has suffered irremedial losses. While contemporary events are but the latest in an ancient iteration, they act on a cumulative heritage of abuse, a progressively exhausted environment.

As social control loosens, goats proliferate. Marginal and degraded lands—the preferred habitat of the goat—magnify. The herds range more freely than in the past, watched more in order to guard against theft than to ensure good husbandry. Poorly restrained, the goats not only feed on the broken biota, they further wreck it. Through their human confederates, they breathe fire before them. The phrygana burns yearly. Worse, the fires are kindled for maximum impact—set in mid-afternoon, on south-facing slopes, with strong winds; after the long drought of summer and prior to the winter rains when the first storms can wash over the blackened slopes and send the soils rushing downward. Instantly, as new growth appears, the herds crowd in and crop it. Even phrygana cannot absorb this kind of abuse indefinitely. The site degrades further; cover abrades. What remains is unpalatable and spiky.

But the fires—like the goats—range more widely still. They spill down the slopes, chewing at marginal farmland and orchards; they infiltrate every mountain crevice, sparing nothing; they pervert the regime with an ecological rule that more resembles that of the Mafia, subverting an honest economy with one driven underground. Social order and environmental order, nature's economy no less than humanity's, crumble into an extra-legal state of relentless feuds, vendettas, indifference, and defiance. Yet Cretans are reluctant to corral the herds, and incredibly the agricultural policies of the European Community actually work to keep many goats in place.

With no clear flight plan for the future the most likely prognosis calls for a controlled crash. The push for full integration into the European Community, and through it into the world market, is probably irresistible. What is needed is a reconstitution of land use on different principles. Too often agricultural subsidies alone soften the blow, at least temporarily, by transferring the costs to the environment. Untended, the biota burns, and in ways that degrade rather than regenerate—the classic Greek fire, fire as weapon.

Real reform demands a change of mind and a change of heart. Crete concentrates problems that are pandemic throughout the Mediterranean, and it could adopt solutions that have been developed elsewhere. Crete could abolish free-ranging goats, as the Balkans have done, or it could substitute sheep for goats, as Cyprus did. It could prohibit construction on recently burned sites, as Italy has decreed; it could instigate cadastral surveys, like France; it could implement a modern firefighting force, like Spain. It could, as has been proposed by its own fire specialists such as Alexander Dimitrakopoulos and Gavriil Xanthopoulos, direct agricultural subsidies into activities that would diminish promiscuous fire, that would reinstate programs of selective harvesting, close grazing, and controlled burning, that would reconstitute traditional practices in ways that could restore a measure of traditional fire protection. It could promote further research like that which suggests some species of phrygana might be gathered and sold for their oils instead of burned for their paltry forage.

The historical perversion of Greek fire lies less with the techniques themselves than with the intentions behind them, not the torch but the hand that holds it. Reform would mean that Cretans—Greeks—would have to surrender not only the goat but the psychology and sociopathy it has symbolized. They would have to escape the cycle of vendetta, dress the land in something other than the black of mourning. They would have to accept that, after millennia, they now rule themselves, that Greek fire is no longer a weapon directed against an alien oppressor but an act of self-immolation.

La Nueva Reconquista (Iberia)

From Galicia to Granada, across the spectrum of Iberia's varied landscapes, fires have spilled out of long-suppressed confinements to propagate a new black legend.

The extensiveness of the burning is astonishing. It afflicts every province, but it strikes with special vehemence in the northwest and southeast, where, respectively, the fuels are most profuse and the climate most droughty. Increasingly it infests every environmental niche, escaping from stubbled fields, burning across grassy mesas, ravaging both relic agriculture and modern afforestation, rekindling wooded *montes.* The fires have laid waste to hillside after hillside of the Serra da Estrela, blackening the symbolic heart of Portugal's forests, roamed across the plateaus behind Valencia and Alicante, and invaded nature reserves like the fabled Doñana National Park. In the summer of 1994, while Americans agonized over 400,000 burned acres and the deaths of fourteen firefighters, Spain experienced twice the burned area and saw a crew of nineteen perish.

No less impressive is the fires' timing. With remarkable precision, the outbreak correlates with the collapse of the dictatorships of Antonio de Oliveira Salazar and Francisco Franco that for so long gripped Iberian society. But with even closer fidelity the fires testify to a reformation in land use more fundamental than any since the reconquest of the peninsula, by fire and sword, from the Moors. Iberia is industrializing. The expressions of that revolution are as much environmental as social, economic, and political. The liberation in political freedoms parallels a liberation of long-suppressed fuels. For this *Nueva Reconquista* firefighting has assumed the status of a paramilitary force; the battlegrounds are forests, parks, and montes; the weapons of choice, torch, air tanker, and computer. The phenomenal wave of incendiarism that has swept Iberia is both an agent for change and a manifestation of the reformation those changes have inspired.

It is not that fire has suddenly appeared where it has never before existed. Iberia has long burned. The mediterranean climate that extends over most of the peninsula is ideal for fire; the indigenous vegetation is heavily salted with pyrophytes, and many of its most emblematic flora, like the cork oak and scrubby *matorral,* exhibit spectacular adaptations to routine fire; the record of anthropogenic fire extends as far back as the hominid presence. The excavations at Torralba suggest fire hunting for elephants, wild cattle, horses, deer, and woolly rhinoceroses 400,000 years ago. The Pyrenees take their name from fire (Greek, *pyr*) and burned so massively in 100 B.C. that ancient chroniclers took note. There is no period of Holocene Iberia without fire of some kind. Fire tempered the landscape much as it forged swords of Toledo steel.

What has changed is the character of fire. Iberia shares the fire histories common to both the Mediterranean Basin and to Europe. But its unique saga is not interchangeable with that of any other place. One reason is Iberia's special geography. It is a subcontinent as much as a peninsula—more like India than Italy—that has acted as a bridge, barrier, and barbican. Tectonically squeezed between Europe and Africa, Iberia rises, falls, crumbles, and rotates, alternately fusing or sundering the two continents. It stands like a biotic castle, surrounded by a saltwater moat on three sides and mountain ramparts on the fourth. The geologic drawbridge at Gibraltar determines whether the Mediterranean Sea fills or dries up. The complex geography of mountains, mesas, littoral, and river valleys harbors species during times of stress, and amply nourishes them during times of climatic calm. Its moisture gradient from temperate northwest to semiarid southeast captures most of the climatic niches of the wider region. In particular the Iberian isthmus has blocked, filtered, and promoted the movement of species, including humans, between North Africa and Western Europe.

But it was possible to mold this raw material in many ways. Throughout the Pleistocene revolution anthropogenic fire was present, almost certainly feeling its way into whatever fuels were available, sometimes coarsely, sometimes with exquisite precision, but ever tempering the biota, encouraging the trend toward sclerophylly that the rigors of Iberia's climatic history and topography had already established. Although the Neolithic revolution arrived later than in the eastern Mediterranean, it quickly began refashioning the landscapes into forms suitable for its exotic cultigens and livestock. The great revolution followed from Iberia's incorporation into the Roman imperium.

Classical agriculture remade the subcontinent. Where it did not rework the land directly, it built roads and through trade and legion shaped the hinterland. New tiles baked in ancient agronomics—olives, vineyards, cereals—joined the shards of past practices such as slash-and-burn farming, transhumant herding, fire hunting for boars and deer, fire foraging for chestnuts and acorns. Through villa and latifundia, the separate pieces of farm, pasture, woods, orchards, fallow, and flock were assembled into picturesque wholes. The result was a complex composite, a landscape analogue to the mosaics that decorated the walls of Roman villas.

Agricultural burning was an integral part of that ancient regime. New lands were cleared by slashing and burning; old lands regenerated regularly by fire. Farmers burned fallow fields and wheat stubble. Shepherds burned mountain pasture and valley meadow, as they followed the seasons and the ripening browse. Where the domestic and the wild met, fires kept wolves and bears away from fields and flocks. Only in the most intensively managed estates was it possible selectively to substitute animal manure for fire, and even then the environment made accidental fire likely if deliberate burning ceased. The persistence of Spanish *quemas* and Portuguese queimadas testified to the ancient economy of Iberia as fully as its relict stone walls, aqueducts, and terraced hillsides. Eventually fire followed the biota into more intensive manipulation, a greater measure of domestication, even of servitude.

Most classical agronomists distrusted fire—Virgil was almost alone in singing its praises, as he was among ancient authors for stoking the *Aeneid* with fire imagery. What agronomists like Cato, Columella, and Theophrastus wanted, ideally, was labor-intensive pruning, weeding, and planting, and fertilizers based on compost and manure. The agrarian ideal was a garden; pastoralism meant animal husbandry, with draft and milch animals treated almost as household pets and herds managed as one would orchards. Infield joined indissolubly with outfield, arable with pasture. Fire belonged in hearth and furnace, not in the field.

But only rarely did the ideal succeed. Instead arable infield *(ager)* and grazed outfield *(saltus)* existed in metastable equilibrium. The linkage connecting them broke constantly. Environmental logic sent flocks into the mountains for summer pasture, then returned them to the valleys or littoral during the winter rainy season. This pattern of seasonal migration, or transhumance, meant that half a year's manure fertilized remote landscapes, not arable. Economic pressures too often promoted flocks instead of tilled fields, particularly when war or disease had depopulated once-arable lands. Cicero's famous exchange with Cato is symptomatic. When asked what was the most profitable thing in the management of one's estate, Cato answered, " 'Good pasturage.' What is the next best? 'Fairly good pasturage.' What is the third best? 'Bad pasturage.' What is the fourth best? 'Tilling the soil.' "

If the strains were real, it was nevertheless possible for large estates like the latifundia to accommodate both, for the diversity of Iberia to accommodate pastoral enclaves by Celts and Basques, and for the agricultural regime of Iberia to reconcile flock with field. Or it could until the breakup of Rome. In the centuries that followed, the conquerors of Iberia were primarily pastoralists. Although they were content to rule, not restructure, neither were they willing to forcibly sustain the dual allegiance advocated by classical agronomics. In particular the Christian reconquest of Iberia from the Moors installed the triumph of herding over farming. The Reconquista was an environmental as much as a political saga.

To this end history conspired with geography. The Mediterranean—Rome's *mare nostrum*—had fused the disparate empire together. In a few critical places Roman power had penetrated inland, notably to Gaul, Dacia, and Iberia; but vigorous though the growth might have been, the provinces survived by grafting onto the Mediterranean rootstock. Iberia's special characteristic was the great plateau, the *meseta,* that bound the encircling littoral together, and connected lowland with mountain. Within the Mediterranean Basin only Turkey featured anything like it. Originally the meseta received influence as power flowed from the outside in. With the Reconquista, however, power flowed from the center out, and that reversal required both novel institutions and a reconstituted landscape.

The Reconquista—perhaps the first Crusade—progressed fitfully over the course of seven centuries. From its base in Asturias and Galicia, the recovery followed climatic gradients, advancing south and east, and only ended in 1492 with the seizure of Granada, the last Moorish stronghold. At this point the politics of the Reconquista toughened; the Inquisition intensified, as Jews and Moors faced conversion or expulsion; and with the exception of Portugal, Iberia's diverse provinces endured increasing control under the consolidated monarchy of Ferdinand and Isabella. Conveniently, as Granada fell, America was discovered, and the Reconquista rapidly relocated to the New World as conquistadores turned from Moorish fortifications to imperial citadels such as Tenochtitlán and Cuzco.

Conveniently, too, the routes of reconquest traced paths for Spanish pastoralism. Herds moved with the troops. Flocks could accompany an army on the march in ways that vegetables and grains could not; the army could quickly occupy seized land; herds of cattle and sheep, in particular, advanced and retreated according to the alternation of summer campaigns and winter truces, with the success or failure of Iberian arms. The ancient habits of transhumance now extended not only over mountain flanks but also across the plains of the meseta. Herds migrated seasonally north and south as well as up and down mountains. The routes of migration thus bound new lands to old, joining León to Salamanca, Logroño to Estremadura, Aragon to La Mancha, constantly extending as the Reconquista progressed. Increasingly formalized, those routes *(cañadas)* inscribed a new agrarian order as fully as Roman roads had centuries earlier. Flung out across the meseta, the cañadas wove a vast web by which the center reached to the perimeter. In short, pastoralism threw a great blanket over the old landscape mosaic.

This did not happen by chance. Lands seized by conquest changed ownership and reorganized. Those lands it did not retain for itself, the Crown dispensed to the nobility, to the church, and, in order to promote settlement, to public or communal collectives. But it was the evolving regime of pastoralism that held the jumble of vacated lands together. The ancestral merino sheep arrived from Africa in the thirteenth century and soon hybridized with local stock. Demand for wool rose as the northern European economies rebuilt and expanded. The prospects were not lost on the monarchy. Hoping to regulate this tidal bore of beasts and to siphon into royal coffers some revenue to replace the tribute lost from now-expelled Moorish vassals, the monarchy asserted its prerogatives and sought control through a regulated monopoly. In 1273 Alfonso X issued to "all the shepherds of Castile" a royal charter of privilege, *El Honrado Consejo de la Mesta de Pastores,* which sought to organize, protect, and extend to the farthest "extremities" the system of seasonal migrations that the Reconquista had set into motion. In 1476 the king himself became the Grand Master of the Mesta. With the merino sheep, the Spanish monarchy formed its most powerful alliance.

The Mesta had its privileges confirmed in 1501, once Iberia had firmly completed the Reconquista; it then survived as an official institution until 1837. Little in Spanish life or land escaped the consequences of its long reach and lengthy rule. The Mesta acted on the Iberian environment much as monarchical absolutism and the Inquisition had on Spanish society, an ecological expression of the fanaticism that is the marvel and burden of Spanish history. Elsewhere the breakup of classical empires had also shattered the fragile balance between farming and pastoralism, but no place else had a landscape equivalent to the meseta or an experience like the Reconquista and no other country evolved anything like the Mesta. When it appeared in Italy, it did so through temporary conquests by the Spanish monarchy.

Something like a pastoral tyranny dominated the countryside. The Mesta's charter granted extraordinary concessions. It oversaw enormous routes of travel, ninety meters wide and hundreds of kilometers long. It brooked no infringement by sedentary farming or trespass by local transhumants on those routes. Farmers had to allow the migrating flocks to graze on the stubble of harvested field—in fact, often found it difficult to exclude the herds from *any* fields. Herds and herders gained almost unrestricted access to woodlands for fuel and forage, thus often subordinating local and communal use to this nationally sanctioned pastoralism; more than 80 percent of communal montes were thus opened to wholesale grazing. The Mesta was allowed, with or without the consent of landowners, to lease whatever land it wanted at whatever rates it chose to pay for however long it desired. Naturally this power inspired a rush of secondary effects, principally forests felled and fired, often initially in the context of war, and then for pasture; local transhumance, too, probed and pushed deeper into the mountains, beyond the grasp of the Mesta. Deforestation did not necessarily mean degradation. Special landscapes, in particular, evolved for swine and oak (Spain's *dehesa,* Portugal's *montada*) that achieved a credible equilibrium. But absolutism corrupted land use as much as it did politics.

The most insidious consequence of the Mesta was the subversion of any other economy, the singularity of land use by grazers and especially by the chartered agents of the imperial *trashumancia.* Spain became the sole economy of Western Europe dominated by pastoralism. Even after the Reconquista ceased, great throngs of cattle, pigs, and especially sheep, like ghost armies, annually reenacted the saga as they marched across the landscape. The country's pastoralism also followed the conquistadores to the New World, where the invaders remade the Mexican plateau into a new meseta and the Valley of Mexico into echoes of Castile and Estremadura. The great exploring *entradas* such as Francisco Vasquez de Coronado and Hernando De Soto marched with flocks of horses, pigs, sheep, and cattle as well as native allies, another reenactment of the Reconquista.

It is impossible to imagine Iberian society without those flocks—the cattle that supplied food, leather, and fighting bulls; the horses that carried conquistador and graced caballero; the mules and burros that long provided transportation in a nation without navigable rivers or canals; the swine that roamed for mast among the oaks and chestnuts; the installment of the ranch as the aristocracy of agriculture. But the dominance of livestock came at a huge price. It unbalanced the Iberian environment as fully as the hidalgo did Iberian society.

Fire was as much a part of this continental transhumance as it was of the Reconquista. It was too deeply embedded in the biota to extinguish; the fire climate was as relentless as the summer sun over Seville. Of course pastoralism and misused fire could damage lands, but pastoralism without fire was almost impossible, and fire necessarily endured. In fact, agricultural burning of all kinds was ubiquitous and essential; only its patterns and scale changed. Swidden long persisted in Galicia, less affected by the rhythms of the Reconquista. Elsewhere over the centuries staff replaced sword, and queima and queimada the torch, and as much as castles and flocks, they claimed the land. Over broad areas pastoral burning replaced field burning or superimposed new fires over those traditionally conducted for field and monte. The Reconquista imposed a new fire regime.

Illicit burning or wildfire in the uncultivated, typically mountainous lands elicited royal edicts of condemnation, like that from Philip II in 1558 in which he forbade grazing on burned montes in Toledo, Andalusia, and Estremadura without orders from the Mesta, a means of subordinating local usage to the prerogatives of the Mesta through control over fire. Where sedentary agriculture continued, the practice of subdividing fields among sons shrank and dispersed landholdings such that it miniaturized agricultural fire to the point that it was indistinguishable from debris burning. Peasants burned pruned branches where they had once fired forests. But everywhere in Iberia queimas and queimadas flared, like candles lit in a great cathedral.

So utterly domesticated was the landscape that wildfire was rare, erupting only where forced by drought, sirocco, or depopulation. Orchards replaced forests; *campesinos* harvested native woods as intensively as they did fields—they tapped pines for pitch, stripped cork oaks for bark, collected branchwood for hearths, fed understories of grass and brush to flocks, harvested selected shrubs for fertilizer, dug up roots for charcoal, and raked litter into bedding for barns. When overgrazed and overburned, forests degenerated into shrubs—heath to the north, matorral on the meseta, maquis to the south; all were in turn grazed and fired in a tight choreography of seasonal and secular exploitation. Burnings proceeded whenever fuels and a dry season made them possible, although the season of choice remained the late fall, when a sunny summer had dried fuels and the winter rains could soon encourage a new growth.

Fires were small because fields were small, pasturage was grazed closely, and forests ruthlessly stripped of fuelwood with even their surface litter raked for use as bedding. War or disease might upset temporarily the demographics of settlement and thus the ability of a particular place to enforce its will on a balky biota, but the larger pattern persisted. It endured century after century, as ruling class succeeded ruling class, as religions came and went, through Inquisition and Enlightenment, through the medieval maximum and the Little Ice Age. For all their ills, the Reconquista and Mesta had imposed an environmental order on Iberia and asserted control over its fire practices. When that power receded, wildfires would sprout up like woody weeds in an abandoned pasture.

It ended only with the halting, often bloody process by which Iberia began to modernize, events that intensified throughout the nineteenth century. Wars jolted Spain out of its lethargy, stripped away most of its overseas empire, and forcibly rejoined it to Europe and the new economy then emerging. Population grew, forcing more land into more intensive cultivation and accelerating, in many places, a degradation that culminated in soil erosion and outright desertification. The power of the Mesta receded, its privileges officially abolished in 1837, although a pastoral economy remained firmly entrenched in Iberian society. Liberalism promoted a market economy, a codification of law, the infusion of scientific and technical expertise such as mining, forestry, and hydrology, and on the example of northern Europe, a reconstitution of land to further those interests.

The results were often at cross-purposes. In what amounted to an Iberian enclosure movement, the great montes of the nobility, church, and monarchy were subject to confiscation and sale *(desamortizacion),* exposing formerly "public" or communal lands to a new level of use, including the conversion of forests to agriculture and pasture. Liberal economists intended the full liquidation of the public montes by 1854 despite critics who argued for the preservation of a communal-based rather than an individual-centered economy. Probably 4.5 million hectares vanished in what one critic called the "abyss" of desamortizacion. In almost the same breath, however, Spain established a forestry school (1848), hired professors trained in Germany, and began to pass forestry codes analogous to those in France. Even as desamortizacion, revolution, and civil wars eroded the remaining public wildlands, new legislation was passed to stabilize the montes (1863), promote reforestation (1877), stymie erosion, and of course prevent damage by fires.

Behind this was a gradual shift from rural to urban life, and from agriculture to industry that became more apparent through the early decades of the twentieth century. Environmental decay and social disintegration were inextricable. On the eve of civil war, legislation laid down a new program of public forestry (*Patrimonio Forestal del Estado,* 1935–36). The war, however, left Spain's economy in shambles, the nation boycotted, its European trading partners at war, and even saw a temporary increase in the population of rural Spain as urbanites joined campesinos in the fields to grow more food.

Much of the landscape, groaning from centuries of abuse, was also a wreck. Although the most notorious breakdowns testified to overgrazing, or more broadly to an imbalance in favor of pastoralism, nearly everywhere centuries of obsessive usage, like ever more burdensome taxes levied on the peasantry, had exhausted the land. In some places it was broken like an overwrought spring. Elsewhere it lay prostrate, stricken by a kind of biotic bleeding from soil erosion and the loss of nutrients. But with four decades of relative peace, planned industrialization gradually reconstructed both land and society.

Industry pulled and agriculture pushed. An industrial triangle developed in the northeast, anchored at Madrid, Bilbao, and Barcelona. To it flowed, in increasing numbers, a rural populace, particularly from the most densely agricultural regions such as Galicia and Andalusia. A similar phenomenon occurred in Lisbon, with the population following the rivers to Lisbon and Porto. Additional émigrés continued farther, to Germany or to other nations eager for labor. In part, the exodus reflected the attractions of a new life in a wage economy; in part, it highlighted the progressive marginality of traditional agriculture.

On the better lands, such as those in Andalusia, ownership consolidated farms into ever-larger aggregates, industrial latifundia, where mechanization replaced hand labor. Overwrought lands further accented the deteriorating character of peasant agriculture, uncompetitive in a world market and unattractive to a generation nourished on rising expectations. But regardless of specific cause, the collective outcome has been a decline in rural population—this in absolute numbers. Equally significant, the age structure of rural society shifted such that only the very young and the elderly remained. The human husbandry that had suppressed the pyrophytic fuels of Iberia faltered.

This transformation has come quickly, so rapidly that Iberians have the look of being immigrants to their own lands. The extended family resembles the classic structure of an immigrant family to America—the grandparents still clinging to the old ways; the first generation (born after the civil war) caught between two cultures; the second, the grandchildren, almost completely assimilated into the new world of metropolis, factory, and international popular culture of blue jeans and rock music. If anything, integration into the European Community will accelerate the trends as all the European nations struggle to reduce their massive subsidies to agriculture.

The landscape reflects these events, in many respects the most momentous geographic change since the Reconquista. Virtually every dimension of Iberian industrialization has influenced the vegetative mosaic; each tile has been removed and replaced, as it were, often with new materials and in patterns much coarser than those that preceded it. What had affected one tile before now influences a larger area, and what had sustained fire grudgingly now erupts into flame. Chemical fertilizers mean that fewer flocks are necessary as a source of manure, that herds need not be integrated into field cultivation, that many plants (such as *Ulex*), long checked by heavy harvesting for use as fertilizer, now flourish in riotous profusion. Mechanized wheat fields replace open range. The diminution of herds amid a global glut of agriculture allows matorral to blossom into woody weedlots the size of mountains. Shrubs and trees propagate beyond hedge and woodlot. Electrification has shut down the endless gathering of fuelwood for hearth and furnace. Fields have shrunk to kitchen gardens. Some land simply lies untended. Railroads have replaced the classic cañadas for the seasonal movement of herds; routes once manicured by the twice-yearly migration of enormous flocks now bristle with revanchist brush. The proliferation of scrub parallels the abandonment of agricultural land; the inflow of exotic plants matches the outflow of indigenous people. Elsewhere exurbanites claim abandoned farmsteads, like hermit crabs seeking new shells, not to farm the land but to use it as a home from which to commute to the city or for recreation; they do not subject the land to the intense regimen of traditional agriculture. Transhumant tourists, herded into flocks by watchful guides, replace transhumant pastoralists.

The rural exodus has made a shambles of the old regimen. Fuels and fires are no longer in sync, and neither operates within the constraints that formerly bound it. Fuels flourish and quemas and queimadas blossom with the seasons but no longer with the built-in firewalls of rural society—without the close cropping of fuels, without an abundance of able-bodied campesinos to control them, without the mores that once governed old fire practices. Those who remain are typically the elderly, who continue to burn but who often lack the capacity to contain, in this new landscape, the fires they initiate. Quemas and queimadas spill over fields, incinerate hedges, cross woodlands, invade long-suffering scrub. The boundaries blur. Any place not subject to meticulous cultivation is at risk. Any fuels not otherwise harvested are available for burning. Fires feed on untouched sites like ants gathering crumbs from a picnic.

But the reformation of land use has also been calculated. Its most spectacular expression has built on the 1940 forestry plan that committed the Franco regime to wholesale reforestation, as much as 100,000 hectares a year. To help support the scheme, Spain established a system of *montes ordenados,* redefining the dimensions and purposes of public land. Both programs peaked in the late 1950s, aptly expressed in the 1957 forest law. Industrial forestry expanded as well, capturing many otherwise marginal or abandoned rural lands. Mediterranean pines spearheaded the afforestation effort, although eucalypts soon reinforced and then overtook them. The major planting occurred in the humid northwest and arid southeast, the former an economic success, the latter a failure, though one for which alternatives are very limited. Much as the latifundia and villa reorganized the landscapes that predated Roman agriculture, just as the *rancheria* and *hacienda* reordered much of the landscape during the Reconquista, so now forest plantations (and other public lands) shape large portions of New Iberia. Revealingly the word for human resettlement, *repoblacion,* is also used to describe reforestation.

Reform brought resentment, particularly when it was imposed by an autocratic regime that came to power amid the horrors of civil war, and when reforms abolished a familiar (if dysfunctional) landscape. Because of the decades it demands and because it controls public access, forestry in particular tends to be a state duty. A protest against one is a protest against the other; both often come in the form of fire. (It did not help public perception that Augusto Pinochet, dictator of Chile, copied exactly the Franco forestry program and applied it even more ruthlessly, an ambition enforced by the most aggressive firefighting brigade in Latin America.) Groves of eucalypts, in particular, are biological deserts, not only because in dry lands their grasping roots soak up water like addicts, but also because they purge a site of indigenous flora and fauna, a planted ecosystem unable to coexist with local villages except through the cycles of a global economy. Large afforestation projects can be eerily empty. The death of Franco, the restoration of the monarchy, and a revised constitution (1978) that granted in principle greater power to Spain's seventeen autonomous regions all confirmed a new political order.

The new environmental order has shown similar adjustments. Afforestation has been discontinued except on sites savaged by soil erosion and selected farmlands in accordance with the European Community's Common Agricultural Policy. Instead, attention has shifted to the reservation of parks and natural spaces, as codified in a 1989 law, places in which the traditional vegetation can flourish. Either way a landscape once given over to pastoralism and labor-intensive agriculture is being reconstituted for very different purposes to suit a very different society. Portugal has told a similar story, compounded by the sudden loss of its overseas colonies and the political turmoil that abrupt severance occasioned. But here, too, the underlying themes—industrialization, emigration to urban centers, the erosion of rural society, a confused response to land reformation as symbolized by afforestation, especially when it involves the compulsory trans-

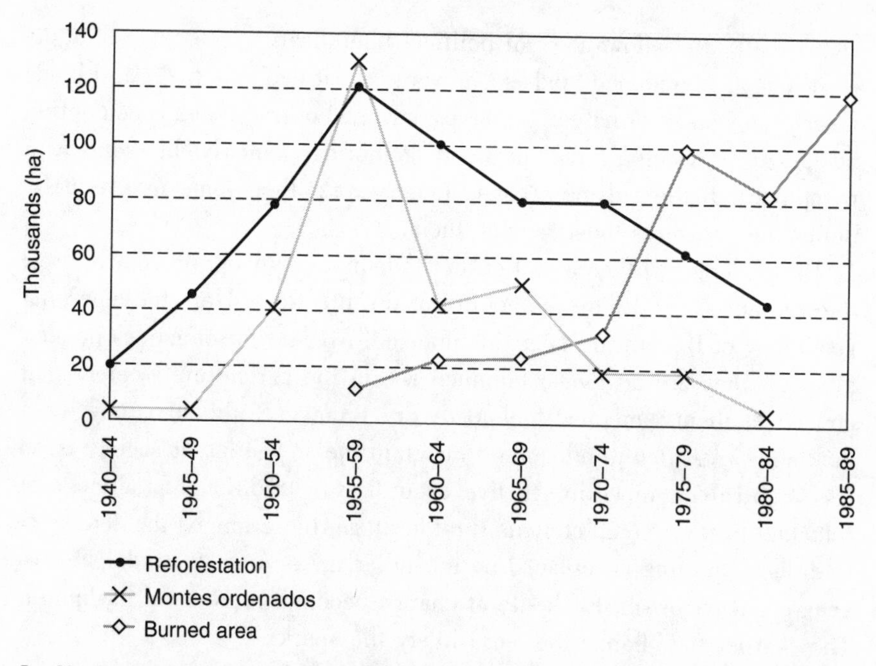

La Nueva Reconquista: reforestation and the return of fire; the record for Spain, 1940–89. The process begins with afforestation and the reservation of montes ordenados and concludes with a rising curve of wildfire burning.

formation of village commons—are identical. Both express themselves in fire. Throughout Iberia as the grasp of peasant agriculture loosens, fire slips through society's fingers.

The new geography of Iberia has thus shaped a novel geography of fire. The evolving patterns of vegetation delineate patterns of fuels. In general more fuels exist than before, not less, and the more pyrophytic species have replaced the less fire-prone. (An interesting gloss is the eucalypt, which suppresses the indigenous understory and thereby reduces the threat of crown fire that plagues closely packed pines. The eucalypt, however, produces its own abundant fuels against which indigenous methods of control are often helpless.) Afforestation is only a minor part of the problem, although the plotted curve of burned lands parallels, with uncanny fealty and allowances for political liberalization, the curve of afforested lands. Abandoned lands soon overgrow with volatile fuels, and lands reserved for parks or nature preserves, stripped of traditional consumption by grazing, burning, and firewood gathering, quickly blossom with pyrophytes. If the old methods no longer work, then some new, equally deliberate measures must replace them.

But no longer is the control of fuels adequate, by itself, to control fires; the burden is equally on the control of fire practices. Here the emerging geography of Iberian fire is again confused. Ancient fire practices persist, although they are no longer confined within the agricultural society that conceived them, and new fuel loads are promoted, although they often lack a social structure adequate to contain them. The larger society is no less ambivalent, unwilling to live according to traditional practices yet reluctant to sever completely its rural heritage. Increasingly, the demise of traditional burning is replaced by incendiarism as protest, vandalism, revenge, and terrorism. Each site of change becomes a potential flashpoint. The sharper the change the more likely the spark.

As traditional fire management has broken down, a modern fire management system, one better suited to an industrialized society, has built up. Spain has moved boldly to pass legislation to substitute for the lost mores of rural life, to fund a program of scientific research and technology transfer to replace lost or irrelevant fire knowledge, and to create an infrastructure to fight wildfire. The Cortes enacted a Forest Fire Act in 1968, and ICONA (the national bureau for natural resource management) approved a "national program against forest fires" in 1988.

The outcome has been impressive. Spain's research in fire ecology now matches, or exceeds, France's; it has adapted fire-danger rating systems and fire behavior models from the United States and elsewhere, and has coded its landscapes into computer-driven geographic information systems; it has installed lookout towers, laid out systems of fuelbreaks, fielded engines, helicopters, and CL-215 airtankers, and fought fires; it has argued for the necessity of substituting some form of controlled grazing and controlled burning for traditional practices lost to contemporary society. It has identified clearly the conflicts, deliberate or inadvertent, that have resulted in Spain's fire problem. It has become a leader throughout the Mediterranean, and by means of aggressive training programs it has projected that influence throughout Latin America. In 1990 Portugal, long a backwater in fire studies, passed new legislation for forestry and fire and hosted an impressive International Conference on Forest Fire Research at the University of Coimbra.

In its impacts industrialization is a virtual Nueva Reconquista, a "new reconquest." In New Iberia tourists replace the itinerant laborers and transhumant herders of earlier eras. Fuelbreaks substitute for the cañadas of migrating flocks. Unruly eucalypts invest even the walls of the Alhambra, challenging the vision of disciplined garden that the citadel had once broadcast. Firefighters battle incendiaries, not so much along a single frontier but as a scattering of confused confrontations and allegiances, a guerrilla war over the environment. *Torres de vigilancia* rather than *castillos*—fire lookouts rather than castles—watch over a restless people and a volatile land. Conflicting values over proper land use replace religion as an inspiring ideology. Flotillas of air tankers substitute for mounted chivalry. The state is reallocating lands, quietly seized by the forces of a global economy, to a society that is once again migrating across the countryside.

There exists a curious symmetry in all this. The conflict focuses primarily on the two regions that, like the poles of a bar magnet, held the centuries of the Reconquista in their fields of force. The two places of greatest population influx during the Reconquista, Galicia and Andalusia, now support the greatest outmigration. Fire loads have accordingly concentrated in the northwest and the southeast for different specific reasons but from related general causes. But Galicia, symbolic heart of the Reconquista, shows perhaps the most dramatic reformation, a cameo of the Nueva Reconquista.

Here in Green Spain (*donde la lluvia es arte,* "where rain is an art") fuels are abundant, unlike the southeast where abuse and aridity have stripped mountains to rocky ribs. On the average the northwest sustains roughly half the fire load of Spain. Between 1961 and 1986 45 percent of Galician forest and shrublands burned, and the process is accelerating. Between 1970 and 1990 33 percent of such lands burned, along with 24 percent of afforested lands. The most recent statistics calculate that

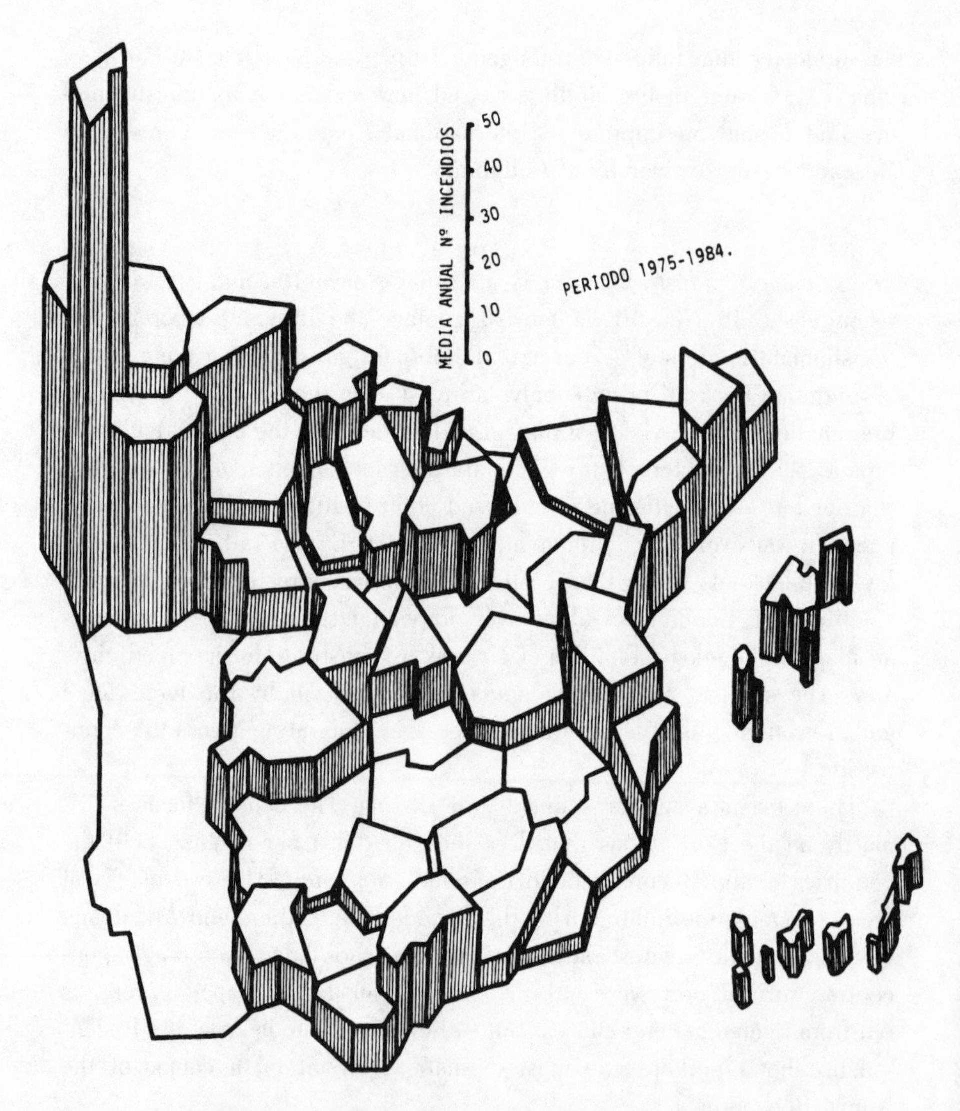

Wildfire's terrain: fires by province in Spain, 1975–84.

Galicia, with 5.8 percent of the landmass of Iberia, features 37 percent of its fires. Similarly the mountains of northern Portugal—the regions of Minho and Beira—with which Galicia has ancient ties have that nation's heaviest fire loads. The liberation of fire mimics closely incendiarism and the liberation of fuels through depopulation. Since the mid-1970s, as drought and political agitation have compounded rural flight and exurban reclamation, the burned area has increased alarmingly. Even lightning has reappeared like an apparition, kindling a 30,000-hectare fire in 1979.

The keep of the Galician castle remains Santiago de Compostela. During the Reconquista, its cathedral housed the putative remains of Santiago Matamoros—St. James the Moor Slayer—the central religious icon of the Christian advance. Santiago became the patron saint of Spain; the road to his relics, *El Camino de Santiago,* the most popular pilgrimage route of medieval Christendom. It could well remain so to anyone interested in the contemporary saga of the Iberian landscape. All the factors that have fashioned the Iberian scene, that have released long-suppressed fuels, here collect as though drawn by a new *camino* of fire. In Galicia every new firefighting technique and apparatus, from ground engines to air tankers, has had its first test. Here, too, ICONA promulgated its model fire plan, announced in 1990.

The great cathedral caps the city, itself perched on a swollen hill. At the base of the city flows a stream and beyond it stretches a narrow valley surrounded by a battlement of higher hills. Forests drape their slopes, bristling like a stockade. The scene would be almost unrecognizable to a medieval pilgrim, who would have seen pine groves tapped for pitch, flocks grazing where hillsides of eucalypts now thrive, the hills swarming with a peasantry gathering, agisting, cutting, pruning, harvesting, and otherwise disciplining the landscape into intensive cultivation. In the more remote reaches the reformed monastic orders would be slashing and burning open new lands not yet broken to plow and hoof. Now that process is rapidly reversing. The towers that guard the entrance to the valley broadcast television and radio, and the rooftops of the city sprout antennas, the cruciform icons of a new era.

There is no Santiago Matafuegos to lead the Nueva Reconquista against its enemies, but the reformed iconography of Santiago's cathedral does support its message in subtle ways. At the conclusion to their travels medieval pilgrims thrilled to the sight of the cathedral's great *torres*—twin towers—capped with iron crosses. Today those torres have, beside their crosses, other metallic spires. Lightning rods reach above all else, and their cables dangle incongruously down the flanks of granite to ground. New Iberia must be shielded against fire.

But it is farther south, at Castanheira de Pêra in Portugal, that the symbolism of the new era comes together with remarkable clarity. Here planners propose to combine tourism with forestry. Some rural lands they wish to convert to golf courses and spas, and farmsteads to rental condominiums. Others—the sweeping hillsides—they have planted to trees, a less seasonally driven industrial base. Expansive afforestation by even-aged plantations, however, demands expansive fire protection. Today hillside follows burned hillside. Where crown fires have decimated the native pines, eucalypts seize the sites. Between the retreating forces of traditional agriculture and the probing advances of an urban and industrial society lies a frontier of fire, a scorched earth. The airstrip intended to bring in tourists will begin as an air tanker base from which to fight fires.

Red Skies of Irkutsk (Russia)

The elements for a conflagration are slowly edging into position, like planets aligning in some cosmic tide. Geography, as always, remains favorable, the immense dialectic between a fire-sculpted biota and a fire-prone climate. The taiga is the largest contiguous forest on the planet, fluffed with conifers and rank with understory. Against it the continental climate of central Eurasia ensures that there is drought or dry lightning or sweeping winds or all of them synchronized, somewhere, every year. The land bubbles fire as if some subterranean coals seeped flame to the surface wherever a weakened crust allowed. But now history is conflating geography.

What immense space had diluted, history is distilling. Large fires often erupt during times of rapid transition, not merely as metaphor but as fact, and the rapid restructuring of the former Soviet Union promises to make its boundless taiga the scene of major fires, the next battleground in humanity's war on nature. As the Russians debate how, in the twilight of the

Cold War, to reorganize their military, so they must debate how to reorient their forest firefighting forces, the largest in the world, as they lurch into a market economy. The quickening pace may transfigure *leso pozhar,* the garden-variety forest fire, into *oghnennaya stikhiya,* the elemental fire, the conflagration.

The boreal forest is a fire forest, subject to a syncopated rhythm of fires—routine surface fires on the order of decades, stand-replacing crown fires by the century. Fires both great and small have hammered together an ecological architecture not unlike Russian Orthodox cathedrals, at once monumental and ornate. The age of forest stands dominated by one species or another typically dates from their last serious fire. The composition of stands—the relative dominance of fir, spruce, pine, larch—announces their typical regimen of fire. Restrict fire and spruce rules. Increase fire and pine dominates the western taiga, larch the eastern. The more fire, the more pine or larch. Studies of isolated islands amid the swamplands of western Siberia suggest that surface fires ripple through the pines every seven to ten years. Fir trees claim a midrange, shifting like glaciers with the tidal pull of changing climate. Everywhere patches of birch and aspen decorate once-burned sites. Unless revived by fire, they fade into oblivion within forty or fifty years. Somewhere in this fiery center, on some Siberian Sinai, it is believed that the genus *Pinus* emerged tens of millions of years ago.

Taiga history is actually a complex ecological fugue between fire and fuel. Some fuels result from past fires, and very large conflagrations can initiate a cycle of reburns, each scavenging on the residue of the last, that can persist for decades. Most fuels, however, result from aging, disease, insects, windstorms, and the ax. Different sites host different understories according to their genetic reserves, their particular hydrology, and the peculiarities of their soils; these, in turn, help control the pattern of underburning, and the likelihood that surface fire will escalate into the crowns. These fuel cycles—*rhythms* would be a more accurate term—lay down the basic tempo of fires.

But climate also imposes a rhythm of its own, far more atonal. Prolonged wet spells crowd fire out of sites otherwise disposed to burn, encouraging fuels to stockpile. By contrast, droughts impose new stress, leaching away moisture from soil, understory, and canopy, all of which expands the domain of available fuel. The immensity of the Eurasian landmass—blocked to the south by towering mountains and to the north by a frozen Arctic Sea—stamps the taiga with a continental climate well contrived to bring seasonal dryness almost everywhere and outright drought somewhere. Chronic droughts, in turn, can make available even bogs or organic soils normally shielded against combustion by their sogginess. In 1915 swamp fires in western Siberia spewed forth a stagnant smoke cloud the size of Western Europe. In 1972 drought and dry lightning conspired to besiege Moscow with stubborn bog fires that defied all efforts at control. Between them, fuels and drought make some part of wildland Russia available for burning every year.

But fuels are, or once were, organisms, and if fire is to be integral to the taiga, it must serve biological ends, as of course it does. It performs the same generic duties here as everywhere, restructuring the biotic architecture of ecosystems, recharging their dynamics, recycling nutrients and organisms. For some subbiotas fire is mandatory; without it they would disappear or wither into insignificance. For all, it is simply there, an environmental presence not so different from frosts, windstorms, seasonal changes, beetles, or the other stresses and strains inherent in a boreal existence. The taiga, however, demands more.

Summer is short; winter long with dormancy. The taiga must crowd its complex growth and decay into a brief, furious explosion. Decomposition, in particular, proceeds slowly, ponderous as wood fungus. Unless some way can be found to accelerate decomposition, more and more nutrients will disappear into tree boles or into inert stockpiles of logs and litter that will be removed from the pathways of active organisms who, like other long-suffering Russian consumers, join ever-lengthening queues for ever-shrinking inventories of goods. The situation is even more acute in permafrost. The zone available for active growth is shallow; the available nutrients and water scarce. If hoarding continues, the taiga suffers famine. For new growth to flourish, for biodiversity to thrive, these various storehouses must be opened for access. Fire does exactly this. Perhaps more critically it controls the depth of thaw in permafrost by removing the insulating layers of raw humus and by blackening the surface so that it can absorb more solar radiation and accelerate warming. Without fire the taiga could not survive on permafrost.

In this model of nature's economy, fire becomes an invisible hand, seeking out available fuels, transferring critical materials along pathways to where they are most needed, balancing supply and demand. Interrupt that idealized, laissez-faire flow and the system slows, its geography becomes lumpy with maldistributions. Eventually it may simply collapse, the ecological equivalent of an economy in which national wealth is buried in socks or stuffed into mattresses. This, of course, is precisely what the Soviets had done.

It is a difficult land for humans to inhabit. But without fire it would be almost unlivable; and humans like other foragers of the boreal forest are drawn to those sites where fire has brought the taiga to life. They gather in the greatest numbers at those sites where fire is most prominent. The difference of course is that humans can add to the bequeathed fire load and reorder its geography, using their indispensable fires not only to warm and cook but also to reshape this inhospitable biota to better suit their needs. Until very recently they have exercised that option whenever possible.

Anthropogenic fire has traced the wanderlust of Siberian peoples, etching out routes of travel, sustaining settlements, delineating the dominion of human habitation as surely as blazed trees or a surveyor's stakes. Hunters burned to entice moose and elk into special sites baited with fresh, fire-flushed grasses. Trappers burned along traplines to keep open pathways through taiga and muskeg, and to promote the grasses and forbs that mice and other creatures sought, to be preyed on in turn by the foxes and lynx pursued by humans. They fired marshes and sedges around ponds to stimulate waterfowl. Fishermen burned, deliberately or accidentally, along waterways, their torches serving as a kind of flaming ax to chop back the grim forest. At night they fished by torchlight, and drew prey like moths to a candle. Likely, from time to time, travelers also fired windfalls and other dense concentrations of impenetrable woods, remaking them into sites that would later tolerate travel and attract moose. Foragers knew that berries and certain mushrooms flourished best on lands well burned. Along common corridors of traffic—roads, rivers, and later railroads—fires sprouted like weeds, blown from campfires or smoking fires, eating into the taiga like glowing moths in wool. Natural openings—*polye*—became points of accessibility into the trackless taiga, and fire both sustained and widened them into refugia vital for human existence. What Arthur Adams wrote from Manchuria in 1870 could well extend over the subcontinent. There were a "prodigious number of those charred and blackened trees that strew the ground in every direction," he observed, the outcome of "wandering shooting and fishing parties of Manchu tartars, who always fire the scrub and burn down the trees, to clear the land and make it yield good pasturage."

So, also, with agriculturalists. Peasants practiced slash-and-burn agriculture, fired fields to stimulate the early growth of pasture grasses and increase soil warmth, burned or reburned sites to keep abundant the fields of lavender fireweed that bees favored, pruned and promoted berry patches with fire every few years. Where domesticated livestock partially replaced wildlife as a food source, and cultivated fields the wild rangelands, there was no reduction in the mandatory fires. Without fire, taiga would overrun pasture or the grasses would degenerate into unpalatability. A protective belt of spring fires also helped cleanse the land of low shrubs that housed that scourge of Siberian mammals the tick, a notorious vector for encephalitis. Resin tappers of pine kept their understory clear of excessive fuels, often with careful burning. Most deliberate firing was restricted to the spring, when moisture or snow could contain its spread into the surrounding woods. Anton Chekhov spoke nonchalantly of the May burning—"serpentine fires"—that accompanied his travels to Siberia. "The fiery snakes creep unhurriedly, now breaking into segments, now flaring up again." Sparks, smoke clouds, and an "eerie illumination" among birches highlighted every thoroughfare.

But inevitably escapes occurred, and the desire for a hot fire often forced peasants to withhold their burns until later in May when the fires could penetrate into the taiga like searching fingers, feeling out points of vulnerability, slowly pushing back the border between field and forest. During droughts, such fires could rage uncontrollably and reshape the landscape wholesale for centuries. Logging—sometimes integrated with slash-and-burn cultivation, sometimes independent of it—recklessly lit-tered the taiga with surplus fuel that only fire could dispose of, if not with deliberation, then as wildfire. Especially after the Trans-Siberian Railroad opened the interior, settlement advanced eastward like a wedge driven into a log, widening the ancient crack between forest and steppe.

This accelerated settlement, in turn, accelerated fire. Settlement became a flaming front, fanned by the violent winds of European expansionism. In a famous prophecy a Buryat shaman announced a vision in which the dark fir forests were burned and replaced by birch and light pine, a premonition of how the white-skinned Russians would succeed the dark-skinned Mongols. And indeed the conquest of Siberia came with fire as well as sword.

The effects were evident for anyone who wished to look. V. B. Shostakovitch, after noting the mandatory burning required for habitation, lamented that "such fires being repeated from year to year, most of the new Russian settlements within the taiga are surrounded with burned out forests. This sad picture is to be observed over millions of acres." Fridtjof Nansen concurred. "It was strange," he noted in 1914, "here, as everywhere in Siberia, how seldom one saw really big trees; the forest seemed often to consist of nothing but young trees; this is not because they are felled, but rather because they are wantonly burnt; and there is no end to these fires, one sees signs of them everywhere." Wildfire threatened wooden villages as well as surrounding woodlands, accentuating the consciousness of fire. So often did towns burn that they almost constituted a form of urban swidden. So pervasive was fire that S. V. Maksimov likened "forested and wooden Rus' " to "an inextinguishable bonfire which, never going out completely, first weakens, then flames up with such monstrous force that any idea of fighting it vanishes: an entire sea of flames is spread by a whirlwind of fire from one end of our unfortunate land to the other, and destroys without a trace forests, planted fields, villages, settlements and towns."

Still, until recent decades, the dominion of anthropogenic fire was restricted. It branded a matrix within which lightning fire had to operate. Eventually the fire-plastered mosaic that was Siberia became more mobile, flexing with the movements of humans, distorted like a balloon squeezed into new shapes. Outside the pale fire of human settlement, anthropogenic and natural ignitions intermingled in ways that are difficult to disentangle. For the true taiga, it appears that natural fires, supplemented by some human burning, were probably ample for the sparse human population that prowled this lonely land, a fire species more opportunistic than calculating. But it is equally true that the human population of Siberia was restricted *because* fire could not remake this environment as readily as it could others. Compared with grasslands or shrublands, the boreal forest responds slowly, compensating for its ponderous pace by its enormous expansiveness. It substitutes space for time. The parameters of fire geography count for more than those of fire frequency.

These traditional practices have endured. It is common folk wisdom that berries grow best on periodically fire-pruned shrubs, that certain favored mushrooms erupt profusely after the right combination of fire and rain, that large game migrate into the flush browse regenerating from severe burns—that much of what humans require to live in the taiga is possible only because of fire. If nature fails to supply adequate ignition, then humans must. The grandparents of those now charged with administering the forest knew this truth and exercised their stewardship, where possible, with fire as well as with ax. In a sense what the ax took from the taiga, fire helped restore.

Settled areas and their penumbra still burn regularly. Early burning occurs, as it must, around nearly all agricultural settlements. Each spring traditional fires redefine the boundary between the wild and the domesticated, delineating their relative dominions as surely as wolves marked off territorial perimeters with urine. Without routine fire, fuel stacks up like biotic cordwood, compounding on itself as though it were a usurious debt, which in a sense it is. Without intervention the relentless taiga will reclaim the lands wrested from it by fire. So, even when condemned or prohibited by statute, the burning continues, sub rosa if necessary. It is probable that many fires now attributed for statistical purposes to "unknown causes" or "cigarettes" are really examples of traditional burning that officials either do not understand or choose not to recognize.

There were of course areas that humans wished to spare from routine fire. Fur trapping, especially for sable, required older forests. While sable would be smoked out of tree cavities, it was not hunted by fire drives or seduced by fire-baited traps; sable hunts, moreover, occurred during the winter, when broadcast burning was impossible. Also fundamental is the curious fire ecology of the Siberian pine, whose nuts furnish vital food to squirrels, bears, boars, and humans, the common kernel of taiga food chains. And even the prevalent pastoralists—reindeer herders—apparently shunned burning outside special pastures such as the *mar'*. Fire quickly boosts grasses but depresses, often for decades, the lichens that reindeer prefer for winter range. Spring range required fire use; winter range, fire control. The more normal association between pastoralism and broadcast burning—see, by way of contrast, the nearby Mongols and Manchus who burned grasses annually—was here sundered. It is probable that controlled browsing by larger herds of reindeer could substitute for some controlled burning, at least with respect to modulating surface fuels. Otherwise herds gravitated between treated grasslands and wild taiga as seasons and nutrient cycles allowed.

But these attempts at fire restriction took on whole new dimensions with the arrival of professional forestry and industrial logging. Fire loads piled up like logs at a mill, threatening ever greater holocausts. Professional forestry aggravated the situation by seeking to banish not only logging fires but all fires, and so revealed its origins in German silviculture, implacably hostile to fire as a putative enemy of humus. These proscriptions extended even to logging residue, so that the growing and harvesting of timber no longer resembled its agricultural cognate, slash-and-burn cultivation. In Russian forestry, burning followed slashing only indifferently, most often as wildfire.

In much of European Russia, where agriculture intensified, foresters could plausibly argue their case, exiling wildfire to Siberia. After all, the heartland of Western Europe had more or less accomplished just this task. But England, France, the Low Countries, and the German states could count on a benign environment not subject to dry lightning, drought, or foehn winds; on a landscape so intensively managed that it resembled a garden; on a dense population long disciplined against careless burning—all circumstances within which fire could be contained. Accordingly, the more progressive elements of Russian society sought, on European models, to contain the fires that pervaded Russian life. In 1893 members of the nobility and professions, with royal patronage, established the United

Russian Fire Association, a focus for the volunteer fire organizations that had sprung up over the previous two decades.

None of these conditions applied, however, to Siberia. To control fire would require a gargantuan investment, a larger cause than mere fire control. It meant not only controlling settlement—this at a time when Stalinist Five-Year Plans called for a "great transformation of nature" and targeted Siberia for massive hydroelectric dams, industrial expansion, and new cities—it also demanded control over those fires that lightning kindled with supreme indifference to the centralized edicts of the socialist state. The command economy would also have to command nature. Fire suppression had to go beyond the ideology of received forestry. To be inclusive it demanded the mobilization of total war. All-out fire control came, in fact, only with World War II and the Cold War that glowed in its ashes.

Geopolitics underwrote the spectacular advance of forest firefighting into the deep taiga of Siberia. Aerial fire control came to the Soviet Union, as to the United States, with World War II. It blossomed in Siberia, as in Alaska, with the deepening Cold War, a paramilitary force that would advance geopolitical goals—to keep the skies clear of obscuring smoke, to firm up an administrative presence in distant landscapes, to encourage the exploitation of natural resources. Amid the vastness of Siberia, ground forces are inadequate, often poorly motivated, overly reliant on tracked machinery that looks like something out of a mechanical Cretaceous Era. Only aerial fire protection could serve the strategic interests of a command economy. By 1990 some 8,500 firefighters, cross-trained as smokejumpers and helirappelers, staffed twenty-two aerial fire centers across the USSR.

But, as Francis Bacon once observed, nature to be commanded must be obeyed. This monumental exercise in fire control could not succeed by sheer intimidation, technological power, extortion, or exile. Nature would reciprocate. Fire defined a new, often deadly dialectic. Increasingly the ancient minuet between fire and climate became an atonal dervish between fire and culture. With remarkable consistency the great Russian fires of the century have coincided with periods of both climatic drought and social unrest. The conflagrations of 1915, 1921, and 1936–39, for example, occurred amid world war, civil war, and internal purges, and resulted not only from a hostile environment but also from a fragmented social order unable to control either its own or nature's fires. The Great Terror was expressed in taiga fires as well as show trials. Of course Russia is vast, fires common, and political turmoil frequent; but this hall of flame registers something more than coincidence—it reveals a serious uneasiness, even a contradiction, in how Soviet society related to its natural surroundings. The attempt to suppress Siberian fire was an environmental equivalent to the postwar seizure of Eastern Europe. It could be held in thrall for only so long.

What geopolitics gave it eventually took away. First perestroika, then the August revolution of 1991, then the dissolution of the Soviet Union promised to redefine those informing causes. Some of them—those most closely identified with the Cold War—have already vanished. Others are only measurable according to the calculus of a market economy that is still tentative and alien. Until the Russians install an economy based not only on the market but also on money, no one will know the economic value of the taiga. Still other causes, such as granting serious protection to parks and nature preserves, must serve goals outside the stated parameters of the market, however it evolves. In effect the institutions for fire management—those essential not only for firefighting but also for fire research—are being disestablished. Beyond that is the question of suitable means and ends. To reconstitute that establishment with any meaning the Russians must reconsider how their land will be used. That will determine why and how much they should protect it from wildfire, and which institutions are the appropriate vehicles to accomplish those ends. The dialectic continues.

All in all, Russia boasts the most interesting fire scene in the contemporary world. The scale is enormous, the consequences vital, the velocity of change swift, and fire inexpungable. Some one billion hectares of forest land may be at stake, probably 750 million hectares actually forested, perhaps 25–30 percent of the world's total. Pressures mount to convert this forest into cash, preferably into hard currency. With much of the tropical forest slowly shutting down, either from exhaustion or legislation, global attention must necessarily shift to the opening boreal forests of Siberia. With traditional political controls loosening but market controls not yet in place, the temptation will be strong to cut now and let burn later. It is an easy matter to accelerate logging, particularly for export, and to choke off fire protection, or to shift the burden of firefighting from aerial attack in remote regions to ground-based methods restricted to those areas of heaviest exploitation. After all, the "real" value of those services and the values of the resources they protect are unclear. But with or without humans, fires are abundant, and firefighting the only form of forest insurance available. The magnificent fire heritage of the taiga will force the Russians to make decisions. The August revolution may inspire a biotic revolution of equal magnitude.

That reformation has already begun. Much of Russia's vaunted forest land is no longer forested. The taiga is far less trackless than it used to be. And fires seem to be increasing in severity. The number of fires appears roughly constant, but the size of burns has increased dramatically. After decades of decline, at least since World War II, the amount of burned area is rising even more sharply than costs.

Why? No one knows. One explanation is that the figures since glasnost are honest and those before are not. Certainly commissars routinely falsified the official numbers, often downgrading burned areas by one or even two orders of magnitude. A burn that firefighters held to 1,000 hectares in Yakutia might shrink to 100 hectares by the time reports reached the central air base in Pushkino. (The 6 million hectares that burned in Chita in 1987 metamorphosed in official statistics to a mere 94,000 hectares.) One legacy of this statistical corruption is that, with a few exceptions, the fire agencies do not have a competent data-set upon which to base their plans, unless they kept a double set of books, which some did.

But there are legitimate reasons for the increase, too, which relate to environmental conditions. After a parallel decline in historic burned area, Canada also experienced a massive reversal during the 1980s, so much so that it has had to reconsider its rationale for fire protection. While the numbers differ, the statistical curves for Canada and the USSR are uncannily similar. For this, global climate fluctuations are partly to blame, and so is the logic of fire control with its capacity to accumulate forest fuel.

Aggressive fire suppression broke the ancient dialectic between fires and fuels. Following the millennial fires that laid siege to Moscow in 1972, the Brezhnev regime's immediate reaction was to airlift 1,200 Sibe-

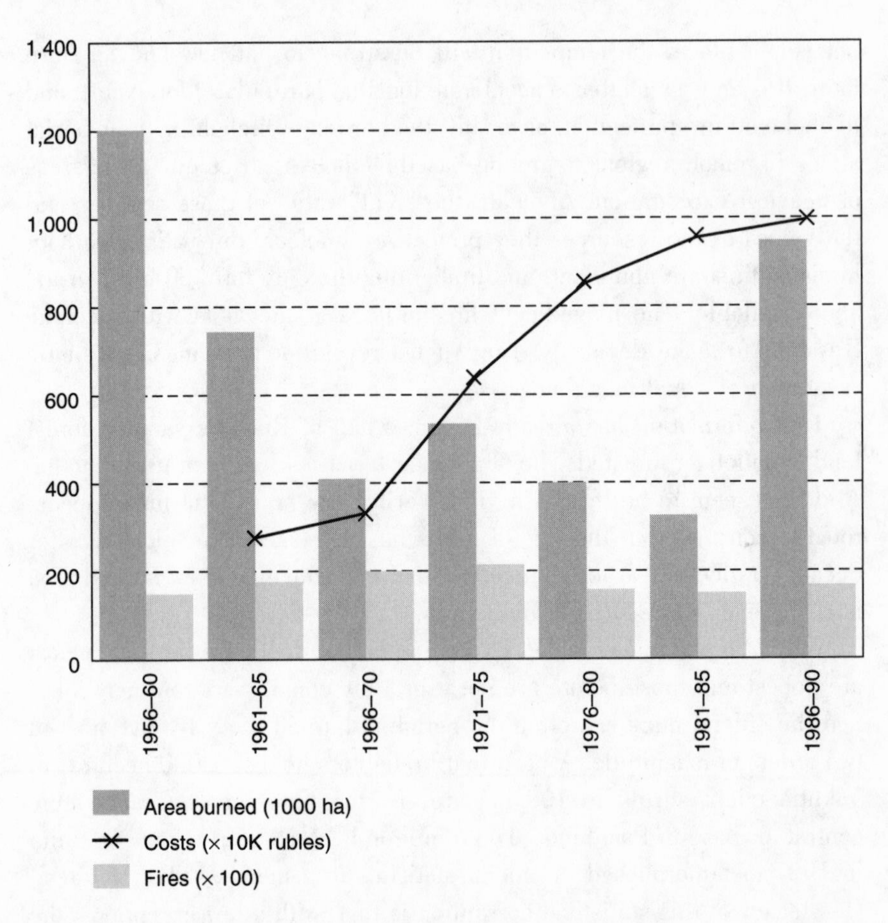

Red Skies: the success, cost, and breakdown of fire protection in the Soviet Union. The actual numbers are almost certainly wrong; reports of burned area were routinely downgraded as they moved from field to Moscow in order to meet production goals, often as much as a hundred times; but the trends are very likely correct.

rian firefighters to break the fiery blockade. The longer response followed the Brezhnevian doctrine that "the more that is forbidden, the better." All burning—all legally sanctioned fires, that is—ceased. Traditional burning around fields and villages disappeared, or went underground (literally). Fuels built up; what had been buffers, annually burned, now became fuses, stuffed with combustibles and ready to carry fire from village to

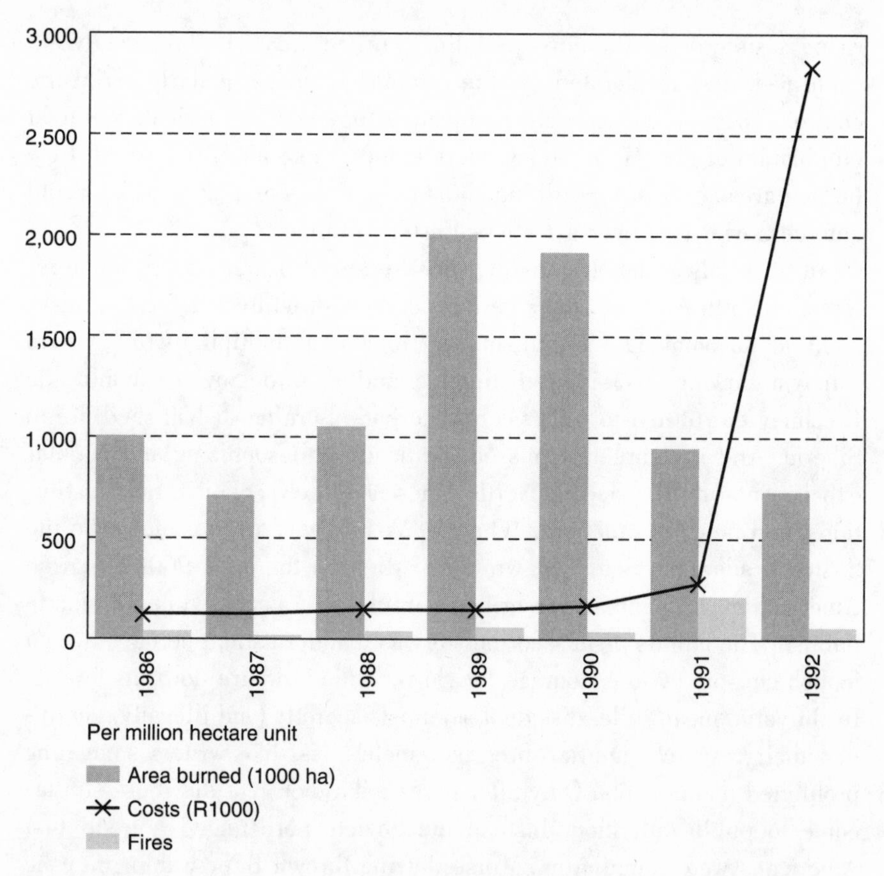

Revolution and inflation. Fire management cannot escape the reordering of Russian society, but the drought-powered fires in European Russia during the summer of 1992 staved off a systemic collapse. Aerial fire protection, in particular, has been vulnerable to rising costs of gas, otherwise a source of hard currency.

taiga and back again. When fires broke out, they burned more fiercely. Not until gargantuan fires burned millions of hectares east of Lake Baikal in 1987 was this absurd edict quietly shelved, not abolished but like so much Soviet law conveniently ignored. After the August revolution, controlled burning was again sanctioned in Transbaikalia.

The 1972 fires were a watershed. Plot the fundamental statistics governing the Russian fire establishment, and the years surrounding the Moscow conflagration mark a major intersection, a time when the curve of rising costs crossed the curve of falling burned area. The fires catalyzed, confirmed, and accelerated the fire establishment, particularly aerial fire control. They posed as a clear objective, however chimerical, the total elimination of fire. More money, more planes, more air bases would drive burned areas ever downward, much as more troops and more tanks would enforce Soviet rule over a restless Eastern Europe.

In an eerily symmetrical way, those years also announced major reforms in North America. But where America restructured its fire establishment to accommodate pluralism, to promote a multiplicity of fires to satisfy a growing diversity of distinctive land uses, the Soviet Union of the Brezhnev era turned to repression. It forbade burning. It banished fire to Siberia. And it completely missed the heady philosophizing and popular experimentation that led in North America to new policies, new institutions, and new fire practices. When V. V. Furyaev, a fire ecologist in the Soviet Academy of Sciences, wrote an article in the late 1960s based on American experiments in prescribed burning, he was denied permission to publish. The editors dismissed the story as disinformation, perhaps even a foolish conspiracy to encourage the Soviet Union to burn down its forests. In Buryatia, meanwhile, dissident scientists secretly (and illegally) continued their research, igniting fires in remote sites, like writers smuggling prohibited manuscripts. Only after perestroika took root did those studies come to public attention. Instead the Soviets perpetuated a trend that Americans were repudiating. Worse, having thrown Br'er Rabbit into the brier patch, they then went in after him.

Now a second crossroads has appeared. Those statistical lines are intersecting once again. Fire starts are slowly increasing, investment in real money (adjusted for inflation) is dropping, and the amount of area burned is reaching record levels. Having been artificially depressed after the edicts that banned burning, fire is reasserting itself. Simultaneously the costs of aerial fire protection are tracing out a logistic curve, now apparently approaching an asymptote; the first-order infrastructure is built out; a point of diminishing returns has passed; it is more likely that old aerial bases will be disbanded than new ones added. Aerial fire protection will have to do more with less. Already Avialesookhrana has shrunk its staffing to six thousand, and only the 1992 fires that threatened Moscow prevented a free fall into disintegration.

But it is the burned area that most shocks. It is as if, once perestroika lifted the dead hand of suppression, fires are making up for those lost decades, as though they are intent on accomplishing with cataclysmic haste what ham-handed repression had denied them. In fact the revival of fire is many times larger than the official statistics claim. Satellite imagery of the May 1987 fires suggests that 12–14 million hectares burned in Transbaikalia and southern Yakutia, probably the largest wildland fire complex in the earth's last fifty years—this after almost half a century of paramilitary fire control.

The official response was denial—and the Russians continue to deny that anything occurred. Even the Academy of Sciences has refused to sponsor any research according to the Orwellian logic that it cannot study what does not exist. The apparent scenario is that normal spring burning met a sustained outburst of dry winds, the *burya;* thousands of fires set around hundreds of villages suddenly became a hurricane of flame. The full extent of casualties is unknown, but considering tinder-dry villages of log and clapboard, they were likely extensive. No firefighting organization could cope with such conditions, certainly not one caught up in the uncertainties of perestroika.

Predictably, apparatchiks urged better technology, more planes, more firefighters, as though they were suppressing an uprising in Prague or Budapest. The more savvy observers, however, realized that the failure was systemic. Because spring burning was essential but illegal, it could not be regulated; because Russian firefighting had a similar structure to the Russian military, it suffered from similar breakdowns, its firefight roughly equivalent to the misadventure in Afghanistan; because the old order had exhausted itself, its stagnation of purpose, its bureaucratic sclerosis, affected even the capacity to manage the overall fire problem. With Western attention turned to the fire at Chernobyl, however, the Soviets could bury the biotic meltdown that occurred east of Lake Baikal. But if they lied to the outside world, they could no longer lie to themselves. Eventually the August revolution came to the taiga.

Now the Russians confront an environmental dilemma not unlike and not unrelated to their political crisis. They need their forests to help finance the move to a market economy but they may not be able to pay the costs necessary to protect those resources long enough to get them to market or to rebuild, not simply strip-mine, them. Their desire for increased logging coincides with a likely increase in large-fire potential, while their own success in fire control will probably help stoke small fires into larger conflagrations. If global climate warms, as some fear, the boreal forest will burn with a ferocity not witnessed since the medieval maximum. Having engaged the forest, the Russians will have a very difficult time disengaging. For many reasons, they cannot. The Russians have a taiga by the tail.

The worst-case scenario calls for Siberia to become a boreal Brazil, though that probably will not happen. An infrastructure exists in Russia for forestry and fire protection that does not exist in Brazil. The Russian problem is one of institutional readjustment, not of inventing an apparatus for research and management. The Russian fire establishment at least has a reasonably comprehensive system for fire surveillance and response. Moreover, Siberia, unlike Amazonia, has an indigenous fire load, a background count quite independent of any human agency, that will ensure that the problem remains one of *fire* management, not simply of people management. The issue will not be solved by social policy alone but by a proactive fire policy. The strains are severe, however, as much intellectual as operational.

Here, too, there is reason for restrained optimism. The Russians have a folk attachment to their forest not apparently true of people in Latin America and Africa who are zealously clearing off their woodlands. Russian firefighters share this heritage. Most grew up in villages made of forest materials. Their diet is supplemented by forest products; when not attacking flames, firefighters are foraging for mushrooms, berries, pine nuts, and medicinal roots. They often jump into a fire without tool handles, preferring to make them from materials on the spot. They feed off fish caught in streams. They literally live off the taiga. Sometimes they die from it; in recent years bears have killed two of them. All in all the older Russkies, for whom firefighting is a career, offer a striking contrast to American crews, overstocked with suburban students working as seasonals. When fire season ends, the men retire to the taiga to hunt for three or four weeks.

Avialesookhrana, the aerial fire protection services, is the most progressive element in the Russian fire establishment. Russians manage to patrol a landmass the size of North America with forty-year-old aircraft, secondhand military hardware, and boots that only survive three or four jumps—no small accomplishment in a country where even tourist hotels lack normal plumbing, soap, or curtains. Yet even before the August revolution the directors and their staffs believed they could make a new future—this in a country whose citizens have seemed ready by and large to flush everything since the October Revolution down the toilet, if they could only find a toilet that worked. They have looked to America for help.

The United States has something to offer them, at little direct cost to itself. From American history Russians can learn how to reorient their bloated military establishment to civilian purposes. They could divert some of their national service obligations into forest jobs, a kind of Civilian Conservation Corps. They could study how to transfer military hardware to fire control, as the United States did so successfully after the Korean War. But it is the record of the past twenty years that speaks most directly.

In those two decades the U.S. fire establishment experienced its own perestroika—the outcome of redefined missions and a reallocation of public lands, of rising costs, of an assertion of biocentric values. American fire protection evolved beyond a simple (and unrealistic) policy of all-out fire control, an aberration of the command economy for public lands laid down during the 1930s, to a pluralistic program of fire management, one in which fire is adjusted to meet ecological and economic diversity. Those experiences are exactly germane to the Russians. They need to learn how to withdraw from certain regions without leaving a scorched earth behind, how to substitute controlled underburning for random wildfire, how to negotiate the contradictory fire programs demanded of natural areas, especially those such as Lake Baikal subject to international "watches."

The USSR missed this fire climax, as it missed so much else of a "normal" political evolution, because of the dreadful hiatus imposed by Stalin. The debate about appropriate fire practices has been an environmental rite of passage common to every industrializing nation. On the average it seems to last about sixty years, from the first public statements to the adoption of a rough national strategy that balances fire use with fire control. Having avoided this controversy, the Russians must now confront a rebellious environment and a virulent environmentalism without the benefit of suitable inoculations. The shock promises to be severe. The good news is that they can learn from others; the bad is that they may have pushed and entrenched a particular strategy too far to allow for easy reconstruction. The prospects are for a lot more fire rather than much less. The question may boil down to whether Siberia's size can compensate for its distorted history.

The most relevant analogue may be Alaska. As America's Siberia, Alaska experienced a peculiar evolution out of sync with national trends. Debates long cultivated elsewhere were deferred, then accelerated and condensed after statehood and the crisis over land ownership that culminated in the Alaska National Interest Lands and Conservation Act. Suddenly the Bureau of Land Management, which had predicated its administration of the interior on aerial fire control, found it necessary to redefine its fire-protection mission as the public lands were divided among multiple agencies and various ends. Both the American taiga and Alaskan fire management survived the trauma. If Brazil warns of a dark future, Alaska promises a bright one. Besides, what the Russians need most is knowledge. They want to learn.

The United States has something to gain in return. Soviet science has much to teach about fire ecology, and the Russian landscape offers new arenas for investigation into the fundamentals of one of the earth's most elemental processes. If, as many environmentalists fear, large ecological changes anywhere induce changes everywhere, then the restructuring of Siberia has global consequences. The taiga holds tremendous carbon stocks, an atmospheric sponge against certain greenhouse gases.

So with the fire community leading and their governments following, contact is leading to cooperation. What the threat of nuclear holocaust could not do, the prospect of forest fires could. Exchanges of official delegations have led to formal programs for the exchange of personnel during fire season. A mutual aid agreement now allows for assistance on active fires via the Bering Strait. The U.S. Forest Service has installed an e-mail system so that the Russians can communicate with colleagues without relying on opportunistic couriers. Books are being translated; contracts for research signed. The Russian Federation has joined the International Boreal Forest Research Association, a circumpolar consortium focused on tundra and taiga. Thanks to the indefatigable Johann Goldammer a major international symposium on the fire ecology of northern Eurasia has joined Russian fire specialists with colleagues in Europe, China, and North America. A simultaneous field campaign to measure fire-generated emissions integrated Russia into an international survey that has sent experimenters to Brazil, Canada, the United States, and Africa. Even nature has helped. The great drought of summer 1992 brought large fires back to European Russia, particularly the Moscow *oblast.* Alarmed, the Russian government granted special status to fire protection and allocated several billion rubles to upgrade its capabilities. The fires rejuvenated the Russian fire establishment, not unlike the way fires can renew the taiga.

Not least, the U.S. fire establishment, too, could learn something from its Russian counterpart. Soviet technology often resembles Australian fauna, the inverted product of isolation and a separatist evolution. In particular, firefighting in the former Soviet Union increasingly orbited around a strategy of helicopter attack, with firefighters rappelling out of MY-8 helicopters, then laying down explosive cord, which is blown to make firelines. The United States has not learned how to make either technique work. (Ironically, Russia, where labor is cheap, opts for high-tech fireline construction; the United States, where labor is expensive, prefers labor-intensive handline.) Any nation that can spend more than $130 million *not* putting out fires in Yellowstone National Park has something to learn from an organization that combats wildfire over one sixth of the world's landmass, and manages to do it with Antonov-2 aircraft, explosive cord, and bagfuls of Siberian pine nuts.

Exchanges, however, will not come easily or without irony or perhaps even tragedy. The old apparatus remains, unable to enforce its will but sufficiently strong to encumber the will of reformers. Lenin still looms over every fire classroom. And there is the story of Misha the Bear, Russia's counterpart to Smokey, a cautionary tale from the taiga. The story began when a fire crew at a satellite base in Bratsk brought back a bear cub. Under their care Misha grew, and grew, and grew some more. This, however, was no celebrity creature, no poster bear, no cartoon. One day, before it was understood how serious the situation had grown, Misha ran wild in the village. The fire crew had to hunt him down and kill him.

Thus an exchange will involve something more than the trading of videos, parachute designs, and T-shirts. What begins as dialogue must end as dialectic. There are many from which to choose—North American fire management and Siberian fire; Soviet past and Russian future; communism's legacy and consumerism's promise; Smokey and Misha. And behind them all lies that most fundamental dialectic by which a prehuman First Nature has evolved into an anthropomorphized Second Nature, a process that inevitably involves fire, nature's foremost dialectician for fusing opposites and disintegrating syntheses. But then, as the Russians well know, fire is often first in wild nature, as it has always been second nature to humanity.

Nataraja (India)

In the center dances Shiva, a drum in one hand and a torch in the other, while all around flames inscribe an endless cycle of fire.

This—the *nataraja,* the "Lord of the Dance"—is more than one of Hinduism's favored icons. It is a near-perfect symbol of Indian fire history. The drum represents the rhythm of life; the torch, death; the wheel of flame, the mandala of birth, death, and rebirth that fire epitomizes and makes possible. In this confrontation of opposites the dance replaces the dialectic; Shiva holds, not reconciles, both drum and torch. Considered ecologically the nataraja thus expresses in graphic language the great polarity of India, the annual alternation of wet and dry seasons by which the monsoon, with faint transition, imposes its opposing principles on the subcontinent. India's biota, like Shiva, dance to their peculiar rhythm, while fire turns the timeless wheel of the world.

Perhaps nowhere else have the natural and the cultural parameters of fire converged so closely and so clearly. Human society and Indian biota resemble each other with uncanny fidelity. They share common origins, display a similar syncretism, organize themselves along related principles. Such has been their interaction over millennia, that the geography of one reveals the geography of the other. The mosaic of peoples is interdependent with the mosaic of landscapes, not only as a reflection of those lands but as an active shaper of them. Indian geography is thus an expression of Indian history, but that history has a distinctive character, of which the nataraja is synecdoche, a timeless cycle that begins and ends with fire.

The cycle originated with the passage of India as a fragment of Gondwana into a violent merger with Eurasia. The journey northward, through the fiery tropics; the violence of the great Deccan basalt flows and of the immense collision with Asia; the installment of seasonality in the form of the monsoon—all this purged the subcontinent of much of its Gondwana biota, and tempered the rest to drought and fire. The populating of India came instead by influx from outside lands, followed by varying degrees of assimilation. Here, in the choreography of the nataraja, east met west, Eurasia confronted Gondwana, wet paired with dry, life danced with death.

What endemics remained were, like India's tribal peoples, scattered or crowded into hilly enclaves. Only 6.5 percent of India's flowering plants are endemic, compared with 85 percent in Madagascar and 60 percent in Australia. The residual biota thrived most fully to the south; peninsular India holds a third of the subcontinent's endemic flora. Some species, Asian in character, entered from the northeast. A diffuse array emigrated from the eastern Mediterranean, the steppes, and even Siberia, the Himalayas serving less as a barrier than a corridor. More recently weeds, largely European, have established themselves. The composition of India's biota thus recapitulates the composition of its human population—the tribal peoples, their origins obscured; the Dravidians who persevered on the Deccan plateau and to the south; the Southeast Asians, migrating through Assam and Bengal; the Aryans, Huns, Turks, Persians, Pathans, Mongols, and others, entering from the northwest; and Arabs and Europeans, mostly Portuguese and British, arriving by sea.

The geographic ensemble that emerged from this vast convergence was

both familiar and unique. Of course there were broad divisions, Asians here, Dravidians here. Of course there were mosaics of field, grassland, and forest, in part because of human influence. But even beyond such matters, this syncretic biota assumed the character of something like a caste society. It is probable that this was no accident. The organization of Indian society impressed itself on the land, with ever greater force and intricacy. Tribal people gathered on disease-ridden hills, better adapted to malaria and other ills. They then reworked those hills in ways that conferred on them a biotic identity. It is no accident that the species most commonly found in such areas are those most abundantly exploited by the human inhabitants, and are often those best adapted to fire. European weeds, like forts and factories, gathered into specially disturbed sites, then spread along corridors of travel or secondary disturbance. The intricate division of Indian society by caste ensured that different peoples did particular things at particular times, and this was reflected in the landscape of India, not only between regions but within areas that different groups exploited at different times in different ways for different purposes.

The intensity of the monsoon assured—demanded—a place for fire.

The sharper the gradient, the more vigorous the potential for burning. Some of the wettest places on earth, such as the Shillong Hills, could paradoxically experience fire and even fire-degraded landscapes. The biota, already adapted to rough handling by India's passage north, responded to fire readily. The flora and fauna that humans introduced, or that migrated into India coincidental with them, also had to be fire-hardened because humans added to and often dominated the spectrum of environmental disturbances.

They certainly exploited fire. Explorers and ethnographers reported the habit among southern tribal groups (and in the Andaman Islands) of routinely carrying firesticks, a practice relatively rare outside of Australia and a few other regions. Probably Alfred Radcliffe-Brown's peroration on fire and the Andaman Islanders could stand for most tribal peoples on the subcontinent. Fire, he concluded, "may be said to be the one object on which the society most of all depends for its well-being." A veteran conservator of forests, G. F. Pearson, noted that even the Ghonds, a long-enduring tribe of India's central forests, "never go into the jungle now, where tigers are supposed to live, without setting it on fire before them, so as to see their way." Almost certainly India's tribal peoples used their firesticks as Australia's Aborigines did. The prevalence of anthropogenic burning in the tropical north of Australia, where the Asian monsoon also dictates wet and dry seasons, is another likely analogue.

But something more than aboriginal fire practices from India's "tribal" peoples shaped the land; agriculture required fire for clearing, converting, and fertilizing. In India, as throughout monsoonal Asia, slash-and-burn agriculture *(jhum)* became dominant outside of floodplains, ensuring that routine fire would visit even remote sites. Where insufficient forest fallow existed, alternatives were found in *rab* cultivation by carrying wood to the site for burning, or mixing it with other refuse and manure prior to conversion into ash. Some peoples fired the hills "with almost religious fervor," observed one disbelieving Briton, in the hope that the ash would wash down to waiting fields. By all these means (and others) a subcontinent of extreme wetness switched, when the polarity reversed, into a land of ubiquitous fire. The nataraja's drum became a torch.

The arrival of the Vedic Aryans was an event of special interest. Beyond their role in confirming hierarchy as an informing principle of Indian society, beyond their heroic literature, beyond their infusion of Indo-European language and customs into the subcontinent, they introduced two items of special consequence to Indian fire history. They imported livestock, and they installed Agni, the god of fire, as first among the pantheon of Vedic deities. Fire and livestock interacted like a self-reinforcing dynamo. Together flame and hoof reshaped the landscape into grasslands and savannas sufficient to sustain the herds. Where jhum was also practiced, its abandoned fallow could be made to evolve into grass and browse through repeated burning. Without fire the process of reducing jungle and reordering landscapes was slow if not prohibitive.

It is no accident that the Mahabharata, part of the Hindu canon, describes the burning of the Khundava forest. It has been argued further that the story is an allegory of Vedic colonization. It begins when a Brahman appears to Krishna and Vamuna, then enjoying the forest. They grant his plea for alms, and he immediately shows himself as Agni and requests that he be allowed to feed himself on the forest. They grant this desire, too; Agni rewards them with a chariot and weapons; and together they consume the Khundava and its creatures. The city of Delhi rises from the site today. The Brahman, presiding over his fire ceremony, was in fact an important pioneer into new lands, provoking by broadcast and ceremonial fire a new order.

Thus the special status granted to Agni went beyond coincidence. Agni was the originating god, and it is to Agni that the Rig Veda opens its invocation; Agni of the two heads, one harmful, one helpful; Agni of the three arms, the manifestation of fire in the heavens as the sun, in the sky as lightning, and on the earth as flame; Agni, the medium between the gods and humanity, the mediator between humans and the earth; Agni, the Indian avatar of the hearth god fundamental to other Indo-European peoples, best known through the vestal fire of Rome. Soon, however, Agni was supplemented by Indra, the king of the gods, and eventually absorbed into that bewildering genealogy of deities and heroes, as overgrown as jungle fallow, that is the wonder and curse of Hindu theology.

But the special status that Agni lost within a proliferating Hindu pantheon he retained through rite. For the Vedic Aryans, the fire ceremony remained at the core of ritual existence. It was to Agni that they sacrificed, and through Agni, as burnt offerings, that sacrifices to other deities became possible. Fire accompanied birth, marriage, and death, if possible flame from the same fire serving all through the liturgical life cycle. Agni was thus both means and end, beginning and end, a continuous ring around the affairs of the world.

> *Agni, the all-knower, the first one*
> *Looked out over the beginning of the dawns,*
> *Out over the days,*
> *And out in many ways alone, the rays of the sun,*
> *He spread over sky and earth.*

Through the centuries the ceremony mutated, and Agni's unique standing declined before its many challenges. Buddhism confronted it directly, demanding a less violent and extravagant practice, preferring useful gifts (*dana,* or "donations") in place of burnt offerings. At Gaya the Buddha, perhaps inspired by the fires that annually burned along the flanks of the Vindhya Mountains, identified fire as a central metaphor of life. "Everything, brethren, is on fire." Passions and desires afflicted human life as flames did the land. They had to be quenched, the Buddha declared, just as the fire ceremony had to be replaced by a less extravagant rite. Nirvana literally meant extinguishing, the blowing out of fire. Hinduism responded by tempering the fire ceremony, relocating it to indoor temples, and granting it a more symbolic, less consumptive role.

Fire remained fundamental, however, as it does yet today. The *puja,* the central ritual of Hindu life, revolves around a fire that stands for the gods, carries sacrifice to them, and purifies the supplicant. Fire begins the day, as it does the world. It ends life in the form of cremation, as the world will end upon Vishnu's final return. Until then fire powers the cycling of birth and death that is the essence of the nataraja.

It is no surprise to learn that, for India, the spiritual interacts with the practical and that what organizes society also organizes nature. The installment of Agni and the Vedic fire ceremony, and the way this acted on Hindu society, had its parallel in the way by which Aryan fire worked on the Indian environment. Fire ordered the landscape as caste did people. The sacrifice to Agni took the form of burning India's forests, or rather of reworking them in somewhat newer ways to support an economy dependent on livestock. The slashed-and-burned Ghats of Karnataka were thus the environmental equivalent to the corpse-burning ghats at Benares. Interestingly the Buddhist revulsion against the fire ceremony had its counterpart in a reaction against the destruction of trees and animals, particularly through fire. The Buddhist king Ashoka the Great thus decreed that forest fires should not be lit "unnecessarily" or with the intention of killing or sacrificing living beings.

The new fire practices folded into the old, much as immigrant peoples and ideas enfolded into India's caste-layered society and its mosaic-wrought landscapes. By the time Enlightenment Europeans began studying India, fire was so prevalent that it merged seamlessly with the natural history of the subcontinent. Writing retrospectively in 1928, E. O. Shebbeare recalls that "every forest that would burn was burnt almost every year." Worse, the fires were chronic throughout the dry season, seizing whatever cured fuel presented itself. Joseph Hooker describes how, during his descent from the Himalayas in the early 1850s, he saw the plains of Bengal immersed in smoke, the product of fires "raging in the Terai forest" and elsewhere, and observes particles of grass charcoal descending like black snow around him. F. B. Bradley-Birt marveled how the "hills round Gobindpur form a wonderful line of light every night during the hot weather," the outcome of native-set fires that smolder for days, and "creep on in zigzag lines from end to end of the hills, invisible by day, but standing out clear and distinct, a brilliant line of light, by night." Benjamin Heyne explains that the "hills here are all on fire, and present a spectacle, the magnificence of which is easier conceived than described." Less enchanted, Inspector-General Ribbentrop fumes in a treatise published in 1900 that the profusion of fire was matched by a "most marvellous, now almost incredible, apathy and disbelief in the destructiveness of forest fires."

A summary of fire causes for the Ghumsur Forest in Orissa tabulated by "Mr. S. Cox," the district forest officer, nicely captures the spectacle, and the disbelieving outrage with which the British witnessed it.

> All the State forests on the borders of the taluk are subject to fires crossing from the numerous surrounding zamindari forests.
>
> The latter, if they are in a condition to burn, are always burnt,

> and the boundary lines are so extensive and run over such difficult country that it is out of the question for us at present to protect them all.
>
> Then in the large hill forests frequented by the Khonds the jungle is fired as a matter of course to facilitate tracking and for other well-known objects.
>
> In the lower hills and more accessible country bamboo cutters and permit-holders generally are responsible for a great deal of the mischief. Wherever a hill is frequented for bamboos there are always constant fires.
>
> Other causes are the practice of smoking out bees for honey—a very common origin of fire—of burning under mango and mohwa trees to clear a floor for the falling fruit and flowers; the roasting of Bauhinia seed; the burning of under-growth round villages and cultivation which might harbour tigers and panthers—this will probably prove one of our most serious obstacles to restocking the sal forests; and the spread of fire from banjar lands under clearance for cultivation . . .
>
> The long list of causes is almost complete if to the above are added the burning of forest by graziers, and for driving out game or finding a wounded animal.

Not least perplexing (and infuriating) was the fact that out of 53 cases of illegal fire investigated within the protected forests, "no less than 27 were caused by the protective staff itself." The native staff recognized, if their baffled masters did not, that the proper use of fire was the best protection against its misuse.

It was in fact the British who did not understand. It was their belief in fire's necessary destructiveness that was, within the context of India, incredible. The indigenous people knew how fire supported jhum cultivation, converted organic residues into fertilizer, kept woodlands and prairies in grass, assisted hunting, cleansed soil of pathogens, and supported foraging for flowers, bees, tubers, and herbs. Fire sustained metallurgy. Fire kept tigers away from villages and opened sites that might otherwise hide cobras. Fire structured the intricate ensemble of biomes that was made by, and that in turn made possible, Indian society. Alone among the elements fire illuminated the complex choreography that bound life with death, the human with the natural. Fire framed the nataraja.

The dance missed beats as British rule extended over more and more of Greater India. The British raj imposed not only imperialism but also industrialism. Britain linked India with lands beyond the reach of monsoon winds, connected it with economic cycles greater than the rhythms of annual growth and decay, and shrank the encircling fire into the combustion chamber of steam engines. The tempo of the nataraja picked up. A ceaseless cycle wobbled, then spun uncertainly into a spiral.

British influence extended piecemeal, as opportunity and necessity presented themselves. Change became serious—and reform deliberate—after the Revolt of 1857 when the Crown replaced the British East India Company as the governing authority. Britain then applied to colonial India the same processes that had restructured Britain over the preceding century. Industrial capitalism and a global market began redesigning the Indian economy. Land reform, or at least the rationalization of land ownership, exploitation, and tax collection inspired a kind of enclosure movement or revenue "settlement" that gradually spread over the newly acquired lands. "Forest settlement" was a part of this process, and quickly brought European-style forestry into conflict with traditional, communal exploitation of Indian woodlands.

The new ruling caste brought their laws, their language and literature, and their sciences. Agronomists sought to modernize Indian agriculture, as political theorists sought to modernize Indian government. Hydraulic engineers erected dams, dug canals, and designed irrigation works. Mining engineers explored for geologic wealth. Cartographic engineers surveyed the subcontinent, imposing a mathematical order on the land, even measuring the anomalous gravity of the Himalayas. Above all civil engineers laid out the grid that would be the means and symbol of Indian industrialization, the railroad. From 32 kilometers laid down by 1853, the system exploded to 7,670 kilometers by 1870, and then continued to grow. Each reform demanded others, however, if it was to succeed. The railroad, for example, was inextricably dependent on wood—for construction, particularly of ties ("sleepers"), for fuel, for cargo. The rationalization of India through the railroad required the rationalization of India's forests.

Indian forestry became one of the great sagas of British rule, however improbable its origins. Britain, after all, had no tradition of forestry and precious little of anything that could be called a forest. But it was clear that the reconstruction of India was doomed without some deliberate intervention. Without forests railroads would run down, agriculture would suffer from drought and flood, soil would degrade, and a timber economy based on the export of teak would collapse. Even by the mid-nineteenth century it was clear that economic and political forces were, like an acid, dissolving the grout that held together the Indian mosaic. If something did not reglue them, nothing would remain but a pile of broken tiles. Besides, the rationalization of the "jungle" (as the uncultivated wildlands were called) was an ideal symbol of liberal reform. If India's jungle could be reordered according to scientific principles, so could the rest of India.

Britain went to the heartland of European forestry for help. In 1856 it appointed Dietrich Brandis as conservator of forests for Burma. A botanist subsequently educated in forestry in the grand European manner, Brandis was the archetype of the transnational forester, Humboldtean in ambition, an indefatigable agent of empire, a Clive of natural resource conservation in Greater India. Two years later Brandis became inspector-general of forests for all of British India, a dominion that grew dramatically not only as Britain added more provinces to its Indian domain but also as the practice of reserving forests proceeded in conjunction with the reorganization of the Indian landscape through revenue settlement.

Brandis pushed for the establishment of the Indian Forest Service, achieved in 1864, one of the compelling institutions of British rule and the centerpiece for forestry throughout the British Empire. Cadets received formal instruction in Franco-German forestry at Nancy in Alsace-Lorraine, then served field apprenticeship in India. From there they might proceed to Sierra Leone, Cape Colony, or Tasmania. This was the same regimen experienced by the founders of American forestry, men such as Gifford Pinchot and Henry Graves. In 1906 the facility relocated to Cooper's Hill at Oxford, and later a separate school and research institution was established for India at Dehra Dun. The Indian Forest Service,

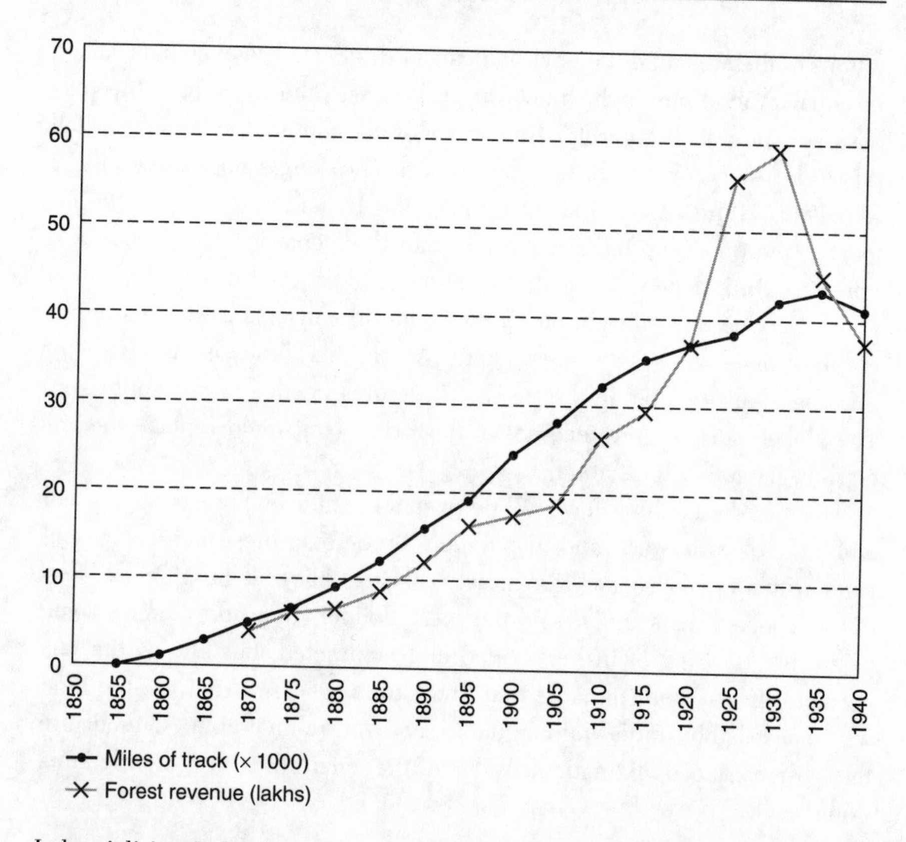

Industrializing India: forestry and railroads, 1853–1940. Much of commercial forestry went to supplying the needs of railway construction and operation, and, of course, the sprawling rails made prime forests accessible to logging.

meanwhile, became a part of the civil service and after critical conferences in the early 1870s assumed its modern form. On the recommendations of the conferees the IFS in 1875 launched the *Indian Forester,* for fifty years probably the premier forestry journal in the world.

Enthusiastic foresters—Sir David Hutchins reminded them that they were "soldiers of the State, and something more"—entered into the reconstruction of India, attempting to regulate timber harvesting, to control traditional forest uses by pastoralists and villagers, to regenerate felled or degraded woodlands, and to control fire. They as much as anyone pioneered the shock encounter between Britain and India, between the institutions of the West and the environments of the East. The encounter mixed in equal proportions high drama, absurdity, grit, the irony of noble purpose and practical stupidity. Rudyard Kipling captured something of all this in his story "In the Rukh," a sequel to *The Jungle Books.* "Of all the wheels of public service that turn under the Indian Government," he intoned, "there is none more important than the Department of Woods and Forests." On it depended the reforestation of India. And it is to the Indian Forest Service that Mowgli, now grown but still conversant with his brothers the wolves, goes as a forest guard. Among his duties is "to give sure warning of all the fires in the *rukh.*" Those fires needed to be suppressed. The globe-encircling fire engines of the British raj would replace the encircling flames of the nataraj.

Here was something new. While over the centuries forests had ebbed and flowed with wars and population pressures, fires had come and gone with the monsoons. Fire practices had changed, but fire had endured. Some years Shiva's drum beat louder than the torch, some years not; the ring of fire expanded and contracted, but always the circle held. It was unimaginable that fire could cease. Without fire the land was inaccessible, India uninhabitable, and life unknowable. Without fire the cosmos faced extinction. Without the encircling fire the nataraja would end.

The pioneers of Indian forestry, E. O. Shebbeare recalled, saw fire as "their chief, almost their only enemy." The extravagance of fire that seeped, simmered, probed, flared, and raged annually throughout India made a shambles of any presumption to reorder those forests along European models. Fires infested the land like malaria or packs of wild dogs. But the challenge went beyond their damage to pasture and woods, beyond the wanton sacrifice of India's immense wealth of forests. Those fires appeared as an environmental superstition, a taunt that mocked the possibility of remaking India in ways that would serve Britain and serve to legitimate British rule. Britain could justify redirecting India's forests to new purposes only if those purposes had higher standing, if they were part and parcel of a more rational order. It had to remake India's "irregular" forests—its tangled "jungles"—into "rational" institutions. It could harvest forests only if it demonstrated how to regenerate and protect them according to some larger principles.

So, in addition to the compelling economic reasons that linked forest to rail, and to the political logic that demanded the subordination of rural villages to a central, industrial authority, the British added the symbolism of science to their justification for fire control. The power to control village life resided in the power to control forest and range, and that depended on the power to control fire. Because Britain's claim to impose a modern ecological rule on India relied on its sanction by scientific silviculture, the British had to oppose "primitive" practices with a "rational" agriculture and a scientific forestry. In European agronomy the divide between the primitive and the modern was fire. Fire had to go.

The experiment began in 1863 when Brandis urged Colonel G. F. Pearson of the Central Provinces to try to stop the burning. No one believed it was really possible. "Most Foresters and every Civil Officer in the country," Pearson observed, "scouted the idea." Edward Stebbing recalled matter-of-factly that in every province "the officers of the Department had to commence the work of introducing fire conservancy for the protection of the forests in the face of an actively hostile population more or less supported by the district officials, and especially by the Indian officials, who quite frankly regarded the new policy of fire conservancy as an oppression of the people." Even forest officers, Stebbing noted, however much they approved of fire control in principle, "were openly sceptical" of its practical possibility. Had his attempt failed, Pearson affirmed, "any progress in fire protection elsewhere would have been rendered immeasurably more difficult." Pearson shrewdly selected a site protected by natural barriers, a biotic counterpart to the fortresses at Ranthambore and Jaipur. He then laid out fuelbreaks, sent out patrols, exhorted locals to give up burning, and enjoyed a couple of exceptionally wet seasons. To everyone's astonishment, the experiment succeeded. The Bori Forest became a showcase of fire conservancy. At the Forest Conference of 1871–72, based on these experiences, Pearson declared that "there can be no doubt that the prevention of these forest fires is the very essence and root of all measures of forest conservancy." Brandis added his imprimatur. "There is no possible doubt," he wrote, as to its "immense value and importance." Fire conservancy was, not accidentally, the first topic addressed by the first conference on forest administration.

Not completely, not without considerable debate and second-guessing, but thanks to militant enthusiasm, patience, and favorable weather, this improbable experiment in fire control evolved into a demonstration program, and then into a prototype suitable for dissemination throughout Greater India. At the Forest Conference of 1875 Brandis reaffirmed that for the improvement of Indian forests "there is no measure which equals fire conservancy in importance." It is, he continued, "the most important task of the Forest Department in most provinces of the empire, and for that reason was awarded first place in conference discussions." Pearson's successor, Captain J. C. Doveton, detailed the ways and means of fire conservancy and observed sourly that these measures were necessary only because "nearly the whole body of the population in the vicinity of forest tracts have, or imagine they have, a personal interest in the creation of forest fire." Not least of all because of that hostility, three classes of state forests evolved, each committed to a different level of use and protection.

Once confirmed, the idea spread, promulgated from the top down. As with the native principalities, so with the native forests; more and more were reduced to British rule by fire protection, for to control fire was to control the native populations. Regardless of the legal status of forests, without fire the local populace had no biological access to the resources of those reserves. By 1880–81 the Indian Forest Service had reduced some 11,000 square miles to formal protection; by 1885–86, some 16,000 square miles; and by 1900–01, an astonishing 32,000 square miles that spanned the spectrum of Indian fire regimes, from semiarid savanna to monsoonal forest to bamboo groves and montane conifers. Fire control grew as rapidly as the railroads with which it was indissolubly linked. Fire protection targeted particularly the great timber trees of the subcontinent, sal, teak, chir pine, and commercial bamboo. What emerged was a robust exemplar, an adaptation of European techniques to exotic wildlands and colonial politics.

But skeptics were not easily stilled. Pearson spoke dismayingly that "it is strange how slow even some, who possess very considerable practical acquaintance with the forests, are to recognize" the intrinsic merit of fire exclusion. In the *Report on the Administration of the Forest Department for 1874* B. H. Baden-Powell echoed and scorned that disbelief.

> Strange to say, that, obvious as the evils of fire are, and beyond all question to any one acquainted with even the elements of vegetable physiology, persons have not been found wanting in India, and some even with a show of scientific argument (!), who have written in favor of fires. It is needless to remark that such papers are mostly founded on the fact that forests *do* exist in spite of the fires, and make up the rest by erroneous statements in regard to facts.

On the matter of fire conservancy science admitted no doubt, and neither did colonial administrators bent on imposing a new order on a very old and complex land.

Like a fire in a punky log, however, the matter smoldered on. Soon field men voiced ever greater doubts about the wisdom of "too much fire protection." In wet forests fire protection seemed to retard natural regeneration, and it allowed fuels to accumulate that, once dried, exploded into all-consuming conflagrations. In drier forests, years of seemingly successful protection would be wiped out by massive fires during exceptional years. Exhortations and bribes with goats could not extinguish all the native firebrands who knew from daily experience what burning meant. Villagers refused to resettle or remain on unburned sites for fear that tigers, hiding in the tall grasses, would seize child herders. (Unlike the American or Australian experience, Indian natives would not melt away, vastly outnumbering the ruling caste, and their fires could not be banished into the past or sequestered onto reservations.) Hunting clubs in the Nilgiri Hills noted the deterioration of game where fires had been excluded. In the absence of suitable fire regimes natural regeneration failed in sal, teak, bamboo, pine—and failed consistently, particularly in wetter sites. Field officers began posting querulous memos about increases in diseases, pests, weeds, and other signs of a forest going feral. An agronomic memoir on Indian grasses noted how "an unforeseen result of the policy of non-interference with the vegetation" was the accumulation of dead straw that defiantly withstood "rotting" and eventually had to be burned, an act that quickly yielded a variety of useful results. Forest guards surreptitiously burned surrounding lands, including the lower-grade forests, to improve their chance of fire control on Class I sites. Upon his retirement in 1952 a native Indian forester commented that in his forty-one years of service he had never known a forest to withhold fire for more than three years.

In what might serve as a cameo, an Anglo forester who signed himself "An Aged Junior" described for the *Indian Forester* the puzzling situation in which, through more or less successful fire protection, the forest had acquired a tiger problem. It is apparent that fire had not been random and ravenous, as it appeared to the British, but had been applied to particular sites at particular seasons for particular purposes by particular peoples. Those selective burns had ordered the landscape. Thanks to fire fresh browse appeared at the proper place at the proper time; deer migrated to those sites; tigers followed the deer; and hunters knew where to find rogue tigers. But eliminating fire, or smearing it, affected that land as the abolition of caste would Indian society. Boundaries blurred. The ecological order became confused. Tigers no longer kept to their place—their place being scrambled and overgrown. They began to menace local communities, follow rangers, and generally make themselves "disagreeable." The forest now had "much fire conservancy and many tigers." Whether successful or not, the *attempt* at fire control was sufficient to unbalance the Indian biota. Changing from small fires set annually to large fires that came every three or four years did not preserve the old order. It was not simply fire that India needed but its syncretic order of fire regimes.

It was not so easy to reconcile European principle with Indian reality. Critics argued for a hybrid program in which controlled burning could supplement fire suppression. In 1897 Inspector-General Ribbentrop, Brandis's successor, had to intercede. To protect regeneration and forest humus (the twin obsessions of European forestry)—to say nothing of saving imperial face—he ruled for the further expansion of systematic fire protection. Edicts, however, did not suppress fires, or doubts. By 1902 the debate rekindled within the pages of the *Indian Forester* and the annual reports of the provincial conservators. In 1905 a compromise was proposed by which controlled burning could be brought into working plans.

Meanwhile sub rosa burning in Bengal, Burma, and elsewhere scorched the landscape like a people's rebellion.

In 1907 protest boiled over into a Burmese revolution. In the absence of traditional fire—slash-and-burn cultivation, routine underburning—teak simply refused to regenerate. Fire control had drained away the economic lifeblood of the Asian monsoon forest; foresters had prescribed a deadly cure where there had been no disease. Faced with a choice between excluding fire and excluding fire protection, the inspector-general began withdrawing fire control from prime teak forests. One after another, working circles that had subscribed to fire protection now receded from it—Pyu Chaung and Pyu Kun in 1906; Kan Yutkwin in 1910; Bondaung, Kabaung, and Myaya Binkyaw four years later. By 1914 conservators of sal forests likewise recognized that regeneration "had ceased throughout the fire-protected forests of Assam and Bengal and that no amount of cleanings and weedings would put matters right." They tried to reintroduce fire, but fuels had so changed that it was no longer possible to run benign light fires through the understory; the *taungya* system by which swidden fields were restocked with planted timber trees evolved as a partial compromise. Chir pine, too, was found to be reliant on routine fire, so that nearly everywhere field foresters introduced some form of "early" (that is, spring) burning of grassy understories for fire protection, and integrated regeneration burns into silvicultural cycles. Whatever the causes for the failure of natural regeneration, Shebbeare concluded for an audience of foresters drawn from the British empire, "fire appears to be the only real cure."

By 1926 the cycle of fire practices had come full circle. Imperial resolve retreated before an unscorched earth, the passive disobedience of Indian silviculture. A conservators' conference amended the rules of the Forest Manual to make early burning the general practice and to extend complete protection only to special sites on a temporary basis. With nice irony that new regime included the Central Provinces. Some critics wanted even more. Writing from the Siran Valley, E. A. Greswell noted that "up to 1922 the [chir] forests had been subjected from time immemorial to periodic summer firing," probably burned once every three to four years. The cessation of those fires damaged regeneration and put the forest at risk from wildfire. The reintroduction of fire was "merely re-establishing a modified form of the environment to which the forests owe their origin." Greswell knotted practice to philosophy when he concluded that "we talk glibly about following nature and forget that the nature we are visualising may be an European nature inherited from our training and not an Indian nature." The fire of Europe was not the fire of India.

But by this time Britain, never fully recovered from the wastage of World War I, was receding in imperial power and enthusiasms, its hold on India becoming steadily more tenuous. Protests increased, often focused on forestry and typically assisted by outbreaks of incendiarism. In 1916 and again, with even greater force, in 1921 political protest in Kumaon inspired a wave of woods arson that brought the regional administration to its knees. Administrators openly admitted their helplessness before the protest of incendiarism, another argument in favor of co-opting burning. But such spectacular outbreaks paled beside the relentless insurgency of small firings. Writing in 1925, M. D. Chaturvedi observed that "prosecutions for forest offences, meant as deterrents, only led to incendiarism, which was followed by more persecutions and the vicious circle was complete." Inevitably, grudgingly, concessions followed. Compromises remained compromises, however, the best one could do under troubled circumstances. With few exceptions—but among them some of the best minds in Indian forestry, such as R. S. Troup—foresters continued to insist that fire was intrinsically bad. They saw it, as they did the native elites, as a necessary evil, not as a powerful ally. Fire remained an impermeable divide in the worldview of European agronomy and silviculture. If a system used fire, it was by definition primitive; if it found surrogates for fire, it could qualify as rational.

In India there were few surrogates possible. Where officialdom approved fire it did so reluctantly, with some embarrassment, and only because fire was seemingly part of an ineffable (and exasperating) East. Fire reduced rational plans to a kind of ecological astrology, and the practices of a scientific forestry to a flame-lit puja. Fire persisted as an untouchable caste within the society of silviculture. The Indian Forest Service burned because it was forced to, not because it wanted to. Where fire was used, it was often not sanctioned, and where sanctioned, often not used properly. As British rule met further resistance, that split widened; theory and practice diverged; the landscape was neither old nor new nor some workable compromise between them. The cycle of fire broke.

What had been a circle became a spiral. The process began well before Independence, and it has continued after the British were expelled. What Britain had done with imperial arrogance, independent India claimed it would do with a social conscience; but whatever their sanction the practices continued, and then accelerated; and this acceleration was itself quickly exceeded by a horrific explosion in the subcontinent's population. However incomplete or mismatched, the reforms of the British raj had initiated a population rise that continues its exponential growth to the present day. In 1800 the estimated population of India was 120 million; in 1871, 255 million; in 1950, 350 million, despite the upheaval of partition; in 1990, 890 million. Until the 1970s the numbers of livestock swelled in almost equal proportion. Much of the human increase gathered into cities; a substantial fraction was absorbed by industry, but the rest (over 70 percent) remained on the land, and one way or another, this maelstrom of peoples and beasts sucked down the Indian environment in its vortex.

The upward spiral of human numbers powered a downward spiral in land abuse. Some 16 percent of the world's population crowded into 2 percent of its landmass. India's forests felt the pressures keenly, particularly the *terai* and hill forests that had, because of endemic diseases such as malaria, been shielded from use other than by those tribal peoples who had acquired some degree of immunity. Disease control, the construction of dams and roads, intensive logging, clearing for additional farmland, and a redefinition of reserves to serve the tenets of "social forestry" eroded India's woodlands, and often their soils. The commitment to industrial forestry that British rule had established, the Indian state reaffirmed; previously unexploited indigenous forests were opened by roads, logged, and often replaced by exotics such as eucalypts that provided pulp but little of the other products India's woods had supplied Indian society. Although the Indian constitution stipulated that 25 percent of India should remain forested in some form (and the Forest Law of 1952, 33 percent), the reality was closer to 19 percent, and critics thought even that number too high;

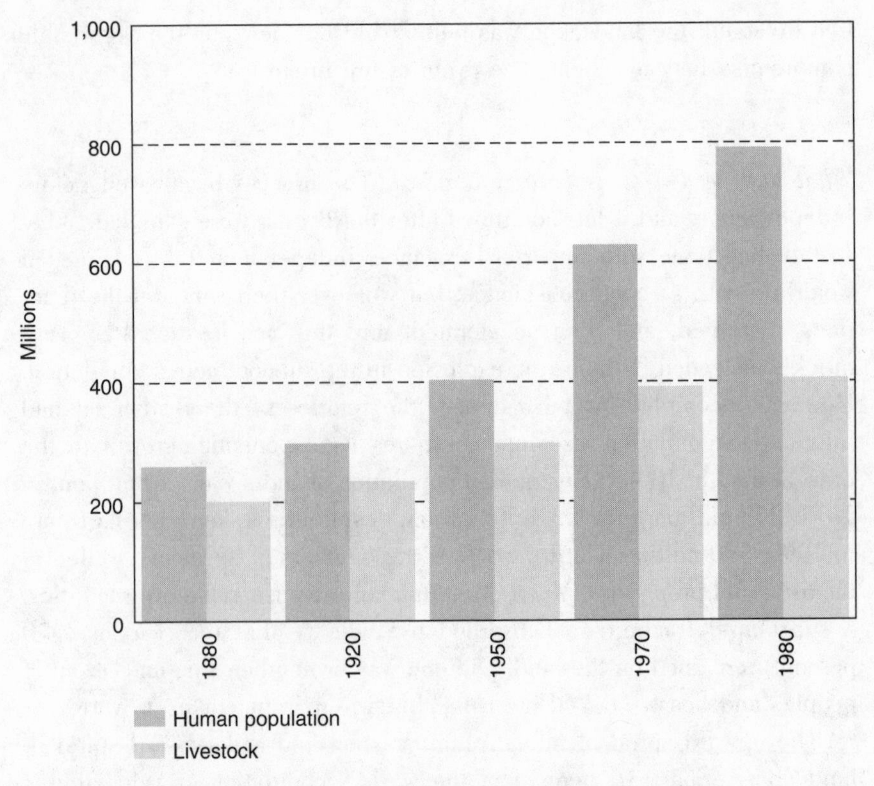

Population spiral: people and livestock, India, from 1880.

much of the reserved jungles were too degraded to classify as productive woodland. Once placed under state care, forests had required the coercive power of the state to survive. As further political unrest threatens the nature of the Indian polity as a secular state, that power promises to recede and to leave India's forests exposed to everyone's grasp and no one's care.

The intensity of use has disturbed the character of Indian fire. There remains plenty of burning, of course. Agricultural fire is common where cotton, sugar cane, and wheat are grown, and among the crop residues of hill farming. Jhum cultivation persists in the northeast, the Ghats, the outer Himalayas, and among tribal peoples in Andhra Pradesh, Orissa, and elsewhere. An estimated 122,000 square kilometers of permanent pasturage is burned annually. Among reserved and protected forests controlled burning assists the regeneration of chir pine, sal, and teak; fuelbreaks are burned early each dry season; particularly where forest plantations are at risk, underburning is practiced to reduce fuels and prevent against wildfire. Altogether this amounts to 5–6 percent of the reserved forest area. Still wildfire, either from "accidental" or incendiary causes, affects an estimated 10,000 square kilometers yearly, as officially reported. Satellite inventories, however, calculate that eighty times this amount burns annually, some 33–99 percent of the protected forests in different states. These numbers do not account for forests subject to less strict regulation. The forest area affected by fire may reach 37 million hectares. Even so, the biomass burned as firewood in villages and urban centers exceeds that of all these other sources combined. Increasingly India's woods are being burned in its stoves.

The quest for a suitable regimen of fire continues. It is pointless to argue for a restoration of traditional practices—the circumstances are too much changed to allow them. What had once rested as forest fallow for thirty years is now slashed and burned in five years, and sometimes as little as two. What formerly experienced small fires that percolated through the jungle over the course of five or six months now suffers from no fire or fire that crowds into short, violent outbursts. Even traditional burning no longer recycles nutrients through a subsistence economy but siphons them off into a global market; where tribals traditionally burned once under *mowah,* they now burn twice, and the harvested flowers do not go to the village but the metropolis. The complex of fires that once fused the human and the natural through the layered intricacies of a shared caste is gone. More and more, India's fire regimes are defined by a global economy in which the forest exists as cellulose and wood is valued as an export commodity; less and less, by the traditional usage of the forest as a medley of usable plants and animals. The beat of pistons, powered by fossil-fuel combustion, replaces the rhythms of seasonal growing, curing, and burning. Artificial fertilizer replaces rab; the tractor and electric pump, the long fallow of jhum; autorickshaws, the bullock cart. Some time around 1980 India crossed an industrial threshold of sorts when deaths from traffic accidents exceeded those from snakebites.

Yet no surrogate complex of fire practices has fully replaced it. To the extent that Indian scientists receive training from Europe or look to European scholarship for guidance, they continue to distrust burning, as though it were still a stigma of primitiveness, a leprosy on the landscape. No one has transformed India's unique experience into a new exemplar for Third World fire powers, a model of nonalignment in the dialectic between those who would base fire management on fire control and those who base it on fire use. India's elite still view fire as an inevitable if necessary evil, like cobras. If it were possible to escape from the endless cycle of fire, to lay down the burden of burning, they would. That would be release, forestry's nirvana.

Instead, with assistance from the U.N. Food and Agriculture Organization (FAO), India launched a "modern forest fire control project" in 1984 that sought to install an integrated fire management system in two demonstration areas, Chandrapur (Maharashtra State) and Haldwani (Uttar Pradesh). The first contains extensive natural and planted teak forests; the second, hills dominated by sal and chir pine. Both projects, that is, intend to apply fire control to support the ambitions of industrial forestry. Incorporated into India's Eighth Five-Year Plan, Phase Two will expand the technologies into ten states and 40,000 square kilometers. Whether the project becomes a latter-day Bori Forest, a misinterpreted experiment; whether it evolves into another showcase of international aid with airtankers and helicopters taking the place of high dams and nuclear reactors; or whether it begins the process of reconciliation between new and old, fusing a uniquely Indian style of fire management, all remains to be seen.

It may be that reconciliation is impossible, that as in the nataraja India must simply hold and live with the opposites. This time, however, fire is not part of a cycle of endlessly reincarnating landscapes, but a spiral, propelling the biota to one extreme or another at an ever-quickening tempo. Without those encircling flames, the boundaries are broken, drum and torch no longer link, their rhythms no longer balance, and the dance must end in either frenzy or exhaustion.

White Darkness (Antarctica)

Fire is so elemental to the earth and so fundamental to human existence that it is difficult to imagine a world without it.

So pervasive is combustion that it seems to constitute a kind of ecological subconsciousness for the planet, found everywhere but everywhere hidden, seemingly manifest only at times of stress. But there it is, asserting the ubiquity and tenacity of terrestrial life, a medium within which air, water, and earth interact. As much as anything it has made the earth habitable and humanity dominant, allowing us to occupy places, to reshape them in our image, and to take our identity from that engaged environment. Remove fire and you fashion a land to which humans have no means either of access or of understanding. The universality of one assures the universality of the other.

But there exists one magnificent exception. Antarctica is a continent without fire. Away from the coast it is also a continent without life. Until fossil fuels allowed combustion to transcend existing ecosystems, Antarctica was a continent without humans. Its ice is a colossal act of biotic negation. That includes the extinction of fire and all that fire means to humans. Antarctica brings to an end one of the earth's great sagas, the heroic cycle of fire. On The Ice fire dies.

Some of Antarctica's properties are common to the high latitudes in which it resides. It is as though their proximity to the earth's poles requires all the geophysical properties to follow their lines of longitude and quickly converge toward some abstract singularity. What elsewhere is diffused, orchestrated, and full of subsidiary plots and rhythms here condenses, not through simple compression but through structural abridgement. Sunlight begins to aggregate. What other regions divide between day and night, winter and summer, the pole combines into one extended cycle. Life obeys the same logic. Ecosystems simplify; biomass replaces biodiversity; fewer species claim greater extents. The magnetic field crowds together into a confused knot. Storm tracks veer away, shedding another measure of environmental variation. Space, time, direction—all undergo a character shift, accelerating toward the poles, that denies the contrasts, the checks and balances, and the syncopated complexity that makes the midlatitudes temperate, complicated, and accessible.

This differentness extends also to the two polar regions themselves. The Arctic is a hole, the Antarctic a hump—the one an ocean enveloped by continents, the other a continent surrounded by oceans. Their antipodal symmetry is nearly perfect, not only in their geographies but also in their histories and in the ways they are understood. The Arctic is confined, shaped, and populated by the lands around it; the frozen Arctic sea draws together, in its icy gyre, the complex world of the circumpolar continents. The Antarctic, by contrast, is almost completely isolated. Its sea is a circumpolar current, circling it like the River Styx. Its pack ice swells outward like an abatis. No isthmus, no sutured range of mountains, no chain of islands link it to the rest of the earth's solid surface. It lacks a terrestrial ecosystem in any meaningful sense. It is exclusive, sui generis, self-referential.

To appreciate the polar regions' distinctive character, resurrect the famous fragment from Archilochus—"the fox knows many things, the hedgehog knows one big thing." The Arctic is a fox, and the Antarctic a hedgehog. Thus a leading commentator on the Arctic—Barry Lopez in *Arctic Dreams*—features the far-ranging, ever-curious fox as an archetypal Arctic creature, and the author himself even mimics the fox in his own endless travels, always sniffing at odd scenes, poking into historical relicts, ever alert to the nuances of native cultures. Polar cold and Arctic ice interact with other things in ways that reveal new patterns. Novelty is ever possible. Surprise is expected.

That is not possible in the Antarctic. Ice—the one big thing—dominates everything else. Life essentially ends at the coastline. There is no way to live off the land; there is no free water, no food, no sustaining ecosystem, no indigenous culture. A journey to the Arctic, though onerous, can be a voyage of discovery. A journey to the Antarctic is a study in things lost, a journey in which one is sustained less by what one finds than by what one brings. Hedgehog Antarctica is a study in ice.

Ice is the beginning and ice is the end of Antarctica. To enter Greater Antarctica is to be drawn into a maelstrom of ice. Antarctica contains 90 percent of the planet's glacial ice, and during the winter when the seas around it congeal into a frozen pack, the size of the Antarctic ice field doubles. Antarctica is literally fused together, as a continent, by ice. The magnitude of the Antarctic ice sheet is enough to deform the planet. The scale of its ice field affects global climate. The dimensions of Antarctic ice control the level of the world ocean. Antarctica—The Ice—is the earth's sink for heat and water, a cold fusion that absorbs rather than emanates, a geophysical underworld.

The Ice is organized, like Dante's inferno, into concentric rings. There is a gradient to the terranes—loose, information-rich, dynamic on the perimeter; fixed, information-poor, invariant at the core. To journey to its source is to pass through an increasingly ice-dominated landscape until land itself vanishes into exclusionary ice. Ice replaces everything that is not ice. Ice confines ice. Ice defines ice. A continent is reduced to a single mineral. In Antarctica more is less.

The patterning of these terranes, however, means that there exist different kinds of ice, each of which can be understood in somewhat different terms. On the outside, orbiting like comets, are the bergs. Here The Ice is most dynamic and varied. Bergs interplay with floes, land ice with sea ice, each obedient to different nuances of current and wind. The berg can interact with ocean and atmosphere, bobbing, spinning, breaking up, even overturning. It erodes into fabulous shapes. Sea life gathers around it. Noises cluster to it. As the sky opens and closes, sunlight glows off the glacial ice in opalescent colors or radiates from its snow-powdered exterior in brilliant blasts of white. White berg shouts an antithesis to black sea.

The domain of the berg extends beyond that of the pack, but the two are generally coextensive. The freezing over of the Southern Ocean—first into tiny clumps of frazil, then into floes, and ultimately into a welded sheet of sea ice—is an annual event. Nearly all of the pack is remade each year. Over the winter the pack grows, prograding outward as skirmishing floes coalesce into a consolidated mass. During the summer the process reverses, retrograding back to the shoreline, sparing only protected enclaves where, for a period of years, sea ice may remain frozen as fast ice. As with the berg, the pack also offers a kinetic art—the motion of individual floes, the stately progression and retrogression of the pack, the syncopated dance with sea, wind, and berg. The Arctic pack persists through many years, constantly torn apart and smashed together into lead and ridge, as the gyre grinds on, giving to its ice a historical character. But the Antarctic pack is remade constantly, and this act of renewal, not the scarred ice, holds the greatest interest.

The Antarctic littoral is a pastiche of ice—here expressed in glacial tongues, thrusting out like swords; there in ice shelves, deltas of glacial ice, often of colossal size, spilling out and anchoring over the surrounding seas; there as fast ice; there, again, as the frozen spray of waves washing against ice or rock. Together they surround the continent with an icy exoskeleton, like an iron band around a wagon wheel. Increasingly the interactions—the contrasts that yield the greatest information and esthetic excitement—occur between different kinds of ice, not between ice and cloud, sea, or wind. The largest shelves, such as the Ross and the Ronne, form ice steppes that introduce the peculiar emptiness and splendor of true icescapes. Shelves of this magnitude owe their shape and dynamics to the properties of ice, not to the interplay of ice with rock or sea. They must be understood as ice alone. Atmospheric effects in which minuscule ice crystals saturate the sky and reflect and refract light into geometric patterns such as halos, arcs, and patches assume special interest because they offer, as the shelf does not, a focus and a point of contrast. They bring perspective to a scene in which the observer otherwise becomes vanishingly small amid the lost horizon of ice.

The floating shelf is a prelude to the frozen sheet. But between them, where mountains allow, lies the fragmented terrane of the glacier. Some glaciers are parochial, the product of local snowfall and ice flow. Others join the interior ice sheets to shelves or to the Antarctic seas. Because they are ice streams they have a complexity of motions and interactions not otherwise possible, and because they often pass through mountains, some with exposed rock, they offer contrasts between earth and ice, much as the pack does between sea and ice. In places such as the Dry Valleys the proportion of land exceeds that of ice, sustaining what passes in Antarctica for an oasis. The glacier, moreover, has clear intellectual and esthetic links with traditions of Romantic travel and painting. In a continent otherwise void of familiar scenes, it offers a comforting analogue, and visitors to it rush like pilgrims to a shrine.

Not so the ice sheets. There are two great sheets, one in West (or Lesser) Antarctica, the other in East (Greater) Antarctica. The West sheet welds an archipelago into a subcontinent, but it rests on a metastable marine foundation. The East sheet—perhaps ten times more vast—is apparently frozen to a continental craton. The West sheet captures storms, displays varied motions, interacts in complicated ways with the mountains it sutures together. The East sheet, however, blocks most storms from the interior, moves with a rhythm akin to the geologic tides by which the world ocean rises and falls, and obliterates any vestige of the rocky landscape beneath it.

Nowhere else on the earth is the land at once so huge and simple. There is only ice. Movement is infinitesimal. Each ice mass is encased in other ice masses almost identical to itself. Little disrupts the surface, which appears as singular and invariant as a Euclidean plane. Few storms ruffle the lapidarian exterior. No ice tectonics break open along hidden fault lines or erupt into ice volcanics. Precipitation falls as showers of diamond dust, minuscule crystals of ice that descend like cosmic house dust. Sunlight reflects off the high-albedo snow. Ice chills the atmosphere into a sheath of air. Atmospheric effects are reduced to optical displays or auroral bands, wholly prescribed by simple geometries of light or ions on ice. There is more of less than anywhere else on the planet. The scene could belong with Charon, the ice moon of Pluto.

It is a subworld, Antarctica, not readily visited or understood. The terranes mirror gradients of energy and information that surround the source regions like force fields. This antipodal vortex does not draw but repels. Leave the coast and there is nothing to live off, nothing to drink, nothing—according to traditional esthetic standards—to appreciate. A journey to the interior is an act of defiance, a thrust into a void. Travel is opposed not so much by active resistance as by a fantastic passivity that leaches away all that is essential to sustain life and thought. The albedo of The Ice reflects information as much as light. The journey becomes a quest pursued in opposition to the natural order, a white whale amid the world's continents. A journey moves beyond wilderness; beyond life; beyond Nature as Other. Here Nature becomes Negation. To encounter the pure icescape of the source regions—the informing principle of The Ice—is to peer into a white darkness.

How do you understand such an environment? Why would you go at all? In point of fact, most visitors don't, or can't, go into the interior. They prefer the margins, where the scenery is relatively accessible. They go to the brink of the alien, admire the palette of ices, but pull back before they pass beyond into The Ice itself. Typical scenes are ensembles of sea, rock, and ices, preferably populated by penguins, seals, killer whales, or if nothing else is available, skuas. They want the alpine or nordic purified by remoteness but still capable of appreciation by triangulation from familiar sites, by a geodesy of traditional esthetics. But this is not true for the interior.

Ice acts in peculiar ways on scenes and consciousness. In small doses, it accents. By removing clutter it can even amplify a scene's message. But when it becomes exclusive, it assumes a totalitarian identity that relates to nothing except itself. It removes all processes and presences other than its own. Ice then functions as a sink, a reducer, and a mirror. The Ice is profoundly passive: it takes rather than gives. The quintessential Antarctic experience is of voids; of things and processes missing; of experiences, sights, sounds, movements, figures, and colors not there. There is no sustaining biota. There is no indigenous culture by which to transfer knowledge or appreciation. There is no obvious moral order, only a terrible solipsism that substitutes soliloquies for dialogues. The Ice replaces everything that is not ice.

Antarctica the awesome derives its intellectual and esthetic power from its combination of the huge with the simple. Intuitively one expects that big places will be filled with many things—objects, images, events, information. But The Ice offers only the purity of bigness. In this ice sink there is no just proportion between what is expended and what is received. The Ice will take all it can, indifferent to nuances of feeling or purpose. What it cannot claim outright it will reduce, then reflect back whatever residue remains, an ice vortex at whose distant bottom lies a mirror.

The reductionism of The Ice is relentless and disturbing. It cannot be hacked through like a jungle or attacked like a hostile tribe or circumvented through the use of native guides who transmit the accumulated wisdom of millennia. Throw something at it and The Ice will absorb the object or idea, rendering it inert. Knowledge is not there to be picked up like rocks or collected like beetles; experiences do not flourish of themselves, needing only to be recorded in journals or painted on canvas; meaning does not radiate from an environment obviously ordered to serve the curiosity and needs of humans. Significance is not found but made by a special dialectic between ice and idea, the latter proposing and the former negating. If it works, the engagement induces a kind of intellectual excitement, the way passing one magnetic field across another generates electricity. But such a dialectic does not simply happen of itself. It must be contrived and force-fed, absorbing more energy than it releases.

Antarctica was, not surprisingly, the last of the continents to be examined by humans. The heroic age of Antarctic exploration came at the conclusion to a grand reconnaissance of the earth, a second great age of discovery, that spanned an era from the circumnavigations of Captain James Cook to Ernest Shackleton's abortive traverse of Antarctica. In a sense, Antarctica announced both the era's beginning and its end. But the techniques of travel, the conventions of art, and the concepts of science cultivated since the Renaissance faltered when they reached The Ice. Effective exploration only came later, when near-miraculous developments in the technology of remote sensing demonstrated publicly by the International Geophysical Year announced a third great age of discovery in which whole planets could be surveyed. Antarctica became a convenient point of departure.

The extraordinary difficulties attendant on understanding Antarctica were only partly attributable to problems of transport. Equally problematic was the apparatus of appreciation, the cultural technologies by which the unknown could be investigated and the unseen made visible. The science, art, and moral philosophy—the robust systems and aggressive rationalism of the Greater Enlightenment—that had carried the second age across all the populated continents and that had mapped and begun to plummet the world's oceans broke down on The Ice. Antarctica was too alien, too much sui generis, to be readily assimilated by a syndrome of thought predicated on the rational order of nature. Besides, those concepts and institutions could deal with abundance, with too much, but not with too little. And while the information content of an ice sheet may, in theory, be unbounded, there is more information encoded in a single penguin than in all the polar plateau.

The intellectual encounter between Enlightenment and Ice resembled the overland expeditions that trudged inland on Nansen sledges. The sledging journey, in fact, stands as a useful metaphor. Whatever you need you take with you. As you traverse The Ice, your supplies dwindle. Nothing by way of food or sustenance is added from the environment. A dialogue with Ice becomes a monologue with self. Life is stripped to the minimum necessary for survival. By the time your travel ends you have only what remains, or what little you have created by a kind of intellectual induction out of this abstracted and minimalist world. Thus artistic conventions are few and stereotyped. The repertoire of literary stories, metaphors, and imagery is small. What ice did to land, it did also to human experience. On The Ice, discovery shed not only its ethnocentricity but much of its anthropocentricity. The Ice took everything else away.

The end of the second age, however, accompanied a wide-reaching intellectual revolution, still in progress, called modernism. The prospect existed that modernism could develop the perceptual equipment by which

The Ice could be assimilated into the culture of Western civilization. Certainly a natural complementarity existed: Antarctica was nature as modernist. The Ice did to landscapes what modernist painters did to the inherited conventions of the pastoral and the sublime. Art became abstract, conceptual, minimalist. The object of art was Art, as the object of ice was Ice. However self-referential Art might become, Ice matched it. When the third age equipped modernism for geographic exploration, it outfitted intellectual culture for encounters with worlds in which neither intelligence nor life existed. It prepared it for lands without an autonomous Other.

Unfortunately Antarctica, the heroic age, and modernism were out of sync. Early modernists lost interest in landscape and geographic discovery just as Antarctica became an object of serious exploration. It may be also that Antarctica and modernism too closely mirrored each other. Modernism worked best when it could negate and reconstitute its inherited culture. But The Ice had already abstracted Antarctica, and to render into modernist idiom a landscape previously reworked along modernist lines by nature was perhaps too much for early modernism. Few artists came. During the third age, photographers Emile Schulthess and Eliot Porter experimented with zest and often startling insights. Painter Sidney Nolan found in Antarctica a frozen version of his native, sun-blasted Australia. But otherwise Antarctica was as remote from modernism as the dark side of the Moon. The essence of The Ice was, it seemed, an esthetic whiteout—obscuring, burying, reducing to a singular nothingness whatever was brought to it. The self-reflexive Ice held up a mirror into which modernism was reluctant to stare.

The environmental movement, however, has rekindled interest in the natural world and charged the search for wild lands with a moral enthusiasm. The very alienness of Antarctica has become a kind of attraction. The remoteness of The Ice suggests an alternate world to that so roughly mauled by humanity. Not surprisingly, advocates for a protected Antarctica have sought links with the attractive features of wild lands elsewhere—with the living world of penguins and whales; with pristine sites easily disrupted by oil spills, garbage, and toxic pollutants; with icebergs rafting like sphinxes, and majestic mountains crowned by storm clouds and ringed with frothy seas. These are, in fact, the least typical Antarctic scenes, those most removed from The Ice. But the politics of international environmentalism has at least redirected the attention of intellectuals other than scientists to Antarctica, and with it has come an interest in images. Artists have returned. They do so less as modernists than as modernizers.

Art comes from art. But there is little precedent for confronting ice qua ice, much less The Ice in all its empty enormity, its reductionism, its inorganic and ineffable simplicity. The difficulties increase as one crosses the geographic gradients ringing Antarctica. The bergs are easiest, their dominion resembling a kinetic sculpture garden. They appear like white Rorschach blobs, alternately empty or luminous, either a tear in the cosmos through which rushes the void, or a flash of hope in a netherworld, an ice opal. Through the domains of pack and glacier, the mind can play on contrasts—ice and sea, ice and earth, ice and sky. The eye refracts off fragments of ice, thrown up like prisms. Before the terranes of shelf and sheet, however, there is little but ice and more ice. The robust clutter of the earth disappears; the vitality of life vanishes.

How to illuminate that white darkness? At best, passage to The Ice can become a kind of esthetic fast that leads to a state of trance from which come new visions. At worst, it results in an intellectual and emotional whiteout. The self dissolves into the self-reference of The Ice. Mostly, people compromise by not allowing ice to control the scene. It becomes one feature or one process among many, shaping and framing but never dominating. It is one thing to abstract from life, another to live in abstraction.

But that is what happens at the source. The connectedness of things thins. The enormity and purity of The Ice erodes away intellectual handholds, polishing the surface to a radiant, terrible reflection. Space becomes vanishingly small, time pauses in frozen hesitation, and the mind disconnects from its referents. Only The Ice remains.

CONTROL

AMERICAN FIRE

Initial Attack: The U.S. Forest Service Fights Fire

Fires are made to put out!

—J. FRANK "SHORTY" MENEELY, FIRE CONTROL OFFICER, FLATHEAD NATIONAL FOREST (1970)

Wrathful but Calm, Austere but Comic, Smokey the Bear will
Illuminate those who would help him; but for those who would hinder or slander him,
HE WILL PUT THEM OUT.

—GARY SNYDER, "SMOKEY THE BEAR SUTRA"

A century ago thoughtful observers saw a United States overrun with fire. Lightning kindled fires by the thousands, particularly in the semi-arid West. But humans accounted for the greater proportion through a sprawl of frontier fire practices, often spearheaded by rapacious logging that ripped through forests like an ax-wielding hurricane and sustained by an unapologetic addiction to agricultural burning. The fires more than compensated for those that had traditionally been set by American aborigines. As late as 1904 an official in northern California observed that the typical fires of the region ("practically the only fires") were light surface burns and noted that

> the people of the region regard forest fires with careless indifference. . . . To the casual observer, and even to shrewd men . . . the fires seem to do little damage. The Indians were accustomed

> to burning the forest over long before the white man came, the object being to improve the hunting by keeping down the undergrowth, which would otherwise shelter the game. The white man has come to think that fire is a part of the forest, and a beneficial part at that. All classes share in this view, and all set fires, sheepmen and cattlemen on the open range, miners, lumbermen, ranchmen, sportsmen, and campers. Only when other property is likely to be endangered does the resident of or the visitor to the mountains become careful about fires, and seldom even then.

In compiling a summary of the nation's forests for the 1880 census, Charles S. Sargent cataloged the damages wrought by fire in numbing detail, a loss he despaired of as "enormous"; the principal source was agricultural burning. Bernhard Fernow, a Prussian-trained forester, thought fires were the "bane of American forests" and dismissed their causes as a case of "bad habits and loose morals." When a committee of the National Academy of Sciences toured the newly established forest reserves for six weeks in 1896, they were never out of the sight of smoke. John Muir lamented that, as extensive as was forest destruction induced by logging, the losses due to fire were ten times greater. And Gifford Pinchot thundered that "like the question of slavery, the question of forest fires may be shelved for a time, at enormous cost in the end, but sooner or later it must be faced."

Solutions were not obvious, however, and European-inspired exemplars were not always agreeable to indigenes. Resident populations considered fires useful and generally benign. Local juries prosecuted only the most flagrant abuses, reluctant to reduce their own access to fire. Active control of wildland fire was considered impossible in a technical sense, indefensible in economic terms, and undesirable on environmental grounds. Conflagrations were regrettable but ephemeral; they would vanish as settlement progressed. Wild fire would go the way of wild lands and wild animals, submerged into a landscape domesticated by agriculture. But the reservation of public lands into parks and forests voided this laissez-faire strategy. A new category of frontiersman, the forester, had to devise new means by which to handle fire, both those fires emanating from humans and those kindled by nature. Aspiring professionals, foresters emulated the example of their European professors and accepted, if only by default, the normative standard of the European landscape.

What they saw, however, had little practical relevance. Coert duBois declared that "American foresters have found that they have a unique fire problem, and that they can get little help in solving it from European foresters. . . . We must work it out for ourselves." Fortunately the coming of professional forestry coincided with a great epoch of American politics, the slow-boiling social revolution known as the Progressive Era that witnessed the transformation of the United States from an agricultural to an industrial nation. The architects of reform turned toward modern science for guidance.

Future chief forester William Greeley put it directly when he declared in 1911 that "firefighting is a matter of scientific management, just as much as silviculture or range improvement." There was nothing peculiar in this manifesto, nothing idiosyncratic to foresters. Rather, it conformed perfectly with the precepts of Progressivism, the belief that scientific knowledge was essential and adequate, that public policy and public lands should be administered by experts trained in scientific management and shielded from political corruption and public whim. The Forest Service quickly committed itself to the formal study of wildland fire. Two decades later Earle Clapp could declare with more bravura than accuracy that "forest fire research apparently originated in the United States, undoubtedly as the direct result of a forest-fire situation which is more serious than in almost any other country."

What forced the issue of fire into public discourse was the reservation of public lands as parks and forests. By excluding settlement, such acts excluded many sources of fires—though not all—while they paradoxically excluded the traditional means of fire control. Administrators had to discover surrogates. They had to create policies to decide which fires to suppress and which, perhaps, to promote. They had to replace centuries of accumulated folk knowledge about fire with scientific data. The first tests came in 1885 and 1886 when New York created its Adirondack Forest Preserve and the U.S. Cavalry took over the administration of Yellowstone National Park, in good part to bring fire control.

But the key lands in question were the vast forest reserves that, beginning in 1891, presidential proclamations carved out of the public domain in the West. The Transfer Act of 1905 delivered these lands to the U.S. Forest Service for administration; modern fire protection as a national enterprise dates from this event. The Forest Service assumed a central institutional and intellectual role in fire programs at all levels of national life—institutionally, by its control of the national forest system and by its promotion of cooperative fire-control programs with the states and industry; intellectually, by sponsoring research and by introducing the standards of professional European forestry into the debate about fire policy. The Forest Service created a national *system* of fire management.

The catalyst came in 1910. In August the Forest Service confronted two literal trials by fire. Parched by a vicious drought, the northern and western landscapes burned an estimated five million acres throughout the national forest system, more than three million in the northern Rockies alone. The great fires traumatized the Forest Service, still reeling from Gifford Pinchot's dismissal as chief forester earlier in the year. Some seventy-nine firefighters died in the line of duty, and the service sank into debt until Congress agreed to honor a deficit funding statute enacted two years earlier. It was the first great crisis for Henry Graves, Pinchot's handpicked successor and another European-educated forester. To add to the embarrassment, the same month (August) that the fires blew up, Graves finished publication in serial fashion of a treatise on fire control. Two future chiefs, William Greeley and Ferdinand Silcox, weathered the fires as officers in the northern region. The fires were the Valley Forge of the Forest Service. Soon afterward Congress passed the Weeks Act (1911), which allowed for national forests to expand by purchase, for federal-state cooperative programs in fire control, and for interstate compacts in support of firefighting.

Yet that same August *Sunset* magazine placed into public view a smoldering debate about fire policy. While professional foresters argued for fire protection by means of aggressive fire suppression, others advocated a program of controlled "light burning" in frontcountry lands and "let burning" in the backcountry. The light burners were an unwieldy amalgam—the state engineer of California, sentimentalist poet Joaquin Miller, timber owners and stockmen, the Southern Pacific Railroad, and novelist Stewart Edward White among them. They argued that fire protection only aggravated fuels, stoking uncontrollable burns; that absolute fire control was technically impossible and fiscally irresponsible; that fire exclusion damaged forests, starved regeneration, and encouraged insect plagues. They proposed instead a rationalization of frontier fire practices organized around controlled burning, what Miller called "the Indian way."

Thus the events of 1910–11 challenged the Forest Service as a fire agency across the board: the 1910 fires, its technical ability to control fire; the light-burning controversy, its capacity to formulate policy; the Weeks Act, its ability to establish a national policy, not just a national forest policy. Surveying the wreckage of 1910, German forestry professor Dr. Deckert noted that "devastating conflagrations of an extent elsewhere unheard of have always been the order of the day in the United States"—that, by contrast, its success with fire prevention constituted "a brilliant vindication of the forestry system of middle Europe"—and he expressed confidence that American foresters would follow their example. Within a decade after the 1910 holocaust, the U.S. Forest Service was trying to do exactly that.

But if, as foresters, they sought to emulate the fire practices of Germany and France, as Americans they sought to reconcile fire management with the tenets of Progressivism. Against folk wisdom, they proposed science; against laissez-faire folk practices, they argued for systematic regulation of burning that would support, not confront, professional forestry; against the self-evident waste of fire—not only what it directly destroyed but those benevolent "forest influences" that it indirectly laid to waste—they conjured up a vision of conservation, the rational, industrially efficient exploitation of natural resources. If forestry was an expression of technology transfer from Europe, firefighting was the pragmatic merger of idealism with reality by means of applied science. On both counts, the Forest Service committed itself to a program of wildland fire research that would support its announced fire policy.

The subsequent history of American fire protection can be dated by five eras. Each has lasted roughly twenty years and for each a characteristic fire problem has informed policy, shaped a distinctive strategy of management, and animated a research agenda. Throughout, the service has been both pushed and pulled—pushed by the wholesale reformation of land use set into motion by industrialization, yet pulled by varied visions of how it could meet the demand for new fire practices. At various times different kinds of fires assumed a commanding role, and around them all the tasks of fire protection were arrayed. At the same time, it is possible to interpret these phases not merely as reactions to particular problem fires but as a means to exploit certain abundances that became available at the same time—"surpluses" of land, money, manpower, equipment, even knowledge.

The two fires—the 1910 holocaust and the light burning proposed for California—defined the early debate about fire policy. Like a pair of scissors, the two sides worked against each other, less a dialogue than a dialectic. The one wanted to found fire protection on fire control, the other on fire use. The advocates of fire control proposed a European model in which foresters could, through rigorous planning and perseverance, gradually strangle fire from the landscape. "Systematic fire protection" would make silviculture possible, and intensive silviculture would weed fire out of the forest. The advocates of fire use proposed an adaptation of frontier fire practices—the ubiquitous use of fire by timber owners, naval stores operators, sheep and cattle herders, landclearing settlers, hunters, and American Indians. Periodic, light surface firing would reduce fuels and fire intensity, domesticating rather than exterminating fire. It would leave the torch in the hands of folk practitioners rather than professional foresters. Each side inspired and goaded the other; each argument or public appeal stimulated an equally vigorous response; light burning and systematic fire protection coevolved, acquiring a sophistication and intensity that neither could have mustered in isolation.

While the controversy had its political epicenter in California, its aftershocks rumbled through the whole country, and while diverse groups rallied around one side or the other, the fledgling U.S. Forest Service led the charge for aggressive fire control. Timing was critical. The 1910 holocaust traumatized the Forest Service, and its coincidence with the popularization of light burning merged in its collective mind as part of a common

WILDLAND FIRE PROTECTION
The U.S. Forest Service Experience

DATE	PROBLEM FIRE	POLICY	FIRE CONTROL		RESEARCH	ABUNDANCE
			Strategic concept	Tactical emphasis		
1910–1929	Frontier fire	Economic policy	Systematic fire protection	Administration	Fire as forestry	Land (Transfer Act) Emergency firefighting funds
1930–1949	Backcountry fire	10 A.M. policy	Hour control	Manpower	Fire as forestry	Emergency conservation programs (e.g., CCC)
1950–1969	Mass fire	10 A.M. policy	Conflagration control	Mechanization	Fire as physics	War surplus equipment
1970–1989	Wilderness fire	Fire by prescription	Prescribed fire	Fire behavior information	Fire effects—biology, economics	Information
1990–present	Intermix fire	Fire by prescription	Integrated fire services	Incident command system	Global change (?)	Rural and volunteer departments

struggle to reform American fire practices. Against the arguments for the light burns proposed for northern California, the Forest Service countered with the horror of the Big Burn in the northern Rockies. For critics to carp over fire policy while conflagrations raged and firefighters died seemed not merely ill-informed but treasonous. Not until the service acquired a chief who had not personally weathered the 1910 fires did it lift its ban on controlled burning.

The fires of 1910 also initiated and shaped fire research. Overall, the period of 1910 to 1920 was one of experimentation, a time not only of testing different methods of fire protection but also of exploring different ways by which scientific knowledge could replace inchoate folklore. Within weeks following the Big Blowup nearly all the major administrators of the Forest Service produced studies for improving methods of fire control. The academic botanist Frederic Clements published his study of fire and lodgepole pine, arguably the beginning of fire ecology. Field foresters experimented with all-out fire suppression, with controlled burning, and with such compromise strategies as "loose herding" and "let burning" in the remote backcountry. As a professional bureau created during the flood tide of Progressivism, the Forest Service was ideologically committed to scientific practices, and when in 1916 the service established its Branch of Research, fire was a formative part of its agenda.

Although the issues were national, the controversy—soon to be a vendetta—concentrated in California. Here Forest Service researchers attacked light burning on two counts. First, they created a model for fire planning and operations, and then they promoted studies of fire effects, some founded on semicontrolled field experiments, others on the self-evident appearance of the landscape. Throughout there was distrust: the experiments were conducted with the understanding that light burning must, necessarily, fail as a methodology of land use. It would fail as policy, as practice, and as knowledge.

The conceptual counterattack was engineered by Coert duBois, district forester for California; the field trials, under his general direction, brought Stuart Bevier Show into national prominence. While duBois had published a model plan in 1911—after abbreviating his honeymoon to fight fire in the Sierras—he completed his masterpiece, *Systematic Fire Protection in the California Forests,* three years later. A brilliant piece of systems analysis inspired by the efficiency studies then popular among industrial engineers, commonly known as Taylorism, the treatise isolated every component of fire behavior and control, assigned each a numerical value, then reassembled the whole to satisfy announced objectives for fire exclusion. Its methodology mirrored closely the motion studies then in vogue that sought to rationalize assembly lines. Light burning could produce nothing of comparable rigor or comprehensiveness; its populist appeals were dismissed as a demagoguery of folkways. Meanwhile Show's studies supplied scientific evidence that light burning was not benign but was in reality a subtle mechanism for forest destruction and soil erosion. A principal tool was the fire report. As reports poured in by the thousands, researchers acquired a formidable data-set that they could process by statistical analysis.

The controlling fire problem, that is, was the existing aggregate of frontier fire practices, perceived by foresters as wasteful, irrational, and politically subversive. Against it they turned to a conceptual model of science strongly refracted through a prism of strict utilitarianism. It was not the purpose of Forest Service science to question policy but to find better ways to implement it, to rebut critics, to improve suppression, to invent useful equipment, to guide expenditures. The political agenda of the Forest Service dictated its scientific agenda. Administrative policy proposed the ends and research sought to supply the means. "We accepted that the route of research was from the general laws and relations to the particular," Show wrote, "seeking to refine and measure the arithmetical values. In no sense did we accept the route of the particular to the general." Administrators and researchers frequently exchanged roles, dramatizing the interchangeability of scientific knowledge and administrative know-how.

The period of administrative experimentation ended with the tenure of William Greeley as chief forester. In 1921 Greeley organized the Mather Field Conference, which assembled in California the best minds of the Forest Service to review and standardize fire policy, lexicon, and techniques. And in 1923 a special panel created by the California Board of Forestry officially condemned light burning; Show, with his new collaborator, Edward I. Kotok, summarized the scientific case against the practice. In 1924 Congress passed the Clarke-McNary Act, a great enlargement of the Weeks Act that further disseminated federal standards for fire protection among the states.

The era reached a climax in 1928. The McSweeney-McNary Act strengthened the statutory authorization for fire research, not only authorizing it by name but also identifying the Forest Service as the principal federal agency to conduct it. Quickly the service established the Shasta Experimental Fire Forest as an administrative model of systematic fire protection, a demonstration unit that boldly challenged light burning in its heartland. Debate ceased. The choices were no longer between fire use and fire control but among various options within fire control. The concept of systematic fire protection quickly evolved into the more sophisticated precepts of hour control, which established standards for response and control based on assessed risk and anticipated fire spread.

In theory—the so-called "economic theory" that in combination with light burning critics had limited fire programs—policy sought to balance the costs expended by fire control with the values of the land at risk. High-value lands deserved greater protection than low-value lands. The resolution to the light-burning controversy had thus supplied both means and end. Suppression through aggressive initial attack became the accepted method of fire protection; and the frontcountry, where suppression could best be applied and most easily justified, demonstrated the purposes to which fire protection could be put.

It was a generation that sought action, a generation that combined European philosophy with American nation-building to create pragmatism; that merged political activism with scientific rationalism to yield Progressivism; that answered Teddy Roosevelt's call for a "life of strenuous endeavor" and made plausible William James's appeal for a "moral equivalent of war." (Revealingly, James published that essay the same month as the Big Blowup.) It was not merely that the times required action; the key personalities of those times demanded action commensurate with their energies and ambitions.

Firefighting in the western wilderness satisfied perfectly the agenda—one as much psychological as political—of Progressivism. It reified, as nothing else did, the Jamesian call for a national conscription of youths to wage a sublimated war on the forces of nature. Ranger Elers Koch marveled at the vigor and esprit of a corps composed, as the Forest Service

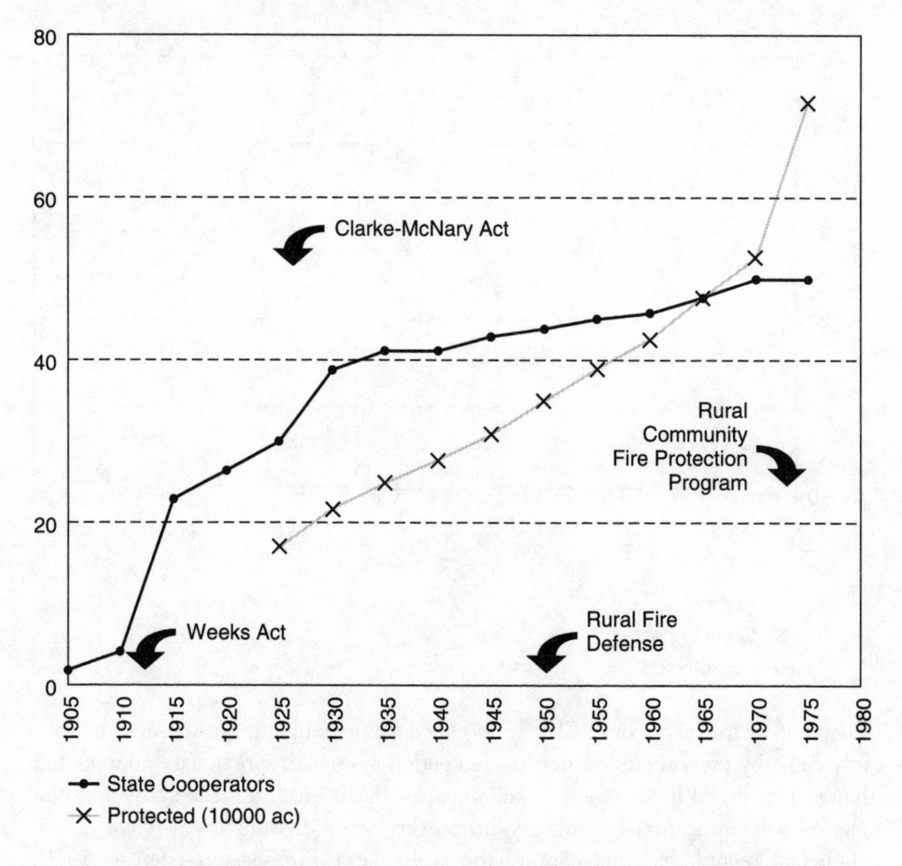

Cooperative fire protection. By the mid-1960s all the states were official participants in cooperative fire programs, and the amount of land under some level of protection escalated. By the 1990s what remained were unincorporated rural (or exurban) landscapes; and these became the thrust of the intermix fire agenda.

initially was, "almost wholly of young men." Even so, the will to act generally transcended the ability to act.

As the Depression darkened the national economy, attention shifted to the country's blighted backcountry—lands in the back reaches of the national forests, and lands, such as those in the South and Lake States, that had been cut over and abandoned. In both cases, fire protection was an investment in the future for which the simple calculus of the "economic theory"

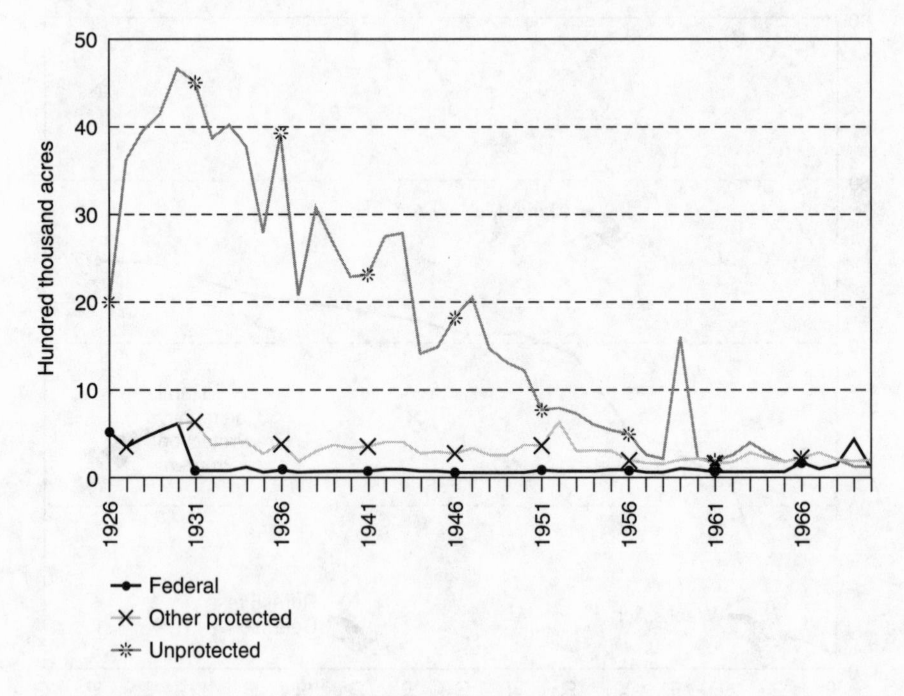

Cooperative fire extinction. Clearly the success in reducing burned area nationally came by creating first-order fire protection systems in previously unprotected lands. This could be done easily and cheaply, but to sustain these systems in the face of worsening fuel hazards required ever larger investments of money. The statistical record does not capture the really large fire seasons such as 1910, 1919, and 1924.

was inadequate. Unless the means at hand multiplied or the value of the lands increased dramatically, it was difficult to justify fire protection for them. Yet something had to be done. The era tested the limits, geographic and financial, toward which systematic fire protection could be pushed.

In the South, folk woods-burning threatened newly acquired national forests and industrial plantations. In the Lake States and the Northeast, fire compromised any attempt to reforest even land abandoned as exhausted or tax-delinquent. Throughout the West, savage droughts led to a near crisis of fire that saw swaths of counties placed under martial law and that culminated in a sequence of spectacular, almost annual conflagrations—the Matilija fire (1932) in Southern California, the Tillamook Burn

(1933) in Oregon, the Selway fires (1934) in the northern Rockies. Either the Forest Service took action or, so it believed, it risked a possible political challenge to its control over the national forests. In fact, the country as a whole appeared in a headlong slide toward an environmental apocalypse; those great fires were to forestry as the Dust Bowl was to agriculture. The national response—the Roosevelt administration's remarkable conservation programs—resolved the issue of how far fire protection could expand.

When the New Deal unleashed its monies and especially the Civilian Conservation Corps (CCC), the architects of hour control already had an ambitious public works program for fire protection on their drawing boards. CCC labor almost overnight created a physical plant for fire control that would otherwise have required decades of normal evolution. Roads, trails, telephone lines, and lookout towers appeared within a handful of seasons. The presence of hundred-man camps available for fire call encouraged a tactical emphasis on manpower, the management of firefighting crews. In 1939 two important prototypes emerged—the smokejumper for initial attack and the forty-man crew for campaign fires. So great was the amplification of means at hand that it pushed the Forest Service into a similar amplification of ends.

In 1935, as an "experiment on a continent scale," Chief Forester Ferdinand Silcox, a veteran of 1910, promulgated the 10:00 A.M. policy, which stipulated control by 10:00 A.M. the day following the report of a fire, or, failing that, control by 10:00 A.M. the day following, ad infinitum. The policy extended equally to all lands; it applied to remote stands of lodgepole pine buried deep in the Rockies, to chaparral-clad sierras, to southern savannas, no less than to lucrative timber sites. The edict shattered the pragmatic equilibrium between the values at risk and the cost of protection. Instead it seemed possible with one bold stroke to break the back of American wildfire.

Fire research rallied in support. Nearly all researchers were foresters first, scientists second, and most had grown up with field-heavy fire research from the early 1920s. One thrust was to improve the concept of hour control, to generalize it into a universal process of fire planning. Some groups, such as that led by Lloyd Hornby in the northern Rockies, funneled the fire reports from new regions into their statistical mills and drafted elaborate blueprints. More daringly, the California tradition of duBois and Show evolved beyond reports by endowing, in 1939, a fire behavior program under Wallace Fons that came complete with labs and wind tunnels that would, it was expected, yield a better typology of fires than that derived from fire reports alone. In the Southeast George Byram boldly transferred similar physical methods to fire phenomena, an enduring achievement made doubly astonishing by his long isolation. Other researchers experimented with smoke detection, hoping to improve the capability of lookouts, to refine the "arithmetical values" for each of the components of the hour control program. A good bit of practical engineering went into equipment development in an attempt to mechanize at least some parts of suppression. More important—awed and overwhelmed by CCC largesse—foresters investigated the principles of crew organization and brought the system to the fireline.

But perhaps the most critical inquiries were those related to fire-danger rating and fire effects. The first was closely related to fire behavior, essentially a forecast under worst daily conditions, and thus served to set the clock, as it were, for hour-control goals. When a rating index was developed in 1935, it quickly became the basis for expending "emergency" presuppression funds under a program intended to increase firefighting readiness in proportion to the rated danger. Not lost on administrators who now had to apply a common policy across the entire country, a fire-danger index also promised a common standard by which different districts and different regions could be compared and their actions judged. Much as fire-danger rating integrated many environmental factors, so it tended toward an institutional integration of fire research. Its advances, however, were incremental, its methodologies those typical of forestry, and its outcome wholly circumscribed by the larger policies it served.

Not so the research on fire effects, studies that the Forest Service could not wholly contain. Centered in the South, these inquiries shifted the regional geography of fire away from the public lands of the West; often conducted by nonforesters, or by foresters outside the employ of the Forest Service, they introduced new methodologies and new data, and placed the debate within a context outside the policy of a single agency. From all sides—from range researchers such as S. W. Greene, wildlife biologists such as H. S. Stoddard, foresters such as H. H. Chapman (dean of the Yale Forestry School)—the evidence accumulated that light burning in the southern pinelands was not devastating and was often ecologically valuable, that a proliferating "rough," or understory, only built up fuels to the point that fires roared out of control with unprecedented intensity. But when its own research confirmed these positions, the Forest Service systematically—shamefully—suppressed the data and sought to influence the publication of any results, even those outside the agency, that contradicted official orthodoxy. To admit the value of controlled burning in the South would, so it feared, unravel the tapestry of political alliances that had brought professional forestry to the region and might allow critics to resurrect light burning, now successfully condemned as anathema. Right-thinking foresters regarded light-burning proposals with the condescension that mathematicians reserved for circle-squarers and physicists for perpetual-motion mechanics.

In fairness, it must be said that the Forest Service faced a vexing, unprecedented task. Alone among the federal conservation bureaus, it had been given responsibilities to regulate and to manage, to administer and to research, a bureaucratic schizophrenia that made conflict of interest a structural inevitability. Researchers accepted the rightness of fire control and looked to science to improve the agency's capacity to apply it; when science, inexorably indeterminant in its conclusions, discovered facts contrary to policy, then the service had no easy way to reconcile those contradictions. It sought to suppress the findings. The momentum of fire protection under the impress of New Deal funding and manpower was too great to resist.

Still, while it might dismiss researchers, it could not defy nature indefinitely. Droughts blasted the South, and fires raged through nominally protected sites. Even by 1932 the service, however reluctantly and silently, allowed its state cooperators to control burn and still qualify for assistance under the Clarke-McNary program. By 1943 it extended that option to the southern national forests. That the new chief, Lyle Watts, who approved the change had not weathered the 1910 fires, that tractor-plows promised to substitute for the CCC crews lost to the army, and that the wartime emergency created a psychological climate in which pragmatic deviations could be tolerated—all were special circumstances that encouraged the sense that controlled underburning was simply another example of southern exceptionalism, a practice not readily transferred or much desired elsewhere.

When, in the midst of World War II, the Forest Service debated the future of fire research, its thrust was conservative. Research still considered itself an adjunct of administration. It sought to keep fire protection firmly within the institutional and intellectual confines of forestry. When the service established a separate Division of Forest Fire Research in late 1948, its objective was likewise restorative rather than innovative. The division would eliminate duplication among regional research stations.

Researchers and administrators still frequently exchanged roles without the slightest agency angst, and they focused on fire control, an administrative problem, rather than on fire, a scientific problem. They dismissed frequent proposals to study fundamentals, the "laws of combustion." Fire-behavior research, they felt, was an ignis fatuus that would lead foresters into a miasma of theoretical science. Instead, administrators urged that fire researchers leave their field plots and statistical compilations and hit the fireline. It is symbolic of the era that two of its outstanding practitioners of fire research, Lloyd Hornby and Harry Gisborne, both died when they did so.

World War II revolutionized Americans' relationship with fire. Watching a conflict in which competing armies hurled newly invented fire weapons at one another—napalm, flamethrowers, incendiary bombs, fire balloons, the atomic bomb—professionals argued that the next war would be a fire war of even vaster proportions, and the public learned to see all fire as hostile fire, as enemy fire. America's two national fire-prevention campaigns—Keep America Green and the Cooperative Fire Prevention Program, best known for its symbol, Smokey the Bear—both emerged directly out of the wartime emergency; almost on cue, too, the Disney Studios released *Bambi,* in which the same villains who kill Bambi's mother also unleash a terrifying fire that threatens Bambi and his father, a powerful antifire message to young children. Before the war most Americans, certainly most rural Americans, at least tolerated fire as an instrument of land use; after it, as fire imagery became more horrific and as the rural population drained away to cities and suburbs, Americans distrusted fire. And once the Soviet Union exploded an atomic bomb in 1949 (two weeks after the Mann Gulch fire killed thirteen smokejumpers in a blowup), fire control became a matter of national security. America entered into a cold war on fire.

Fire protection expanded into new lands. It leaped into Alaska, newly destined for statehood (1959), and extended a network of rural fire defense over that countryside otherwise unclaimed by either federal agencies or urban fire services under the aegis of the Office of Civil Defense (OCD) in the early 1950s. Both grew out of wartime alliances, both responded to the imperatives of Cold War geopolitics, and both relied on the flood of war-surplus equipment released after the Korean War. The Forest Service had priority access to this hardware through the federal excess-equipment program, and it was able to funnel large quantities to its fire cooperators such as the state foresters. Virtually overnight, fire protection mechanized; machinery replaced manpower as an informing abundance. The support given fire protection by the emergency conservation programs of the New Deal was, in effect, replaced by a new alliance with the programs of the national security state. This abundance of hardware, together with the swelling affluence of American society, allowed the 10:00 A.M. policy to flourish.

Animating the new era was the specter of mass fire, a technical term developed to replace loose allusions to "firestorms." Here visions of fire-bombed cities such as Hamburg, Dresden, and Hiroshima merged with the stubborn persistence of massive fires in American wildlands. Shaping both perceptions was the concept of conflagration control. Even more than under the hour-control program, the goal was, through rapid initial attack and various measures for confinement (such as fuelbreaks), to prevent small fires from making the transition to mass fire. The military analogy, of course, was to contain brushfire wars before they could lead to a superpower confrontation and nuclear holocaust. If they could not eliminate the "red menace" altogether, fire agencies would at least contain it. A good expression of this philosophy was the creation in 1961 of a rapid deployment force, the interregional fire-suppression crew ("hotshots"), as a kind of Special Forces that could be dispatched instantly to trouble spots throughout the national forest system.

It was a great era—a golden age—for fire research. The wartime experience suggested how science could be mobilized for national defense and the new funds from OCD and elsewhere pushed fire research far beyond its previous confinement. "There is no reason," exulted Keith Arnold, "why we can't develop fire extinguishing forces as strong and effective in their way as atomic weapons have been in their field." With a new generation for whom World War II, not the 1910 conflagrations, was a point of reference, fire research proceeded on two fronts. One seized on equipment development, the other on fire behavior.

Operation Firestop, a one-year crash program conducted through 1954, announced both. It advertised new institutional relationships, new methodologies for fire research, and a new intensity of purpose. It joined the Forest Service and the California Division of Forestry with civil defense and the military. Through expansive field trials in Southern California, it explored such war-dramatized technologies as air tankers, helicopters, and chemical retardants, refined the mechanics of fire behavior beyond the guesswork encoded in fire reports, and promoted a vision of fire attacking sprawling metropolises. Even as Operation Firestop torched off hillsides at Camp Pendleton and tested hoselays from hovering helicopters, new suburbs crowded into chaparral wildlands, and the hypothetical vision of incendiary attacks on civilian populations seemed frightfully imminent as conflagrations rushed through the streets of Malibu and Bel Air. Against such drama, the statistical methodology of Show and Kotok seemed antiquated, and the adaptations of silvicultural techniques to the study of fire effects quaint. A new-model pulaski might ease backstrain, but a squadron of air tankers could kindle the imagination.

Aerial attack was an old dream, and almost immediately after World War II the Forest Service experimented, inconclusively, with helicopters and bombers. Operation Firestop accelerated the process and bonded to it the problem of mass fire, suggesting that if fire was an attacking enemy, firefighters could counterattack with the same methods that proved so successful during the war. Air superiority became a watchword. As early as 1947 the Forest Service tested a comprehensive aerial program in the northern Rockies, the so-called continental unit, in which detection, suppression, and logistics were all conducted by air. By 1956 air tankers and helitacks, helicopter-led attacks, became fireline realities. To explore other means of converting military surplus to fireline duty, the Forest Service established two equipment-development centers. One, at Arcadia in Southern California, emphasized aircraft, tractors, and engines; the other, in Missoula, Montana, addressed the particular needs of smoke-jumping and hand crews. The service also transferred from Portland, Oregon, an older equipment program that focused on portable radios, and redefined its mission from development to testing.

While gadgets quickly entered the fireline, research into fire fundamentals percolated into practice more slowly. Like the ingredients for mass fire itself, many events had to compound one with another—a chief forester, Richard McArdle, who had a personal interest in fire research; the monumental model dramatized by Operation Firestop; disastrous California fires in 1956, which led to a congressional review of federal fire protection; the military liaison, which seemingly equated fire defense with national defense and took fire research out of the parochial control of forestry. But merge they did. And the merger extended, ambiguously, to popular culture as well. Keith Arnold believed that the political will to endow fire research resulted from fires that threatened to burn the cable to Mount Lowe through which the Rose Bowl game would be broadcast. That snapped the California congressional delegation to attention. The Riverside fire lab was the result.

By 1958 the prospects were such that the Forest Service and OCD requested the National Academy of Sciences–National Research Council to assemble a Committee on Fire Research to coordinate the national effort and, in particular, to advise OCD and the Department of Defense on their investments in fire science. While its charge was panoramic, the committee's agenda sought a more universal description of large fire behavior, one that unified wildland fire and urban fire under the concept of mass fire. That same year the Forest Service, which had plenty of internal reasons for studying fire behavior, launched its own integrated inquiry, a program to create a truly national fire danger rating index, one based on "analytical" principles, to replace the empirical meters that had proliferated, region by region, over the past two decades.

Administrative desire outraced scientific ability, however, and over the coming years a national fire danger rating system (NFDRS) threatened to become the space shuttle of fire research, a technological marvel that somehow failed to fulfill the special needs of either basic science or field operations. What was needed was a more powerful predictor of fire behavior, not fire danger. (It is revealing that the fire behavior module of the NFDRS soon embarked on a separate career, almost an apotheosis, far exceeding that of the system for which it was created. Almost from the moment Richard Rothermel of the Northern Fire Lab first published it in 1972, the model quickly dazzled the entire fire community with the prospects for mathematical forecasts of fire spread. Fire officers seized on the Rothermel model and, to the mixed horror and pride of its creators, projected it far beyond its originating conditions. So intense was the demand that the lab spent much of the next decade refining and elaborating the model into a more universal, computer-based fire behavior system: BEHAVE.)

All this—the flurry of plans, the ambition of new monies—demanded commensurate institutions. Newly installed as head of fire research, Arthur Brown reported, somewhat breathlessly, that the military alliance put his small band of researchers center stage in the great drama of Cold War science, a mobilization that would translate into both prestige and durable institutions. Looking back on the buildup, an incredulous Craig Chandler marveled at the combination of naïveté and ambition that led a new generation of fire researchers to crank out multimillion-dollar proposals to foundations on blue-ink mimeograph machines. Still, it happened. Much as the New Deal had hurled fire protection several stages beyond a normal, incremental evolution, so now the Cold War pushed the Forest Service beyond its homegrown research horizons. Between 1959 and 1963, the service established three laboratories dedicated to fire science—the Southern (in Macon, Georgia), the Western (in Riverside, California), and the Northern (in Missoula, Montana). By the mid-1960s the service had concentrated almost its entire fire-research program in these labs. By 1962, moreover, OCD was sponsoring enough science to hold annual conferences among its contractors. After the Cuban missile crisis, another rush of funding became available for fire-dynamics studies.

The Forest Service had plenty of internal reasons for wanting to better understand fire behavior. Large fires continued, and multiple-fatality fires such as the Mann Gulch, Rattlesnake, and Inaja were a painful reminder that knowledge lagged behind technology. Moreover, the service had plenty of funding of its own to invest in traditional topics. But fire science changed, becoming in some respects as self-contained as the mass fires that claimed its attention. Fire as a physical phenomenon, rather than fire as an administrative problem, became the object of investigation.

Wildland fire had at last left the confines of forestry, but in so doing it also left the traditional means of evaluation by foresters. It became a subject for physics, chemistry, meteorology, and new forms of operations research. It developed experimental models, both physical and mathematical, for fire spread. It invented new measures for the description of free-burning fire. It undertook large-scale experimental tests to discover the essentials of mass-fire synergism, one series blossoming in the early 1960s and another, Project Flambeau, in the mid-1960s. Similarly, it searched out better methods of orchestrating fire research. Fire science brushed against Big Science, and it could no longer remain in the province of backwoods mechanics or tolerate an indiscriminate interchange of researchers and field administrators or dispatch its physicists to dig firelines.

Yet for much of the era wildland fire science seemed remote, overshadowed by the deluge of mechanization. Like its informing subject, mass fire, it started slowly, and, once begun, burned defiantly, almost oblivious to the ambient winds around it. As it became big, moreover, it also became powerful, so much so that its information and formulas could, proponents argued, substitute for fireline crews. By the end of the era, research had created an "abundance" of information that would help shape the coming decades in profound ways. Besides, even before the labs published in volume, fire research confirmed the institutional hegemony of the Forest Service. By the early 1960s, the service dominated virtually all aspects of fire management—its manpower, equipment, research, federal-state cooperative programs, field operations, and policy. But like one of those old-growth ("overmature") forests it sought for harvesting, the service's great size disguised an inner rot. Fire control was dying on its feet.

By 1970 the Forest Service's marvelous hegemony had splintered, and by 1980 an environmental movement had redrawn the political geography of American fire. Increasingly wilderness preservation dominated debate

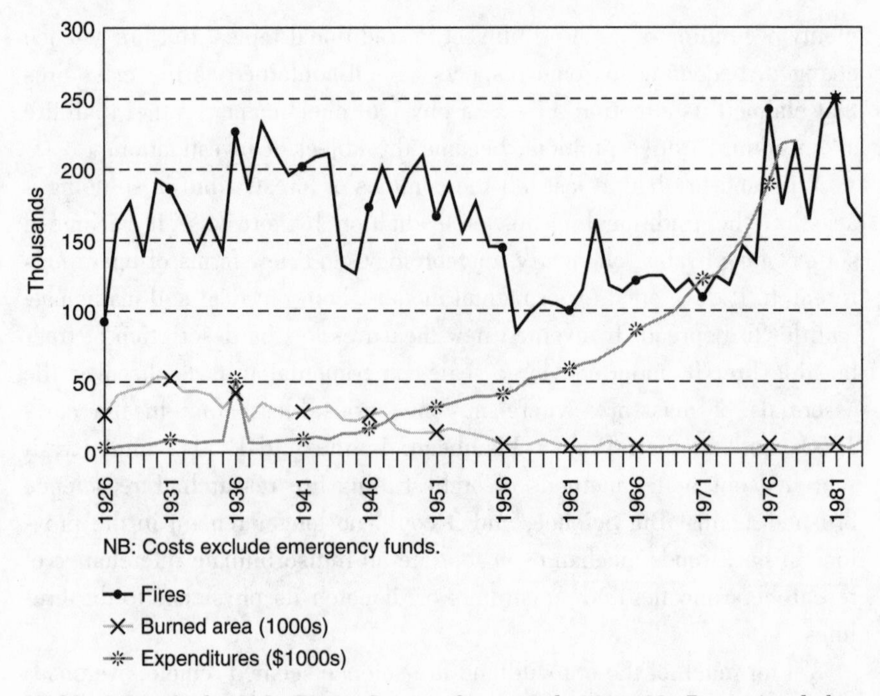

Field of Fire I: the U.S. Forest Service fire record, 1910–83. Protection led to quick results in terms of burned area. Unfortunately, there is no statistical record of controlled burning by which to measure the overall reduction in fire, and even for suppression, the reporting lags.

about land use, deforming it as a black hole warps the geometry of space; the federal land agencies had their organic acts rewritten or fundamentally reinterpreted; and like old chaparral, steadily readying itself to burn, decades of successful fire suppression had created conditions that became more and more insupportable. Fire protection had its costs, and those costs rapidly approached a point of diminishing returns.

The new missions date—for the National Park Service—from the Leopold Report (1963), and for all the federal agencies from the Wilderness Act (1964), which charged that parks should be managed as "vignettes of Primitive America" and wilderness should remain in a state "untrammeled" by human artifice. The statute directly challenged traditional fire control. It argued against the presence of firefighting, particularly where mechanized, and it argued for the introduction of some kind of fire, espe-

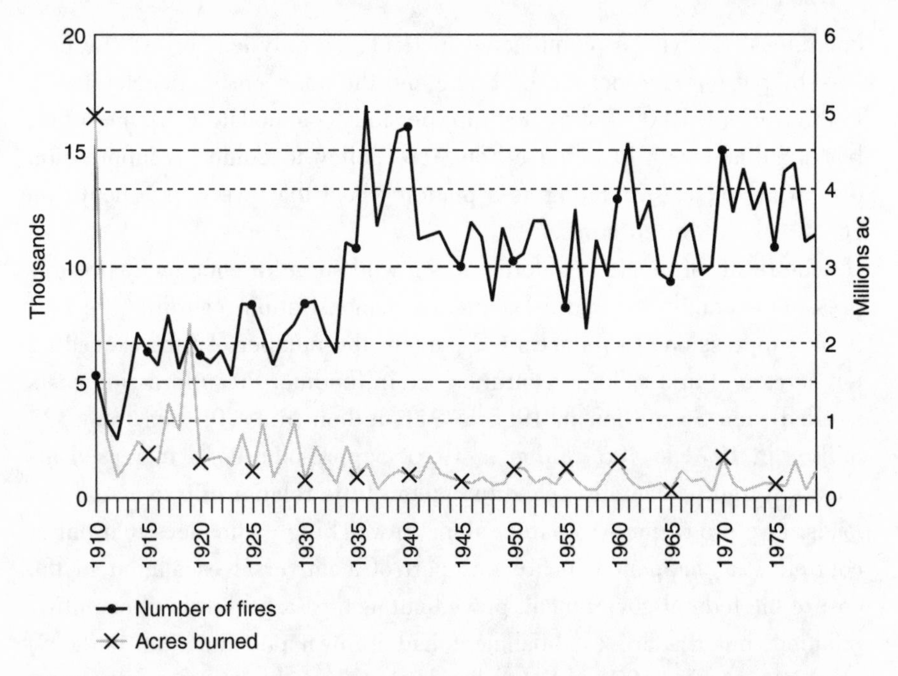

Field of Fire II: America's fires, burned area, and costs for all statistical areas. Note that these costs, while unadjusted for inflation, do not include emergency funds.

cially when natural sources, notably lightning, kindled blazes in official wilderness. But if fire was beneficent in wilderness, then why not elsewhere? Controlled burning could distribute the healing goodness of the wilderness to other, abused lands. Prescribed fires of all kinds in every imaginable setting became the rage. Fire by prescription became a national policy.

A boisterous decade of policy reform, replanning, and field experimentation quickly followed—some painful, some exhilarating. The National Park Service revised its policy in 1967–68, leapfrogging its philosophy ahead of its expertise. Within a decade, after several embarrassments and the nearly catastrophic Ouzel fire at Rocky Mountain National Park, it restored some control over field operatives by reinstating national standards overseen by a Branch of Fire Management. The Forest Service completed its revision in 1978 but came more fully equipped for field operations. A revision promulgated in 1971 had only temporized the process by putting one foot on the brake and the other on the accelerator; it liberalized the 10:00 A.M. policy in order to accommodate wilderness fire, but simultaneously installed a Ten Acre policy to guide presuppression everywhere else, stipulating as a planning goal that every fire should be contained within ten acres.

The result of the 1971 reforms was a wild surge in emergency presuppression expenditures and a fascinating demonstration regarding the limits of suppression. To reduce by 2 percent the number of fires exceeding ten acres required a 90 percent increase in funding. Presuppression costs swelled from $6 million in 1965 to $11 million in 1970, then from $25 million in 1973 to $85 million in 1976. Suppression costs increased almost as rapidly. Overall expenditures bore little relationship to actual fire loads. Fire protection was hardly alone in watching its finances roar out of control; what happened in fire was part of a universal escalation in the cost of the federal government, played out against a backdrop of mounting inflation; but the fire establishment had its own peculiar ("off-budget") mechanisms. The 1978 policy reforms abolished the emergency presuppression fund. However one interpreted the ills of fire suppression, it was increasingly evident that more suppression would not solve them.

Instead, it was hoped that prescribed fire could substitute for customary presuppression practices and fire science for traditional, increasingly suspect methods of fire control. The upshot was, inevitably, less than that. The clamor for burning far exceeded the acres burned; worse, for more than ten damning years the country's most disastrous fires were the result of breakdowns in the management of prescribed fires. Had the climate been hostile, as it was by the mid-1980s, the intellectual sparks thrown off by critics might never have kindled prescribed burning; but for the critical period of transition, favorable weather kept wildfire more or less contained and allowed the experiments to proceed. The resulting reformation was irresistible. In theory, if not in practice, a certain parity developed between fire control and fire use. Painful, self-conscious "fire management" replaced "fire control." A policy of fire by prescription guided practice and left to local discretion the determination of which fires were good and which were bad and how they should be managed.

Still, more was involved than fire alone. It was not merely that fire had to serve land management, but that the means and ends of land management were changing. Two decades of public scrutiny, legislation, and rancorous litigation redefined agency missions and rewrote their statutory authority. Fire protection became a means, not an end, but the ends to which it should be applied were themselves in upheaval. For the Forest Service the critical events were the passage of the National Forest Management Act (1976) and a raft of auxiliary legislation in 1978 that redefined its missions in cooperative forestry and research. For the Bureau of Land Management, the service's principal rival in the Department of the Interior, the complementary events were the institution of the Federal Lands Management Act (1976) and the Alaska National Interest Lands and Conservation Act (1980), which broke its sovereignty over the subcontinent. Even so, the year-by-year, state-by-state congressional fights over wilderness legislation kept the pot boiling. Pummeled by environmental critics on such issues as clearcutting, deficit logging, and its general hostility to wilderness values, forestry lost high ground, and with it fell its presumed authority in fire. All this influenced, in ways both abrupt and subtle, how the Forest Service fought fire.

Two processes were at work, one driving the American fire establishment toward fragmentation, the other toward new forms of integration. Each of the federal agencies sought to create its own autonomous fire organization to serve its particular mission. No single organization could oversee all these needs, any more than could any one policy. But the need to coordinate was equally compelling. One outcome was the Boise Interagency Fire Center (1969), which consolidated the national response to wildfire. Another was the National Wildfire Coordinating Group (1976), which brought together not only the federal agencies but, at Forest Service insistence, also a representative from the National Organization of State Foresters. The NWCG promptly assumed a commanding role in establishing standards for equipment, training, and certification, and helped advise fire research on field needs. The upshot was that, like planets pulled equally by gravitational and centrifugal forces, the American fire community maintained something like its former orbit. If it was unclear how to start fires of the type and on the scale requested by the new order, the Forest Service and its allies could continue to fight them.

In such a scene, inherited wisdom was only marginally useful. But

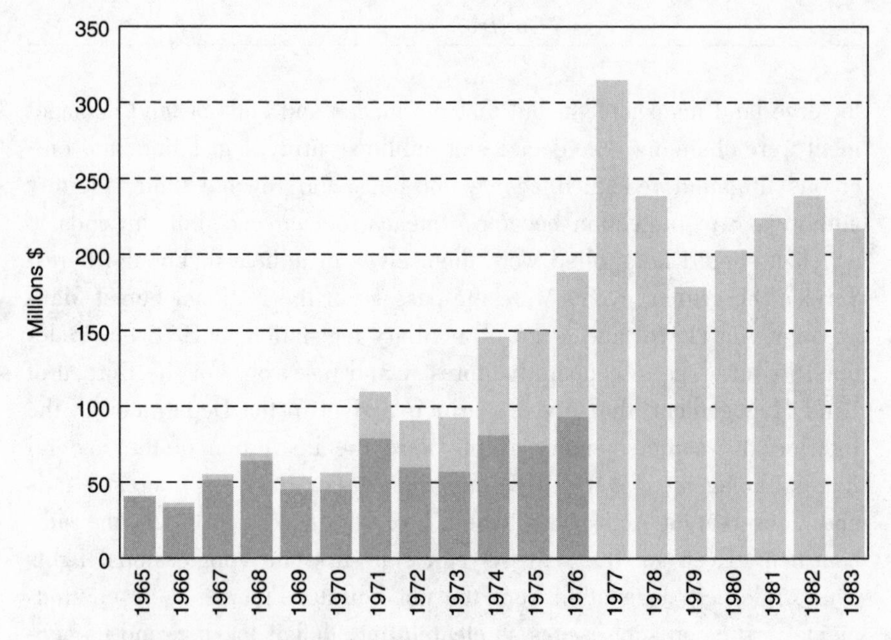
350
300
250
200
150
100
50
0
Millions $
1965
1966
1967
1968
1969
1970
1971
1972
1973
1974
1975
1976
1977
1978
1979
1980
1981
1982
1983
NB: After 1976 numbers not adjusted for inflation.
Deflated $

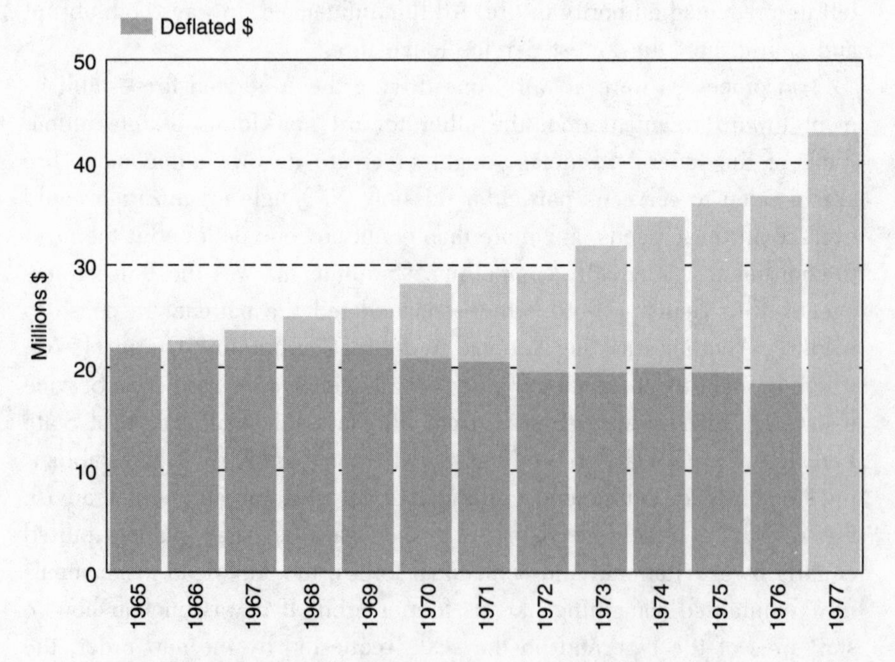
50
40
30
20
10
0
Millions $
1965
1966
1967
1968
1969
1970
1971
1972
1973
1974
1975
1976
1977
NB: 1977 numbers not adjusted for inflation, or for transition quarter.
Deflated $

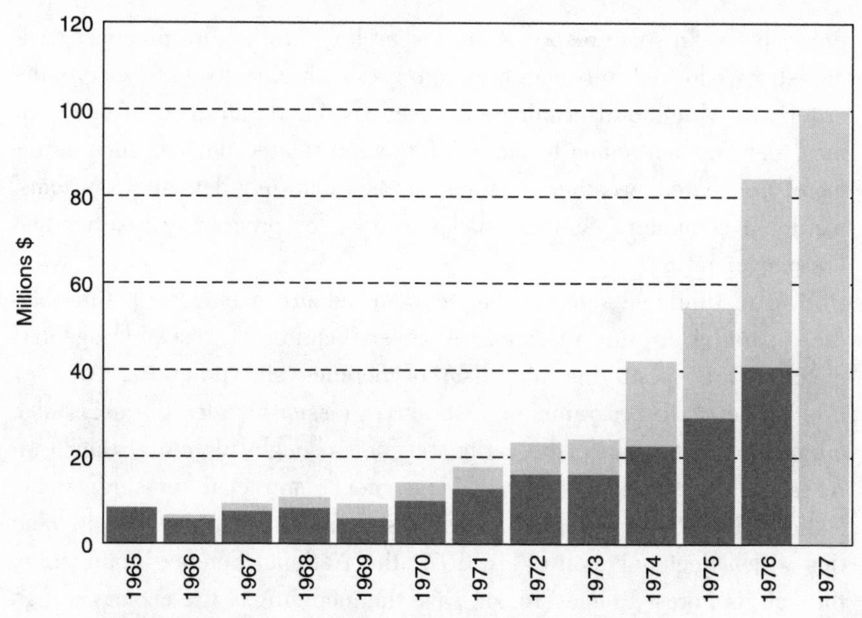

The economics of fire protection at a transitional period: (top left) total expenditures; (bottom left) presuppression expenditures (regular budget); (above) presuppression expenditures (FFF). In real dollars the investment in fire protection was declining, but emergency (off-budget) funds were making up the difference, and then some. The drought years of the late 1980s worsened the situation. In 1994 emergency expenditures exceeded $925 million.

knowledge—information that poured out of the Forest Service labs, fire sciences that thrived outside the service—was at the bold core of the new era of wilderness fire. The revolution in policy and practice had followed from recently discovered knowledge and from innovative ways of seeing old knowledge; and it demanded still more. The demand was insatiable. The strategic emphasis on prescribed fire required a science of fire behavior and, increasingly, of fire ecology. So awesome was the scientific effort, so profuse its publications, that information replaced surplus equipment as a new "abundance." It seemed possible to substitute knowledge for power, to replace traditional forms of fire control with information, to call "controlled" those fires that behaved within the parameters of a monitored prescription. In some respects, fire research controlled fire practices. Administrators looked to research to help establish objectives, to specify the criteria by which to discriminate between good and bad fires. Even equipment development came to mean information-related devices such as remote automated weather stations units, lightning-detection systems, handheld computers, and national networks for processing weather and fire-danger rating.

The institutional reforms that reorganized fire management thus had their parallel in fire research. A diversification of research agendas matched the splintering pluralism of policies and practices. As they evolved separate programs, or distinctive missions under the aegis of a tolerant national policy, the various agencies sought to control their own research. The Bureau of Land Management contracted for studies; the National Park Service fielded a small research organization and cultivated ties with select universities. In 1972 the National Science Foundation, through its Forest Biome Project, gave the modeling of fire ecology a high priority, at long last stimulating academic interest in the field. In the same year the Senate forbade the Forest Service to conduct further research into the military uses of fire, thus divorcing it from a lucrative if compromised connection; the NAS-NRC Committee on Fire Research closed shop soon afterward. Meanwhile the creation of the U.S. Fire Administration in 1974 and the Center for Fire Research under the National Bureau of Standards, while focusing on urban fire and materials testing, nonetheless further severed the liaison with OCD and helped break up a once-impregnable monolith. And not least, the privately endowed Tall Timbers Research Station challenged the authority of forestry as a source of research and policy. From 1962 to 1974 it sponsored annual conferences on fire ecology that, for a while, were the principal, perhaps the only, podium outside Forest Service control for debating fire practices. The symposia brought before the public a cornucopia of counterexamples from elsewhere in the world, argued vociferously for the virtues of prescribed fire, and compelled official research to factor the biology of fire into its agenda.

While debates over policy raged, research flourished, but when the revolution became itself the new establishment, research flagged. All parties had looked expectantly to science to resolve the intractable questions about what they should do and how they should do it. When the Forest

Service adopted its new policy in 1978, fire research was compiling a massive, computer-based fire bibliography (Firebase), sponsoring state-of-knowledge reviews on all aspects of fire effects, and launching an ambitious research, development, and applications program to firm up the foundations of field operations. In the process research into fire ecology had subsumed much of what had gone into fire behavior, and a research program that had flung itself into a high orbit of science-driven inquiry had been returned to more earthbound administrative duties.

Once the euphoria passed, fire research lost its bureaucratic momentum. No less than suppression, it also had limits. There were too many questions it could not answer; it could supplement but not substitute for field operations; the relentless skepticism of real science proved as corrosive to the ideology of wilderness as to an earlier ideology of national forestry. Ironically fire use became as intensive a form of fire management as fire control, substituting information for labor. Because of the way fires were funded, it was not possible to transfer likely savings from fire suppression into prescribed fire. However necessary, prescribed burning was not necessarily cheaper or simpler. It was only, in many circumstances, more appropriate.

Besides, information had its costs, sometimes no more controllable than those of suppression. The research to sustain prescribed fire needed substantial start-up costs. Instead, almost simultaneously with the adoption of their new policies, budgetary restrictions began to slow the rate of growth of the federal agencies. There would be no peace dividend from ending the cold war on fire. Then, during the 1980s, the United States began its dreadful metamorphosis from the world's largest creditor nation to its largest debtor. The funding for fire research collapsed. Instead of more fire ecology, research retooled to examine another dimension of fire effects—economics. In part, this inquiry reflected an effort to maximize the funds allocated to fire; in part, it served to justify agency budgets.

By the late 1980s the mission and financing of fire research had stabilized, but it ceased to command national attention. Its lab reports no longer read like communiqués from revolutionary bunkers. Tremendous energies went not into new research but toward translating existing knowledge into user-friendly formats for the field, of which the BEHAVE project was exemplary. The Southern Lab faded into near obsolescence; the Riverside Lab retooled into systems analysis, regional fire ecology, and the economics of fire budgeting; and the Northern Lab, renamed the Intermountain Fire Science Lab, shielded what remained, like a windbreak thrown around a campfire. New initiatives reworked old themes to better justify existing programs.

As drought recycled large fires back to the West, suppression reasserted itself, and after the Yellowstone Park debacle of 1988, the era of wilderness fire, already receding, rushed to conclusion. A sometimes mindless if exhilarating national mobilization of firefighting resources had greater political clout, certainly greater public appeal as a spectacle for television, than quiet experimentation on intractable questions about an imponderable natural phenomenon. An interagency panel reviewed federal fire policy, which it found appropriate, but set into motion a systematic review of fire plans, which had the effect of weeding out marginal programs and dampening inchoate enthusiasms. Yellowstone Park itself escaped the criticism it deserved—it was too big, too symbolic, to condemn; but it did so by passing along the costs of its malfeasance to lesser places; and as their programs were snuffed out, so was the era. The presumption for prescribed fire, particularly for natural fire, lost its moral drive.

So the era ends in irony. Just when the Yellowstone conflagrations advertised the power of natural fire to the public, an era informed by such fires expired, exhausted of its philosophical energies. And just as fire found a place on the agenda of international environmentalism and Big Science, the fire agencies were ill-equipped to seize the opportunity offered. Instead, the problems of uncontrolled combustion exploded into public consciousness through the ministrations of atmospheric scientists—first, with the nuclear winter controversy, then with alarms over a greenhouse summer and what seemed to be a world fire erupting from such unlikely locales as Borneo, Manchuria, Amazonia, and even burning oil fields in Kuwait.

In striking contrast to its postwar alliance with military research, the Forest Service was conspicuously marginal to nuclear winter research. For global climate-change issues it has mustered a presence, exploiting a useful connection with its ongoing studies of smoke. But it would not, as it once had, dominate federal science on free-burning fire. And if global warming reconnected Forest Service fire research with science, it also disengaged that research from the field-generated needs of the agency, the particular fire problem that promised to inform the coming era.

Even before the Yellowstone conflagration gave wilderness fire a megalomaniacal valedictory, the fire community was packing its operations to the other side of the frontier. Whereas the fire establishment had previously moved into the wilderness but dampened suppression, it now crossed into an exurban landscape crowding against wildlands, one in which wildland fuels and houses mixed like a land-use omelette. The city came to the country. The wild penetrated the urban. In thousands of scattered new settlements a fire mosaic came into definition whose essence was its intermixed nature. In exurban developments that flourished like wood-framed weeds, wildland fire discovered a heady new fuel, and the wildland fire establishment a new problem fire. With the aid of drought the era unleashed a suppression organization that, grown moribund in the soggy years of the early 1980s, roared loudly back to life. The horrific Oakland fire of 1991 announced the new era with imagery as graphic as that with which Yellowstone had closed its predecessor. It was quickly followed by conflagrations in Santa Barbara the next year, and throughout Southern California, most spectacularly around Malibu, in 1993.

The intermix fire redefined, once again, the geography and social significance of American fire. Urbanized sprawl afflicted all the regions of the United States, but everywhere it shifted national attention away from fires in remote sites and onto those that intermingled wildlands and people. It reconnected fire programs to American society not through abstruse concepts such as wilderness but through the realities of property and daily, or seasonal, living. Wilderness fire had the exclusive character of modernist architecture, one distrustful of people; Yellowstone's plans resembled a city designed by Le Corbusier. The intermix fire exhibited a postmodern eclecticism, an admission that people, for all their messy multiculturalism, were a part of the scene.

By the early 1990s the full implications were barely appreciated. Naturally, advocates spoke to its prospects rather than to its future liabilities. Not coincidentally, the intermix fire reinvigorated the political status of wildland fire. The Forest Service watched its liaison with the state foresters renewed, sought out intriguing new alliances, particularly with the National Fire Protection Association (NFPA), and built the case for its fire budget on the new politics and social calculus of the intermix fire. Geographic regions and land agencies that had occupied center stage by virtue of their abundant wilderness were now shunted to the wings. Fire protection followed demographics and became a bicoastal institution. Meanwhile, as fires raged, emergency funding in unbelievable streams poured through old channels, like a debris flow through long-dry arroyos.

The parameters of this new era are yet hazy; they will be sharpened, as against a whetstone, by events that have not yet occurred. But it is possible to make reasonable guesses. Expect a policy of fire by prescription to endure as too flexible to overthrow. Instead, local circumstances will contract the number of practical options open to administrators, so that something on the order of a national consensus will evolve. Expect prescribed fire to persist, too, not as the treatment of choice but rather as one technique among many. No longer will the momentum of reform demand that fire be "restored" to every environment unless proven otherwise; more likely there will emerge a kind of zero-based fire budget in which every practice will have to justify itself. Fuel management, for example, may be better served by building codes and zoning restrictions than by broadcast burning. Expect an integrated fire service to serve as a strategic concept, one given tactical reality by the universal adoption of the Incident Command System (ICS) of the National Interagency Incident Management System. For the requisite "abundance," look to that amorphous population of rural and volunteer fire departments that can be captured by new alliances with states, counties, and the NFPA. If past histories are a guide, expect the era to continue well into the twenty-first century.

A vigorous scenario, this, to restore to political favor a fire establishment battered by two decades of public controversy and inevitable failures of execution. But to its credit the Forest Service looked out as well as in, shunning a budget-hoarding isolationism to project its influence around the world. This vision, too, had a long tradition. Congress granted formal recognition by mandating the service as the leading American agency for international forestry. The internationalization of fire problems—alarm over global warming; the fire-catalyzed clearing of tropical wildlands; the intermix fire problem that also plagued Canada, Australia, and Mediterranean Europe—found the Forest Service as the American contact. American fire engaged the world fire.

There were mutual-assistance treaties signed with Canada, Mexico, and even the Russian Federation to help with fires across the borders, a new twist on the old doctrine of hot pursuit. Through the Peace Corps, the Agency for International Development (AID), and the Disaster Assistance Support Group, the Forest Service organized training courses for Mediterranean countries, Latin America, and even Indonesia; dispatched advisers to wildfires in such locales as Costa Rica, Ghana, and the Galápagos Islands, sites far removed from traditional fire provinces; hosted international conferences; exchanged fire specialists with Russian colleagues in Siberia; and rushed assistance to fire-blasted regions from northern China to Amazonia. A special memorandum of agreement between the United States and Brazil worked to create an infrastructure for fire management.

The dissemination of fire research was even more pervasive. The concepts and data of North American fire research seeped into nearly every landscape on the earth; by 1990 the Rothermel fire model was used on every continent save Antarctica. The Intermountain Fires Sciences Lab had become something of a pilgrimage site for fire researchers around the world, a vital source of diffusion to such disparate lands as China, South Africa, Portugal, and Australia. The American and Canadian researchers plotted out the strategy for a common North American fire danger rating system, one that might also engage Mexico. Forest Service scientists joined field campaigns in South America, Africa, Canada, and Russia under the International Geosphere-Biosphere Program, an attempt to assess the interrelation between fire and global climate. For good or ill North America became for the world what the Forest Service had been for America, an exemplar, a source of expertise and order.

But foreign assistance is also self-help. Understanding fire ecology in Botswana, Siberia, or Brazil will toughen the fabric of knowledge in the United States; a universal model of fire behavior must predict fireline intensity in mallee, taiga, cerrado, and sourveld as well as sawgrass and chaparral; comparisons sharpen appreciation for the relationship between fire practices and society, sensitizing Americans about the extent to which the choices that have informed their fire establishment are not self-evident truths but the product of their cultural history. Not least of all, it is likely that alarm over planetary environmental change, especially climatic, will require a global redefinition of fire practices, of which fires are good and which are bad.

Here is the possible solution to the emerging era's formidable weakness, its uncertain linkage with fire research. The field needs of the intermix fire do not demand large research initiatives; what is required is the transfer of existing wisdom to the local political arena. Zoning boards are as vital as fire coordination centers. Building codes will go further to reducing damage than new prescriptions for underburning. But on the global scene the situation is different. The politics of international environmentalism through such issues as biomass burning may well influence how the American fire community does its work. The paralysis that field operations experienced in the era of wilderness fire, when wilderness values dictated fire practices, may be repeated, this time with evaluation by atmospheric scientists and their publics. Without an aggressive research program, the Forest Service and the fire community at large will be unable to answer those critics. Whether or not it can control fires, the American fire establishment will no longer control fire policy.

The contours of this new era are barely discernible. It may be that the intermix fire will merge with concerns over biomass burning into a larger, amalgamated problem of global fire. Or it may be that global issues, now a subtext, do not break through to the surface until the next era. Either way, because there is little direct connection between field operations and fire research, pressures will mount to downsize fire research, to see no further than the preservation of the next budget. Once the continuity of fire research is broken, however, it will not be easily restored.

But the same might be said for a fire establishment that had grown ever more precarious, its real funds replaced by off-budget borrowings, its capacity to control fire marginalized by its incapacity to control fuels, its credibility corroded by deficit logging, spotted owls, and its damning resistance to wilderness. Gone was a Forest Service composed, as it was at first, "almost wholly of young men." Gone was an American fire establishment dominated by the Forest Service. Gone was the sense of mission that had, year after year, mobilized and motivated thousands of firefighters for most of a century. Like America itself, fire management had fragmented and then gridlocked, unable to match purpose with procedures, equipped only to spend staggering sums of money. This was a bureau hamstrung with contradictions and scarred with self-inflicted injuries, with more than 20 percent of its land base committed to the National Wilderness Preservation System yet with the remainder inscribed by 370,000 miles of roads (more than the interstate highway system). Certainly the old order could not have endured. The new order, however, can not seem to replace it. For all this the uncertain status of the Forest Service is both explanation and symbol.

Experience had shown the Forest Service that, contrary to early beliefs, fire control was not a onetime affair, that wildland fire was inexpungable so long as wildlands existed. If in the early years of the century foresters had declared that industrial forestry and the protection of reserved watersheds were impossible unless surface fires were eliminated, by the 1980s they insisted that land management at large was impossible without recourse to some such burning. Ironically, precisely because its fire-protection mission did not wither away, the service acquired a special source of strength. If professional forestry had brought systematic fire protection to America, it was equally true that fire protection had created the need for foresters.

From the beginning the Forest Service saw science as indispensable to its mission. The agency was charged to conduct science and it was committed ideologically to applied science; its fire practices had encouraged, and in some cases were predicated upon, a knowledge of fire that the agency believed had scientific validation. The relationship that evolved between its missions to manage and to research, however, has been awkward, always balanced on a razor's edge between the demonstration of accepted techniques and the creation of new knowledge, between the values of those who have to act and those who seek to know. If foresters, however, had not committed to the systematic study of wildland fire, it is likely no one would have. The reservation of public lands had rendered folk knowledge inadequate; yet the recession, or suppression, of fire folklore did not by itself ensure that modern fire science would move into the vacuum. The Forest Service put it there.

But having acquired a proprietary right to the study of fire, the Forest

Service discovered that intellectual ownership of wildland fire science was more problematic than political ownership of public wildland. Some science it misinterpreted, some it misused, some it misplaced; but without the service it is unlikely that there would have been any science at all. The outcome was both ironic and cantankerous. The Forest Service had allied itself to scientific research in the expectation that science would merely rationalize and render more efficient existing policies and programs. It discovered instead that the practice of science had its own dynamic and its own ethos, and that genuine science could serve field forestry only if field forestry served it.

Early in this century the U.S. Forest Service founded a national strategy of wildland fire management on the promise of fire suppression, and suppression on the power of initial attack. But there is a metaphorical sense in which that strategy applied not only to individual fires but also to the issue of American fire at large. The Forest Service believed that a vigorous assault on burning of all kinds could quickly restructure American fire regimes and eliminate unwanted fire. "Hit 'em hard and keep 'em small" was the formula of successful fire suppression, a truism of operational logic.

But what the founders envisioned as an exercise in initial attack stretched into an extended attack, one that continued across years; extended attack evolved into an uneasy truce that lingered through several decades; and truce concluded finally with a negotiated settlement in which fire control shared power with fire use. The Forest Service discovered that it had tried to extinguish an eternal flame and found that it was unable—and ultimately unwilling—to do so. Its ancient enemy had become a favored ally.

Coldtrailing

> I'm not interested in stories about the past or any crap of that kind because the woods are burning, boys, you understand? There's a big blaze going on all around.
>
> —WILLY LOMAN, IN *DEATH OF A SALESMAN*

In the early years firelines were often called trails; and they resembled them, paths cut down to mineral soil around which firefighters patrolled. No fire was technically controlled until a fireline surrounded it.

But fires were often fast and firefighting with handtools slow, so a technique evolved by which burned-out sections of the fire could be incorporated into the constructed line without the labor of cutting, digging, scraping, and trenching. The dead, or cold, sections of the fire were treated as de facto fireline, and crews leapfrogged over them from hotline to hotline. Those burned-out sections were known as cold-trails.

In a sense that is the path fire protection itself followed. Before the advent of industrial fire control, fires cycled through the landscape. What burned out today would rekindle tomorrow. The land was a kaleidoscope of burned, burning, regrown, and greening pieces. As seasons and droughts turned the kaleidoscope, the parts reorganized, but according to a kind of conservation principle. While its locale might migrate, fire remained constant.

Fire control broke that cycle. Burning areas were extinguished, and burned-out sites were prevented from reignition. Fire control cut a fireline through history. Burned-out patches were coldtrailed into the new order, absorbed by a larger process of suppression into what was advertised as a path of progress. America needed to reconstruct fire regimes to suit an industrializing society, and the easiest first step was to overthrow the old regimes. There was no thought that what didn't burn today might need to burn tomorrow. There was only a drive to contain that flaming frontier. Systematic fire protection snuffed out fire after fire.

But firefighting alone was not a sufficient philosophy of fire management. It did not solve the problem of how fire and Americans ought to interact. Now as a troubled present hotlines into the future, that coldtrailed past has become a burden. What allowed a rapid expansion in control—the incorporation of burned-out patches into a program of fire suppression—has become a liability. The fire establishment finds itself, shovel in hand, at the end of a path not merely taken but hacked out of history, knowing that it cannot continue but unsure where to turn or what else it can do.

The environmental consequences were visible anywhere anyone cared to look for them. Like billboards along a highway, fire-degraded landscapes were so common they were hardly even seen, just accepted like road striping and NO PASSING signs. But some could not be ignored. The Blue Mountains of Oregon, for instance, have become a cameo of what has gone dreadfully wrong with the stewardship of American ecosystems. Remote, embedded in public lands, and subject to professional management, the national forests of the Blue Mountains should epitomize a near century of conservation, the nurturing of quasi-natural biotas. Instead they advertise a disaster, or worse a tragedy of good intentions and bad practices gone horribly awry. They have become a paradigm of forest health issues, of the complex and often contradictory demands on contemporary fire management, and of perhaps the greatest failure of fire management, the fire that was never lit.

The causes of decay are many and distressingly common. The complicated land practices of the native peoples vanished with the tribes in the latter nineteenth century. Grazing swept into the region like a plague, scouring away the grasses that characterized the lower-elevation mosaic of prairie, forest steppe, and savanna. Farming and scattered settlements broke up the land with roads, plowed fields, and towns. Logging targeted old-growth larch and pine, an assault accelerated dramatically during the post–World War II housing boom. And fire seeped away, like heat from a bowl of soup placed on a predawn stump.

When explorers first encountered the Blue Mountains, they found fire in abundance, and some even ascribed the blueness of the range to its perennial haze of smoke. Lightning and indigenous peoples kept the region simmering with a regular regimen of burning. A likely reconstruction has grasslands fired routinely. Lower-elevation forests of larch and ponderosa pine burned at least every ten to twenty-five years, enough to flush away invading understories of brush, grand fir, and Douglas fir. Upper-elevation forests of lodgepole and whitebark pine, ceanothus, and larch were pocked by patches of crown fire, surface-burned understories, and older growth, with major fires returning between 40 and 150 years. Undoubtedly many of these fires moved upslope from one biome to another, particularly during drought years or as long-smoldering burns crept and flared through a hot summer and prolonged fall. A single fire manifested itself variably as it met diverse pockets of fuel, terrain, winds, and moisture. But although many factors shaped the Blue Mountain forests, fire—hot, cool, simmering, flashing, wild, controlled—bound them together, like grout in a mosaic.

With the suppression of fire those tiles came unglued. Fire exclusion, too, had multiple origins. Settlers broke up the fuels that once carried flame, ranchers sought to hoard winter forage, loggers were eager for timber not tinder, and the administrators of the reserved public lands were committed to fire control as a species of wise conservation. Probably, however, indirect effects counted more than did direct action. The geometry of settlement prevented fires from passing through old corridors, not unlike the way it segregated migratory herds from old ranges. Grazing stripped grasses, the essential fodder for the former fire regime. And the indigenes who had for centuries, if not millennia, restructured the landscape with their burning practices were gone. The aggressive suppression of what remained, lightning fire, was comparatively easy, especially on those frequent-fire sites once dominated by grasses.

It has become much less easy. The pine forests changed composition as fire suppression and selective logging promoted grand fir and Douglas fir reproduction in what became dense thickets. When they reached sapling and pole size, western spruce budworm invaded the fir in what appears, despite extensive spraying with pesticides, to be an unquenchable epidemic. In many stands kill exceeds 60–70 percent. Simultaneously, bark beetles and root disease infested other sites, and mistletoe swarmed over second-growth ponderosa pine. The bugkill pumped up the region's fuels until, like a speculative binge, the forest threatened to crash, burn, and plunge its dependent human economy into depression. Major wildfires broke free during the drought years of the late 1980s, gorging on the fuels piled so profusely throughout the Blues. The forest was a shambles. By 1990 the collapse inspired an official declaration of forest health emergency.

But how to correct it? It was not enough to regulate (or end, where possible) the disrupting practices; it was necessary to create alternatives. The Blue Mountains Natural Resources Institute became a focus for many science issues. The Forest Service sponsored research, fielded interdisciplinary review teams, proposed management strategies, and created demonstration areas. But if observers could agree on causes, they could not agree on solutions, or even a strategy by which to thrash out a solution. Some contributing events were reversible, some not, and some only at costs at which society had to balk.

Since fire exclusion alone had not created the catastrophe, fire restoration would not, unaided, abate it. A fire strategy had to integrate with logging, grazing, recreational use, mandates for air quality, habitat for elk and anadromous salmon, local residents, bureaucratic battles within the agency itself, and the national symbolism and political posturing of environmentalism. Estimates called for a tenfold increase in prescribed burning and the installation of a new fire regime, not simply a rash of onetime restoration burns. A down payment demanded burning on the order of 100,000 acres a year, amid fuels that made the task akin to running through the woods with vials of nitroglycerin. But mostly reform demanded trust and money.

Neither was abundant. The agency whose counsel and oversight had created the mess now asked the public to trust it to set things right. Although its past practices had failed, revised practices, it assured critics, would rectify matters. It asked the public to accept its assessment that selective logging (of a different kind) was essential to remove fuels, that prescribed fire was mandatory because there were no acceptable ecological alternatives, that smoke in towns such as La Grande and in Class I wilderness areas was acceptable, that a major investment would, this time, unlike the last, restore forest health. Paradoxically, the old advocates for fire control found themselves insisting on massive controlled burning as a complement to remedial logging, while fire control's old critics in effect demanded more fire suppression to prevent any further harvesting.

In a context in which the public agenda was mesmerized by the cost and management of human health, in which the United States continued its relentless spiral into trillion-dollar indebtedness, the Forest Service pleaded for major funding to restore ecosystem health in a remote landscape of marginal economic value and of few special environmental characteristics. The money was there for suppression, seemingly endless, conveniently off-budget. The money, new money, for prescribed burning on the unprecedented scale proposed by experts did not exist, or could come only at the expense of other programs. The burning in the Blues had set to boil a witch's brew of American environmentalism.

The worst news, however, is that the Blue Mountains are not isolated. They stand for a cavalcade of biomes, particularly those subject to high-frequency fire regimes, that fire suppression shocked into decay and that prescribed fire has failed to resuscitate. For some the failure was strategic, the inability to imagine the essential need for fire in quasi-natural ecosystems or to create fire-use institutions of equal power to those for fire control. For others the failure was practical, the inability for whatever reason to begin the tedious, unglamorous, unending, often bureaucratic task of burning, patch by patch, year after year. Regardless, the outcome has been identical.

By the 1990s, however, it was no longer necessary to pursue wildland fire into the backcountry. It had come to cities and suburbs, or they had come to it. Cities moved out, fuels moved in, and fires glowed in the cracks between. The process was pandemic, but of course it was heightened to the point of parody in Southern California.

The landscape surrounding the greater Los Angeles basin is notorious for its volatile fuels, rugged topography, recurrent droughts, episodic eruptions of Santa Ana winds, and overall fire-prone intransigence. A mediterranean climate mixes wet winters with long droughty summers, an ideal formula to create fuels and ready them for burning over most of the year. Considered in evolutionary time the region's biomes burned only as their fuel cycles allowed. For chaparral, the fabled brush of California, this meant twenty to twenty-five years of regrowth before a site could carry fire vigorously again. Charcoal-laden varves in the Santa Barbara Channel record fires as far back as two million years, long before any human occupation. In this environment, in brief, fires are inevitable. The fire regime as it exists, however, is an artifact of anthropogenic intent and accident.

Each year, and for long periods of every year, fire can propagate somewhere everywhere. Humans ensured that ignition remained more or less constant. California nourished an intricate mélange of native tribes, none of which, interestingly enough, practiced agriculture. Instead, with fire for plow, rake, and ax, they harvested the native flora and hunted the resident fauna. Fire use was most intense and the fires smallest near settlements, particularly abundant in grasslands, oak savannas, or ecotones of grass and chaparral, precisely those sites most amenable to anthropogenic burning. The geography of fire reflected seasonal usage and the dynamics of fuels. Some sites burned annually; others, as needed. Probably the most frequented mountains had their slopes dappled with chaparral and grass, the signature of an anthropogenic economy.

Colonizing Spaniards arrived in the eighteenth century, and found the native fire regime not to their liking. Their missions brought not only the cross but also Eurasian cultigens, livestock, and habits of agriculture; they sought to convert the land as well as the natives, the one being essential to the other. Although they, too, used fire, they did so according to different purposes, and with different rhythms, and within another very different social context. A controlled burn for one society became a wildfire for the other; burning to hunt rabbits or harvest seeds threatened the winter pastures that herders required. So long as indigenes continued their seasonal routines of migratory burning, they could not be settled onto missions. By the 1790s Spanish officials condemned aboriginal burning and sought to contain it. The fire problem and the social problem were one and the same.

By the time Americans infiltrated into Southern California, the missions had collapsed, the indigenous population languished from disease and contact, livestock ranged widely, and anthropogenic fire continued, although in a somewhat different regimen than before or with different effects. Patch burning persisted wherever possible. But pastoralists transformed grasslands through overgrazing, and enormous herds modified the mountain mixtures of grass and chaparral. After the American acquisition, old missions and *rancherías* evolved into orchards, fields, pastures, and towns. Ranchers ran herds in the shaggy hills. Logging crept into the mountains.

The big change came after the Forest Reserve Act (1891) allowed the president to create forest reserves out of the public domain. At the insistence of irrigation agriculturalists and urbanites—both desperate for secure watersheds in the mountains—Southern California acquired some of the earliest reserves. Programs to control fire and grazing promptly appeared. With each decade the effectiveness of fire programs, largely under the jurisdiction of the U.S. Forest Service, improved, with steady decreases in the total number of acres burned and equally steady increases in fuel loads and the uniformity of chaparral cover. But federal foresters were never alone. State legislation established a Department of Forestry that, in time, contributed powerfully; two regional counties, Los Angeles and Ventura, opted to field their own firefighting forces rather than contract with the state; cities found that their responsibilities included wildland protection for chaparral-encrusted parks, dispersed suburbs, and borderlands. To the attrition of fire that accompanied the disintegration of aboriginal and Hispanic society, the new colonists promoted active fire suppression.

Firefighting on this scale distorted the province's already destabilized fire regimes. A smaller number of large wildfires tended to replace a large number of smaller fires. Comparison with Baja California dramatically illustrates how a landscape dappled with small burns has been supplanted by one in which the burning is concentrated into a few mammoth wildfires. Historically, even large fires at the onset of the twentieth century obeyed a rhythm of smoldering and flare-ups that left a landscape mottled with burned and unburned brush. Those fires shared the complexity of the living systems on which they depended for fuels.

Active suppression changed all this, much as levees and channelizing could eliminate nuisance floods but lead to more frequent large floods. Fire control could, by deferment, contain the wildfire menace for several decades. But so long as fuels grew like a cancer across the sunbaked mountains, so long as Santa Anas blew the aridity of the Mojave Desert over mountain passes, so long as a mediterranean-type climate prolonged a fire season across seven or eight months of the year, fires could not be abolished. Worse, new sources of fuel and ignition continued to kindle fires outside the parameters of any former regime.

Not everyone accepted fire control as necessary or practical. No less a figure than William Mulholland, architect of the Los Angeles water system, refused to send men to battle fires that raged in the mountains in 1908 and again in 1919. Big fires, he insisted, were beyond "the power of man to stop." Their reputed damage to watersheds was "greatly exaggerated." Those big fires were dangerous, and putting them out was, over the long term, no less dangerous. It was better, Mulholland insisted, to "have a fire every year" that burned off a small plot than to wait several years "and have a big one denuding the whole watershed at once." There were plenty of others who agreed, and although the light-burning controversy had its epicenter in northern California, southerners felt the tremors and joined the debate. The greatest check on unrestricted fire exclusion, however, was simply the lack of tools, men, and money. That began to change during the New Deal. And the sense of limits—limits of any kind—appeared to vanish completely with World War II.

With the war Southern California crossed a divide. The Los Angeles region emerged, chrysalislike, as an industrial dynamo. Its population doubled almost by the decade. Old valley fuels disappeared—first into fields and orchards, and then into houses, factories, and asphalt. The exploding population thrust into and around the mountains, pushing with

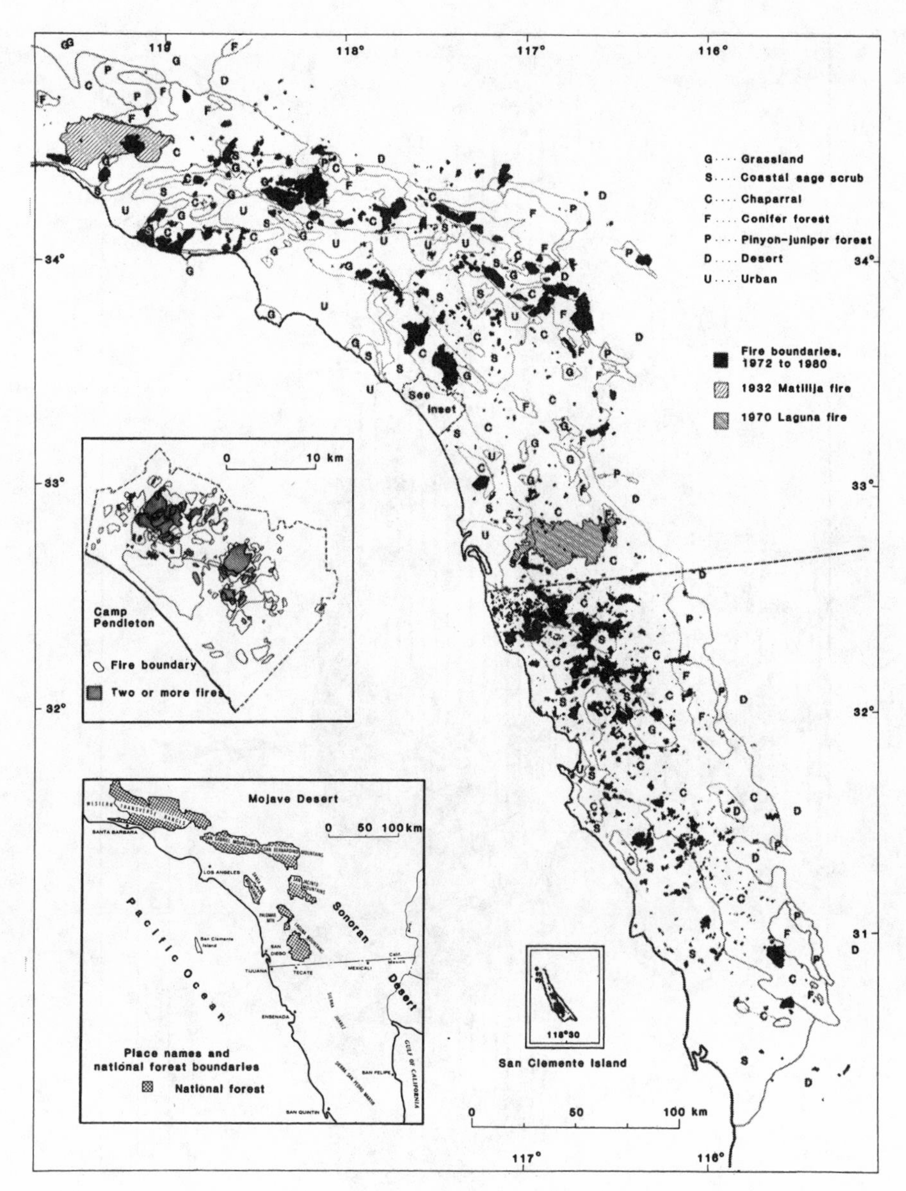

California Dreaming. (a) By comparing the record of fires across lower and upper California, this map highlights the shift in the American-side fire regime from large numbers of small fires to less frequent but more devastating conflagrations. (b) A burnt-out case: Malibu fires between 1919 and 1982. Fire returns with almost astrological regularity. The 1993 burns, not shown, washed over the same fire-plain.

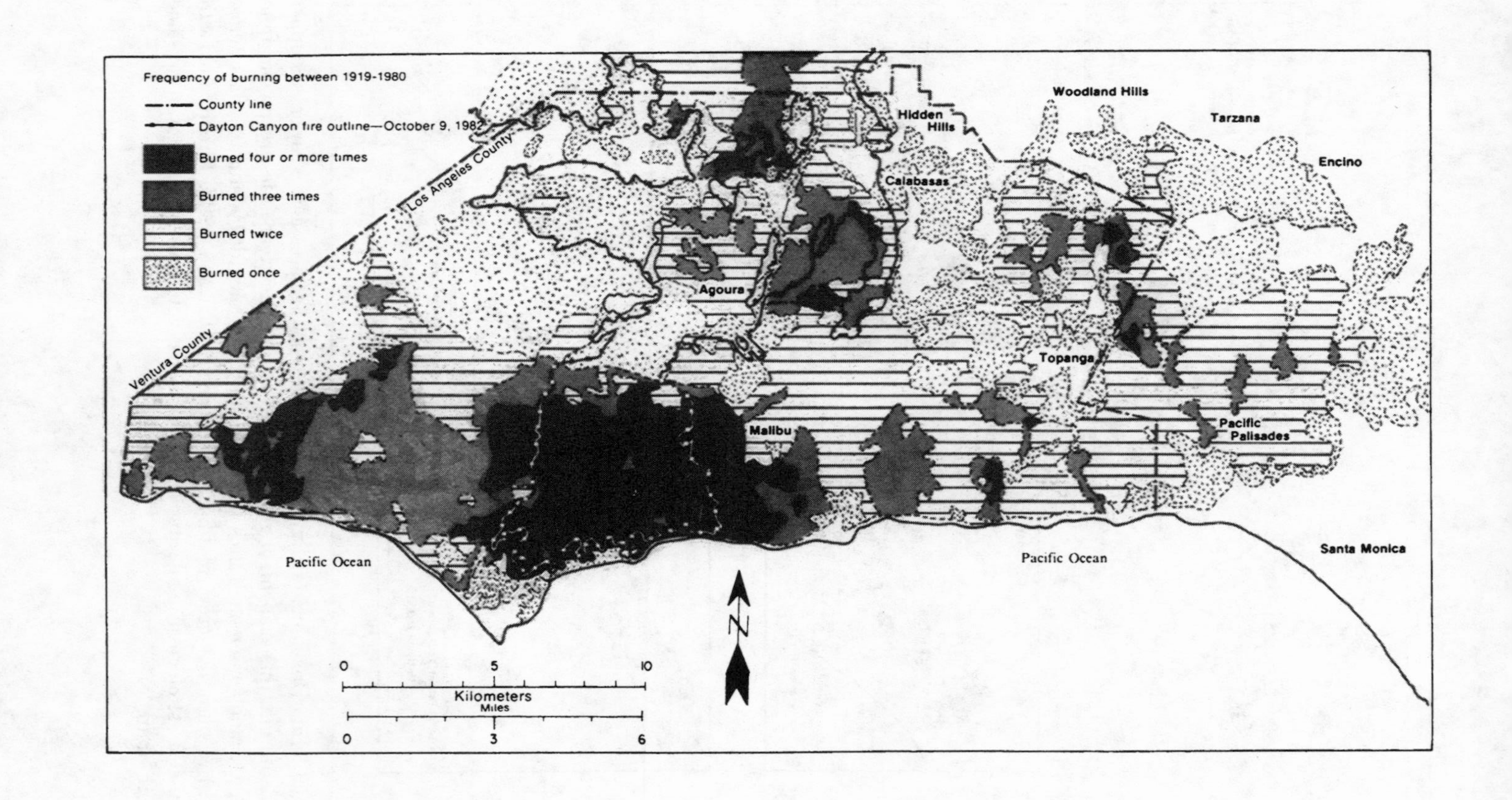
Frequency of burning between 1919-1980
County line
Dayton Canyon fire outline—October 9, 1982
Burned four or more times
Burned three times
Burned twice
Burned once
Ventura County
Los Angeles County
Hidden Hills
Woodland Hills
Tarzana
Encino
Calabasas
Agoura
Topanga
Malibu
Pacific Palisades
Santa Monica
Pacific Ocean
Pacific Ocean
N
0
5
10
Kilometers
Miles
0
3
6

mounting force against forest, park, and reservoir reserves. The center fell apart, if it had in fact ever really existed; a jumble of built landscapes, stocked with immigrant flora, created a postmodern ecosystem ready for deconstruction by earth, air, water, and fire, nature's least forgiving critics. To aging chaparral fuels the new era added houses, often outfitted with highly flammable wooden roofs, and to the old cycle of ignitions it added further sources from powerlines, machinery, children, and arsonists. Conflagrations that had formerly raged in the backcountry or along a tattered rural fringe now, by virtue of this instant geography, burned into suburbs. Residents lived on a fault line of fire as powerful as the San Andreas.

"The city burning is Los Angeles's deepest image of itself," Joan Didion has observed. During the Watts riots "what struck the imagination most indelibly were the fires. For days one could drive the Harbor Freeway and see the city on fire, just as we had always known it would be in the end." It is no accident that the Smokey the Bear campaign had begun in Southern California during World War II amid alarms of a fiery invasion by Japan. But fire as urban catastrophe didn't require riots and wars. It demanded only that the city develop as it had, and that the fuels from suppressed fires build up until they could be released in a fury of natural rage.

As the threat of wildfire escalated, so did the response from fire-protection authorities. There could be no compromise with fire, no détente with prescribed burning. Fuelbreaks transformed ridges into battlements. If on a national scale the postwar era resembled a cold war on fire, the Los Angeles region epitomized its arms race. The incredible intensification of human life in this fire-molded region compelled a similar intensification of fire protection on a scale almost unimaginable elsewhere. The region housed one of two Forest Service equipment-development centers, and one of three fire labs; in the early 1960s, as much as 25 percent of the national investment in fire protection rushed into Southern California, like dry leaves sucked into a whirlwind.

Of course this situation could not endure. The national enthusiasm for prescribed burning and fire control through fuel management broke down the region's long antagonism to any alternative to full suppression. Fire management accepted prescribed fire in principle, and sought to integrate it into operations—this despite some spectacular failures and very tough constraints imposed by air-quality boards. The amount of burning was laughably small compared with the magnitude of the fire-famished landscape. And it was hard-won, given the grotesque condition of the region's fuels. In 1990 wildfire savaged Santa Barbara, consuming 4,900 acres, 427 homes, 221 apartments, 15 businesses, and 10 public buildings; and in 1993, a cluster of wildfires struck Southern California, exploded across 152,000 acres, and burned down 720 buildings. Over a twenty-five-year period the average wildfire in California had doubled in size; between 1920 and 1989 California had lost 3,500 structures to wildfire, while between 1990 and 1993 some 4,500 structures had burned. Californians' feared China syndrome did not have to wait for a nuclear accident: the biotic meltdown had already begun.

But if not denied outright, the causes were personalized. The blame fell on arsonists, and probably some fires did originate from arson, although a year later none was proved. The search for firebugs and psychos, however, diverted attention from the fundamental problem, like telescoping society's problems with crime into the question of drive-by shootings. The power of arson derives from the power of fire to propagate. As an antisocial act arson more resembles a riot than it does an assault; it can thrive only under the proper regimen of packed combustibles and wind-driven corridors; the greatest check on it is to structure that nurturing landscape in such a way that fires cannot spread. Some of the major fires of 1993 started from downed powerlines, another failure of social design, but these were conveniently forgotten in the public angst over arson.

California's arsonists succeed less because they are clever or particularly malignant or more abundant than because residents choose to live in a way that makes arson plausible. In the past, arson was self-limiting because the landscape could burn only so much at any one time, even under extreme conditions. Today there is little to stop the flames except high-technology firefighting and the Pacific Ocean, and they are, respectively, too little and too late. With or without its city the landscape will burn. The issue is whether it burns in small chunks over centuries or in great catastrophes, whether from accident or error or malice, whether under controlled circumstances or in crisis.

Protection is possible, but firefighting has at most a modest role. Postfire studies suggest that a nonwooden roof increases the chances of structural survival from 19 to 70 percent. Clearing flammable shrubs thirty feet around the structure improves the odds from 15 to 90 percent. If, in addition, a firefighter is present—which requires adequate water sources and road access—a house is 99 percent defensible even under extreme conditions. But these are precisely the measures habitually ignored by the public. Still, reforms have come, grudgingly and spottily. Thirty years after the Bel Air conflagration the Los Angeles Fire Department was delicately spot-burning in its revegetated wake. Select counties even banned wooden roofs on new construction.

The past, however, cannot be extirpated. The compound of houses and chaparral, of city and wildland, cannot be wiped away; it remains metastable, capable of exploding with the right spark or the next surge of a Santa Ana. What had once been diffuse pressures that had metamorphosed the Los Angeles basin in the postwar era were now concentrating, like a diesel piston compressing air and fuel. More than anywhere else, Southern California announced the dimensions of the intermix fire era. And at least as much as anywhere else it expressed the problems of reconfiguring a fire establishment set up to put out fires into one that can balance many purposes with many techniques, the environmental equivalent of multiculturalism. The difficulties Los Angeles faces in converting to a post–Cold War economy are not much different than those it faces in converting its fire establishment from its coldtrailed past.

The consequences for fire management were no less complex or profound. If it was difficult to rehabilitate the land, it was often no less difficult to reform the fire establishment. The bureaucracy of fire management had taken on the attributes of American government, and the ends and means of fire management had subsumed, absorbed, and accepted as their own the malaise, social tensions, fixations, denials, and ultimately the character of American society. Wildland fire was in gridlock among nature, Congress, industry, bureaucracy, and environmentalism. Or rather the national gridlock was expressing itself through the medium of fire management in even such remote locales as Big Cypress, the Tonto Rim, and

Little Tujunga Canyon. The one thing the system could do was build fireline. It built fireline even where building line did not solve the fire problem. But then building line was where the money was.

America had the most expensive firefighting operation in the world. Whether, like its Cold War military, this apparatus advanced national security or improved the quality of national life or whether it simply spent money, the capacity of Americans to mobilize against wildfire was a marvel of the modern world. Only Canada could claim anything like it. The American fire establishment could, within days, assemble and ship around the country tens of thousands of firefighters, hundreds of engines and tractors, and the mobile kitchens, powered tool rigs, gasoline tanks, and timekeepers to feed the throng, fuel the machines, sharpen worn pulaskis, and pay both those on the fireline and the mob back in camp their overtime and hazard-duty differentials. The suppression apparatus could flood the sky with scores of air tankers and helicopters and observation planes, a claim to air superiority unmatched by the military of many nations. It could spend $1 million a day fighting a single fire. It could spend $24 million simply "rehabilitating" the landscape after the Foothills fire outside Boise. For a large complex such as the Yellowstone fires of 1988 it could, without any significant effect on the fires or improvement of the ecosystem, expend more than $130 million in nominal suppression. Over the 1994 season it spent an obscene $925 million. Clearly something was broken.

The mechanism for leveraging flame into fortune was the Forest Fires Emergency Act of 1908, which established the practice of paying for fire-suppression bills after the end of the fire season. There was nothing insidious in this concept, any more than the originating legislation for Social Security or Medicare was a subterfuge for national bankruptcy. The act (later known as the Forest Firefighting Fund, or FFF) recognized that it was impossible to guess in advance what a fire season would cost. Even today's long-range weather forecasts are laughably inadequate, and their utility becomes nought when fire budgets are proposed two years in advance of their use. Instead the Forest Service sought to establish a base-level firefighting force targeted for an "average worst year," and then build up its forces with supplemental appropriations as individual seasons dictated. Congress would authorize those expenditures after the fires ended.

The alternative was to transfer funds from elsewhere in the budget, a practice that compelled agencies to shut down operations or else to cease fighting the fire. Before it received similar authorization in 1928 the National Park Service experienced just this choice each summer and had to close down several parks to pay for the 1926 fires in Glacier. The same was true for state forestry programs. Virginia's Forestry Service closed shop in 1931 after exhausting much of its year's appropriations fighting a fire in the Great Dismal Swamp.

The result for the federal agencies, however, was that they programmed for a base funding of infrastructure and relied on the "emergency" funds for actually fighting fires. Perversely, the system rewarded agencies for having fires. It encouraged the agencies to do what they wanted to do, build fireline. Even where there was interest in controlled burning, there were precious few appropriations with which to do it. As federal spending liberalized, so did expenditures under FFF. During the New Deal the development of a fire danger rating system argued for expanding FFF into "presuppression," which meant spending unprogrammed dollars to bolster firefighting resources during times of high fire danger. Fire officers could keep lookouts in their towers a few hours longer, hire an extra crew, pay more overtime for extended patrols. The idea was fine—an ounce of fire prevention should save many pounds of suppression costs—but of course the monies became a kind of slush fund and the Forest Service ended the practice as part of its 1978 reforms. (The Department of the Interior agencies still retain access to an "emergency presuppression" fund.) Instead the Forest Service sought to increase its regular appropriations, for which there could be tougher standards of accountability, and reduce the corrupting reliance on emergency funds.

But regular budgets decreased, and with that decline the basic structure for initial attack decayed in relative terms. FFF money flowed, however, so long as there were fires. The more fires, the wealthier the agencies. Following a 1981 fire at its Merritt Island refuge that killed two firefighters (one the son of a federal judge), the U.S. Fish and Wildlife Service saw its fire appropriations rocket upward. After the Yellowstone disgrace, the Forest Service, which had avoided a meltdown in the surrounding forests, suffered budgetarily, while the National Park Service, which orchestrated the debacle, saw its regular fire funds swell. More and more often agencies paid for fire management with FFF dollars that were conveniently "off-budget." They had little choice.

And they were far from unique. Wildland firefighting simply mirrored the free fall of indebtedness that characterized all of government, or of American society, for that matter. American fire management resembled American medicine—capable of spending unlimited third-party funds for heroic intervention on behalf of dying seniors, but unable to immunize its children. Money had less and less connection to work actually done, to purposes larger than political symbolism. It had become an end instead of a means. The size and trajectory of its budget signaled whether an agency, or a job within that agency, was in or out, up or down, a winner or a loser. None of this was production or investment; it was simple consumption that made a big fire into a kind of bureaucratic potlatch. Why, after all, should fire management invest in the future when it appeared to many that the nation's governing institutions were strip-mining the economy and selling their grandchildren into debt peonage?

There had always been problems matching money and needs where fire was concerned. The public agencies were, in the end, political institutions, not economic ones. Much of what they protected stood, by legislative intention, outside the market economy. When was enough too much, or too little? The situation worsened amid the legislative reformation of the 1970s. New policies could repudiate the supremacy of suppression as a goal, but they could not achieve clear consensus on what new goals should replace it, or how success should be measured, or how new programs should be financed. A Government Accounting Office (GAO) report on the state of federal fire management after the Yellowstone crash criticized the retarded progress made on restoring prescribed fire programs, particularly prescribed natural fire. The GAO estimated that the Park Service had 69 percent of the funding it needed; the Forest Service, probably 1 percent. But the Park Service saw only its costs, not its productivity, rise. What had required ten dollars per hectare before the 1988 season now cost two hundred dollars per hectare. The number of acres burned continued to plunge. The agencies were in fact doing what they were paid to do.

The absence of consensus regarding the purposes of fire management had left a void, and into it rushed money and lawsuits. In Southern California the Forest Service could be sued for violation of air-quality regulations if it burned, and sued for not managing wildlife habitat if it did not. The agency could be sued for escaped fires and sued for not halting wildfires. Following the 1985 Painted Cave fire outside Santa Barbara, insurance companies staggered by hundreds of millions of dollars in claims threatened to sue the Los Padres National Forest for failure to control fuels, that is, for maintaining a hazard. Here was the national gridlock on fire.

The only operation that could count on abundant funds and public support was the active suppression of wildfire. But that, experts agreed, was not solving the problem. It was possible that nothing could. The backlog of untreated fuels and fire-famished biotas was too great. Increasingly managers opted to do nothing, or rather to let nature take its course, which evidently Congress, the courts, environmental groups, and the public could accept. Eventually lightning or an accident would start a fire, leave to nature the responsibility for its environmental consequences, and pump big bucks into the local economy. The strategy was analogous to leaving urban renewal to arsonists and hurricanes, or public health to measles and cholesterol. Bureaucratic (and national) prestige was possible, however, because huge sums could be spent fighting the mess, once it kindled. Photo opportunities abounded. Relief would follow catastrophe, a kind of lottery. No one could be sued, blamed, or voted out of office, or would face financial ruin. A wild fire, if it was big enough, brought morning again to America.

So the fire establishment shouldered its shovels and dug more line. There seemed no way to stop its reliance on a fire-powered deficit without plunging into bureaucratic depression. If FFF funds shriveled, the whole organization would burn to its roots. The intellectual revolution had been won—fire managers accepted prescribed burning in principle; the technical trials had succeeded; within limits fire specialists knew how to burn in different landscapes and for different purposes; policies were sensible if not always practiced; but means and ends were no longer in sync. The fire community began to appreciate that coldtrailing was a metaphor for its own history, and had to decide how to incorporate that burned-out past into a usable future.

There was genuine cause for pride, nonetheless. Those firelines *had* stopped a lot of bad burning, and had helped discipline the fiery consumption of industrial logging. Moreover, the act of suppression revealed, as no other experiment could, the ecological power of fire. George Bush's call for a "thousand points of light" was answered in the pointillist glow of fusees and drip torches, which kindled hopeful fires where throngs of mesquite had throttled prairie, sagebrush had crusted over valleys like scabs, and hardwoods choked out once-flowered fields; where sequoia seedlings withered in shaded humus, and starved wetlands waited for the jolt of fire-released nutrients. But while such spot fires were locally important and symbolically significant, the larger ecological darkness grew. Suppression prevailed, paradoxically to the world's envy.

America fielded a fire establishment that for sheer size and exuberance was unrivaled. It could fight fires with as much muscle and tenacity as any country on the globe. Brazil lacked an infrastructure. Chile could suppress fires only on artificial forestry plantations. Argentina could not cooperate with neighbors. Ghana watched helplessly as a third of its cereal crops burned along with its savannas. Germany and Austria saw fire as a kind of civil disorder, and lacked fire institutions capable of balancing fire use with fire control in the name of land management. Russia could not muster an extended attack if its first-pass aerial assault failed. China worried more about bonfires in Tiananmen Square and burning buses in Shanghai than about conflagrations in Heilongjiang. Indonesia transformed central Borneo into an inferno whose smoke shut down airports as far away as Singapore but could hardly raise a rake against the flames. With United Nations assistance, India launched a "modern forest fire control project" that put air tankers into the sky but left firefighters with bare feet. The United States (particularly when allied with Canada into a North American bloc) dominated fire research. Together they provided the global standard for wildland fire. No one could understand fire on earth without grappling with that strange, marvelous, powerful, confused, fiscally undisciplined but commanding American presence. American firefighting looked a lot like America.

Call it inevitable or call it ironic, but a fireline, if successful, eventually outflanks and encircles its fire. It returns to its point of origin. That, in a sense, is what has happened in America. As long as it retains wildlands those wildlands will burn. Until American society accepts this fact, its wildlands are ungovernable. The time has come to return to where that coldtrailed past has regrown (or overgrown) and begin burning out. In most places this requires some environmental restoration, at least to the point where the landscape can burn again with relative ease and at low cost. But delay can only lead to further ecological deterioration and more ferocious wildfires. The time has come to convert that encircling fireline into a fire cycle.

WILDERNESS FIRE

Vestal Fires and Virgin Lands

They came together, fire and wilderness, like flint striking steel, and threw sparks that kindled a special kind of flame.

The belief grew that Americans had a unique relationship to nature, that wilderness was something vital to our national experience, that natural areas ought to be preserved to express those values and bolster that identity. With the Leopold Report (1963) and the Wilderness Act (1964) beliefs became matters of policy, ideas evolved into an ideology, and wilderness claimed the high ground of a new environmentalism. Inevitably changes in how Americans managed lands affected how they managed fire. By 1970 the question of wilderness fire began to dominate the American fire establishment, a reign that flourished until its self-immolation, like some wildland Wall Street, in the Yellowstone conflagrations of 1988.

In the beginning the issues seemed obvious, and the solutions self-evident. It was held that fire is a wholly natural process and wilderness a completely natural environment; that the two are intrinsically compatible, and have been for geologic eons; that the question of fire management was simply to remove the impediments, all anthropogenic, that inhibited their natural association. To establish wilderness it was necessary only to abolish the human presence, and to promote wilderness fire it only remained to eliminate the intrusions of human fire practices, particularly the obnoxious conduct of high-technology fire suppression. Wilderness fire management was simply a process of restoration. Remove the intrusions and wilderness would thrive; free-burning fire would reestablish its symbiosis with the wild, like lianas intertwining with tropical trees. In this model of nature's economy untrammeled natural processes, like Adam Smith's invisible hand, would seek an ideal equilibrium, balancing supply and demand, fuel and fire. By removing ourselves we would allow nature to preserve a vestal fire on America's virgin lands.

The reality has been more complicated. With wilderness fire we are not dealing with a natural phenomenon and a natural environment but with two hybrids of nature and culture. We are not simply putting a natural process back into a natural landscape but trying to reconcile one hybrid—fire—with another—wilderness. Neither is a fixed idea or set of practices. Both have their own complex and independent histories. They had not been created together, or coevolved over millennia; they were thrown together, violently, like continental plates ramming into each other. That the process of harmonizing the two should be perplexing—institutionally, intellectually, and operationally—goes without saying. Theirs was a shotgun wedding. Separation was inevitable.

Their encounter not only reshaped American cultural geography but also themselves. Fire demonstrated the fallacies of laissez-faire wilderness management, exposed the paradoxes of banishing humans from nature, and revealed the way in which wilderness was an artifact of American society. For its part, wilderness compelled a national obsession with fire control to confront its absurdities, to admit its ecological and economic costs, and to reaffirm that its purpose is to serve land management, not itself.

Their collective consequences—call it wilderness fire—exceeded their individual effects. But what wilderness fire announced, it could not answer. While it forced its originating traditions to question their ends and means, what began with philosophical conviction has concluded with uncertainty; early accomplishments have become future conundrums. It is no longer clear how to rekindle the vestal flame without burning down the sacred grove, and it is not obvious that American society cares enough to commit special attention to the enterprise.

Wilderness is not an immutable order of nature or a universal concept in human societies. Rather, modern-day wilderness is an intellectual construct, and wilderness sites are cultural artifacts, the product of a constantly evolving state of mind interacting with a constantly changing state of nature. It is increasingly clear that wilderness is only one of many forms of nature preservation, that its particular expression in America shows strong cultural biases, that wilderness is, in fact, a peculiar creation of a peculiar people at a peculiar time in their national history. Pretending otherwise is itself a cultural decision—or an exercise in collective self-delusion.

Well before it acquired legislative definition, wilderness had become a fundamental part of a national creation myth—in fact, much of the urgency and sense of moral fervor behind wilderness preservation derived precisely from this association. America's virgin lands took the place of those cultural monuments that Europeans used to express their national saint. Americans lacked cathedrals like that on Mont Saint Michel, Roman coliseums, Parthenons; but they had Niagara Falls, the Grand Canyon, and giant sequoias to serve as surrogates. The encounter between Americans and those natural wonders became our *Iliad,* our *Aeneid,* our *Gilgamesh.* Like them, the story defined an American identity and a national destiny. Like them, too, the story is a tragedy in that the founding hero must leave or die; the pioneer destroys the conditions that make pioneering possible.

This notion represents the latest installment in a venerable intellectual enterprise, the encounter of Old World ideas with New World environments. As they pondered these exotic peoples and places—outside the domain of ancient philosophers, beyond Ptolemy's *mappa mundi,* indifferent to the genealogies of the Old Testament—many savants began to imagine concepts by which to exploit the apparent discrepancies between the two worlds in order to advance assorted political or cultural purposes. The

Noble Savage, the Forest Primeval, the Virgin Land, the Untrammeled Wilderness—all are ultimately parables, moral templates, with which to criticize the decadent civilization of the Old World and to exhort the New World to do better. They represent myths of a past Golden Age of natural and moral order, relocated from a Mediterranean Eden to the American Wilderness. Their population is prelapsarian. Their promise is that America is a chosen land, spared the ravages of the Fall, closer to the Creator. Such ideas are philosophical paradigms and literary conventions, however, not reports on the natural environment.

To this basic formula other values have been grafted. That the land possesses information vital to science; that it preserves biodiversity and affirms biocentric values; that it offers the opportunity to reexperience the awe of western explorers and the hardihood of pioneers; that it is a part of our landed heritage, the raw stuff out of which our civilization has evolved—all presuppose the values and institutions of American civilization; none is inherent in the landscape itself. American wilderness and its meaning were shaped, and continue to be shaped, by the society that defines them. Yet by configuring our conception of wilderness, even to the point of fixing it in legal language, these ideas have assumed the status of management goals. What began as a state of mind has reified into a putative state of nature. Almost all of the paradoxes of wilderness fire derive from this bold, flawed translation.

But it is important to remember that wilderness is only one of many forms of nature protection, that the preservation of the natural world has extended also to endangered species and their habitats, to biosphere reserves established as scientific laboratories, to a vast array of reservations around the world—from Russian *zapovedniks* to African game preserves to the sacred groves of Mount Athos—for which wilderness, as Americans understand that concept, is more or less irrelevant. These sites have sought to define a special relationship between humans and nature but they do so as an expression of different social values, for which the creation story underlying American wilderness is as indifferent a guide as the *Ramayana* would be for managing the Yosemite or the *Kalevala* for the mountains of the Bitterroot-Selway.

What was a strength has now become a weakness. The thirtieth anniversary of the Leopold Report was also the hundredth anniversary of Frederick Jackson Turner's celebrated essay "On the Significance of the Frontier in American History," delivered at the Columbian Exposition that convened in Chicago to celebrate four hundred years of European discovery. Today both the reputation of Columbus and the Western saga lie in tatters. Discovery has been redefined as Encounter, suggesting a greater equivalency between Europeans and Amerindians; the Winning of the West has become a grimy Legacy of Conquest, with the frontier only an exercise in ironic imperialism and environmental dislocation; the politics and scholarship of multiculturalism have, in brief, shredded the intellectual and mythic story behind the wilderness ethos. Shane won't be coming back. Many professional intellectuals regret that he came in the first place. The contemporary wilderness mosaic must somehow express ethnicity and gender as much as forest age classes and pre-Columbian biomes.

This deconstruction takes many forms inspired by many motives, not least its uncertain relevance to a population swollen by immigrants from places outside Europe and the cultural reprivileging of peoples (such as African Americans and American Indians) who stand outside it or who were its putative victims. The aspect most pertinent to fire management concerns the place of pre-Columbian peoples, those who were here before the wilderness. It is one thing if Americans, operating out of European traditions, choose to declare themselves unnatural and deny themselves access to wilderness sites; it is another altogether to extend that declaration to America's indigenes. Even a decade ago the question of "Indian burning" was a quaint appendix to fire management; as "native peoples," American Indians were either part of the natural scene or were perceived as unable or unwilling to influence it in any serious way. That Indians burned merely meant that fire suppression was wrong, and hence the effort to "restore" fire was appropriate. Their fires, like the natives themselves, simply merged into the landscape. They were part of the natural order, nothing less and nothing more.

Of course, this is absurd, and its ecological and historical absurdity has long compromised practical management. More recently the cultural, philosophical, and moral premises that underscore wilderness management have become equally untenable. They can no longer be ignored.

The absurdity begins with the land. How can a park such as Yellowstone, for example, argue for reintroducing wolves because they are an indispensable part of the "original" ecosystem and yet deny the most prominent predator of all, *Homo sapiens*? How can anyone dismiss anthropogenic fire as inconsequential or indistinguishable from lightning fire? Biotas are adapted not merely to fire but to fire regimes, and all the regimes of Holocene America have emerged within the context of anthropogenic burning or that negotiated matrix between lightning and humans. Eliminating that fire does not restore wilderness to a Golden Age but fashions an environment that, in all probability, has never before existed. Anthropogenic fire was as vital a process as floods, droughts, and epidemics in shaping pre-Columbian landscapes. That process cannot be dismissed without disrupting the systems under protection. This is eminently obvious for biomes such as the tallgrass prairie, but in different ways it is true everywhere, from the Brazilian cerrado to the Swedish heden to the French maquis. Only in America is the role of indigenous peoples regarded as trivial or irrelevant. That is a judgment rendered on philosophical and political, not ecological, grounds.

Instead natural areas must grant to humans ecological roles commensurate with their numbers and powers, allowing them biotic citizenship as fully as bison, elk, and coyotes. Surely humans, even hunters and gatherers who after all preyed on wildlife, fished, foraged for grasses, nuts, fruits, and tubers, and of course set fires—much more than those peoples who cleared, slashed and burned, and herded livestock—had as much ecological impact as snail darters, pupfish, spotted owls, and blackfooted ferrets. How, on biological grounds, can those consequences be dismissed? Or if they can be excluded as ecologically trivial, then why not those endangered others?

The dismissal is no less absurd historically. American explorers did not encounter a raw nature, as later explorers would on the Antarctic ice sheets or the Viking space probes would on sandy Mars; they encountered a landscape already populated and, within the limits allowed by their technology, remade to better suit those inhabitants. Everywhere people mediated the encounter with the wild. Frontiersmen relied on Indian guides, learned Indian skills, and often took Indian wives. They seized Indian fields for their own, and grazed livestock where Indians hunted wildlife. To the dangers posed by grizzlies, blizzards, and prairie fires, they faced the much greater hazards of hostile humans. Successful frontiersmen learned to live with the indigenes, to live like them, or to overcome them. Those were the vital wilderness survival skills.

All this is gone from today's wilderness. Re-creating the vegetation at the time of European discovery or preserving select natural processes does not re-create the historic wilderness experience because the most critical element, the encounter with humans, many hostile, all alien, is gone. It was those native peoples who made the wilderness "wild," which is to say, exotic, unpredictable, dangerous, exciting, and wondrous to those for whom it was not already home. Similarly dismissing the things those peoples did, including burning, only sustains a landscape that is historically incomplete. But to accept as valid those indigenous peoples only leads to other questions. One must ask which native practices to preserve, from which native people, and from which times in their complex histories? The past is multiple, and history a negotiated compromise between the records of those pasts and the needs of the present.

Not least, the calculated dismissal of pre-Columbian fire is absurd on social and political grounds, as well. Multiculturalism simply won't allow that pre-European landscape to be emptied. Stripping American Indians of the power to shape their environment with fire is tantamount to dismissing their humanity. Our capacity to manipulate fire is a species monopoly, unique to humans but also universal to all humans. Deny someone the right to fire and you deny him that claim to humanity.

There is more at issue, however, than crude political correctness. By forcing a reexamination of the creation myth behind wilderness, multiculturalism has compelled American society to reconstruct the moral universe within which it manages wilderness. The decisions we make regarding fire practices are, inescapably, moral acts, which is to say, they emanate from ideas about who we are and how we should behave. This influences even the way we manage fire because, although as a species we are genetically equipped to handle fire, we do not come programmed knowing how to use it. Decisions about what kind of fire to apply and withhold have to rely on a larger cultural context, and they have to seek some basis for legitimacy. Along with the power that came with fire, humanity assumed a responsibility to use it and to justify that use.

Initially it was a sufficient principle of wilderness and park management to renounce as inappropriate certain practices, particularly when they relied on guns or mechanization; the wholesale slaughter of elk or fireline construction with bulldozers, for example. But long-term administration demanded some positive goal, as well. To this a wilderness ideology proposed a simple solution. The "state of nature" conveyed a moral order, one ultimately descended from Nature's God. Wild nature—that is, nature unpolluted by humans—was the purest expression of that natural order, and hence the highest moral condition. The state of nature might consist of biotic objects and landscapes, arranged according to some pattern, or it might occur as an ensemble of processes. Typically this condition existed in the past, but it might also be re-created in the future. Efforts to preserve or restore wilderness thus were sanctioned by an order of nature that transcended humanity and that located the source of judgment outside human society.

In reality those decisions are our own. We occupy an existential earth that can assume many forms, has known many pasts, and can evolve toward many futures, none of which is privileged. Likewise human fire practices are profoundly relative, and their suitability depends on criteria that are as much a part of our cultural context as they are of the natural landscape. Removing anthropogenic fire from legal wilderness does not create pre-Columbian landscapes. More probably it helps degrade those biotas. To claim, as some have, that nature is amoral does not change the fact that, for humans, how to act toward wilderness (or any other land) must originate from human consciousness and human will. The decision to allow a patch of land to exist "untrammeled" by humans is not an instinctive part of our genetic code. It is rather a deliberate decision made by a particular society. It is part of the peculiar power of fire that it exposes these circumstances, like a torch thrust into a dark room.

If humans cannot be excluded from wilderness on ecological grounds, if there is no historical validity to the expulsion of pre-Columbian peoples, if the philosophical arguments for a state of nature are specious, if social and political pressures for multiculturalism are redefining the national character and its expression in nature preservation, if the moral privilege claimed by laissez-faire wilderness management is untenable, then it would seem that the question of wilderness fire is at best muddled, and at worst cynical. One might satirize National Park Service goals, in particular, for their uncanny fidelity to the prime directive issued to *Star Trek*'s starship *Enterprise*. (Curiously both the National Park Service and Gene Roddenberry's sci-fi television program began their careers in the same year, 1967.) The quirky quality of American wilderness management might aptly be characterized as Starfleet's mission to Earth.

Of course humans can destroy the biodiversity of a site, and do so often. But it is also true that much of the earth's biodiversity has resulted from the presence of humans as peculiar intruders, for whom fire is a preferred enabling device. What is destroyed today is not raw nature encountering *Homo sapiens* for the first time but landscapes shaped in long association with assorted human practices. The destruction results from the removal of those practices as much as from the removal of the vegetation. (Thus the Swedish national park Dalby Söderskog boasted 208 species in 1925 and only 122 in 1970, having been exempted from the cutting, grazing, burning, and plowing that it had known since the retreat of the ice sheets.)

The issue is not whether humans are present but the character of human presence. Shutting an area off from industrial exploitation does not, by itself, ensure the protection of the species or ecosystem that existed previously. An emphasis on natural processes may do no better unless those processes also include select anthropogenic activities, fire among them. Humans cannot be abstracted from the earth without effect on the natural system, some biomes becoming enriched, and some impoverished. Even where it can be done for several decades, we have learned painfully that fire exclusion is an error. The next phase is to admit that abolishing all anthropogenic burning is as much an error as the attempt to abolish all fires.

The problem is more one of confusion than hypocrisy, and part of the dilemma is that the cultural precepts, both written and unwritten, that have sustained the enthusiasm for wilderness have changed. Other ecological values have clambered to supremacy, such as ecological integrity and particularly biodiversity, which seem to resonate better with the politics of cultural diversity. Preserved lands, that is, exist for many purposes, not all of which synchronize. The assumption that one set of practices will equally serve all these purposes is untrue. Take it as symbolic that Roderick Nash, rather than further revise his *Wilderness and the American Mind,* wrote *The Rights of Nature,* a broad survey of biocentric philosophies. That is where nature preservation is going, and where fire management in wilderness and parks must follow.

Despite its ambivalences, the legacy of wilderness fire has been immense—and reciprocal. At first such fires granted wilderness management a dramatic iconography of what allowing natural processes to range freely meant. Towering smoke columns testified to the power of untrammeled nature, a burnt offering ascending to Nature's Creator. But fire didn't remain only a symbol; it couldn't be finessed in the field as readily as in philosophies. Eventually wildfire boiled off the philosophical froth and left managers coughing in a pall of doubtful practices. In the end fire has compelled a reconception of wilderness.

But if fire forced wilderness management to change, so, too, has wilderness compelled changes in fire management. Wilderness fire chartered a new era in the administration of American fire. It broke the supremacy of fire suppression among fire practices, and chastened the autonomy of fire protection within public agencies, forcing fire into the service of land management and especially wilderness values; it validated prescribed burning, placing fire use into parity with fire control at least in principle; it shattered the hegemony of the Forest Service as a fire agency, and broke through the boundaries of forestry as a controlling profession over wildland fire; it inspired an efflorescence of research, particularly in fire behavior, ecology, and applications; it redefined the meaning of control to include information, and rewrote policy accordingly, allowing the power of prescriptions to take the place of traditional fireline forces; it devised reasons and techniques for dealing with natural fires in wilderness areas; and, based on this cloudburst of change, it inspired a massive exercise in replanning. If you conceive of American fire history in terms of "problem fires," each of which dominates the scene for roughly twenty years, then wilderness fire clearly qualifies. It informed an age.

With the clarity of hindsight, it is obvious that the fire establishment was primed for reform—already overextended and misshapen by the late 1960s. Fire protection was squeezed by the pressure of diminishing returns and the failure of existing methods to suppress an irreducible quantum of large fires. It could not adequately control fuels, and hence wildfires. It could not satisfy ecological objectives, nor could it reconcile its heavy-handed mechanization with wilderness values. It had to change. But it was not obvious in which directions fire protection would proceed or by what means. The wilderness movement furnished those means and ends.

The era's most spectacular achievement is probably its vindication of prescribed burning. If fire was essential for wilderness areas, then it could also be good for other, less pristine environments. This intellectual conviction merged with ancillary considerations—such as a buildup of fuels in some environments and an accelerated awareness about the potentials for prescribed fire in the service of biocentric goals, spearheaded by the Tall Timbers Fire Ecology Conferences. But fuels had built up implacably in some areas for decades without leading to the almost universal adoption of prescribed fire as a solution. Similarly, the range of applications for prescribed fire might have expanded only slowly, site by site, purpose by purpose, without becoming a generic solution to fire management's problems or claiming status as its defining policy.

Instead prescribed fire became identified with wilderness fire. It was not practical issues such as fuels that led to the fervor for prescribed fire; it was conviction about the value of prescribed burning, inspired by a wilderness ideology, that encouraged a search for legitimate uses. It was as if distributing prescribed fire became a surrogate for distributing the saving wilderness. No site was too abused for redemption by prescribed burning, no reason too implausible to justify its application; every dimension of wildland management had, it seemed, a record of fire in its past and cause for fire in its future.

In the flames of prescribed fire, techniques fused with ideas. Some practices, such as fuel reduction and habitat maintenance, channeled prescribed burning into areas of traditional concern to foresters and helped connect old concerns with new goals. But others escaped. Once ignited, those fires broke through the institutional and intellectual control lines forestry had delineated. Other agencies and professionals demanded to share in decisions regarding fire practices and policies. Wildland fire no longer remained the unique purview of forestry or the Forest Service. It is probably no accident that wilderness fire, prescribed burning, and the evolution of collegial institutions for fire management, from the National Interagency Fire Center to the National Wildfire Coordinating Group, occurred simultaneously.

To counter its accomplishments, however, the era of wilderness fire created an equally impressive array of breakdowns, operational dilemmas, and intellectual paradoxes. Wilderness fire management was an innovation. It had to be used in order to be understood, and understood before it could be used properly. There were, accordingly, plenty of failures in the field—prescribed fires that fizzled, prescriptions that suppressed what they should have promoted and encouraged what they were intended to depress, controlled burns that went wild. For nearly a decade, from the mid-1970s to the mid-1980s, most of the spectacular failures of fire management involved prescribed burns that went bad.

One does not have to be a Hegelian to see a kind of dialectic at work here. At first wilderness fire was defined and promoted in terms of the problems it solved; eventually it was shunned because of the problems it created. It began as an alternative to failed practices and a repugnant philosophy. It concluded as itself a hopeless quagmire, its bright flames reduced to a bureaucratic landscape of smoldering guidelines and ashy punditry.

In retrospect it is apparent, moreover, that nature favored the experiment. After the 1978 reforms in Forest Service policy and Park Service guidelines, the American West experienced a cycle of relatively benign fire seasons. Within that climatic nursery, prescribed fire, and especially wilderness fire, grew to a kind of maturity. Then drought returned, wildfires raged, and prescribed burning as a privileged practice ended in Yellowstone's 1988 *Götterdämmerung*. Had the reformation begun six or eight years earlier or been delayed an equivalent time it is likely that it would have failed or evolved in dramatically different ways.

Then came the fires of '88. The Yellowstone fires were more than a biotic spectacle: they were a cultural epiphany like the stock market crash of '29 or the landing of Apollo 11 on the Moon. They were as much an expression of American society as the conflagrations that raged that same year in Borneo, Amazonia, and Thrace were of Indonesian, Brazilian, and Greek societies, respectively. The fires stated boldly and confusedly that Americans sought a special relationship with nature but that they were unable to match goals with methods, or institutions with values.

Those fires ended the era of wilderness fire. It is important to acknowledge that the era was already fading, and would have assumed a new avatar with or without Yellowstone. Even as the conflagrations raged, America's quest for fire sent it hiking out of the backcountry and into exurbia. What American civilization packed into the wilderness, it now packed out. Revealingly those immense fires posed no new questions to the fire community, nor offered any new answers. The conundrums that had emerged over the purposes of wilderness fire and the suitable means by which to attain them remained.

But as a consequence of the fires, federal fire policy was reevaluated, and fire plans nationwide had to be resubmitted. Wilderness fire had to rebuild itself, park by park, wilderness by wilderness. And it would have to do so in a context in which industrialization was compelling anthropogenic fire to compete not only with lightning but also with fossil-fuel combustion, and in which global environmentalism was condemning all fires, wild or prescribed, as the flaming matches set to kindle a global apocalypse. Probably wilderness fire will never fully recover. Almost certainly it will never again assume the commanding role it had claimed for nearly two decades.

Wilderness fire management will have to reconstruct itself without the moral enthusiasm and philosophic privilege it possessed when it had led rather than defended its reformation. Revolution had become Establishment. The wave that carried wilderness fire had crashed on shore some time ago, and advocates now stood in the sinking sands of its backwash. What John Dewey had observed about great philosophical questions—that they were never solved in any technical sense but simply "got over" while society moved on to other matters—applied to wilderness fire. Wilderness had become one form of wildland management among many; prescribed burning, one fire practice among others; and wilderness fire, one fire management task among dozens, with no particular claim to priority.

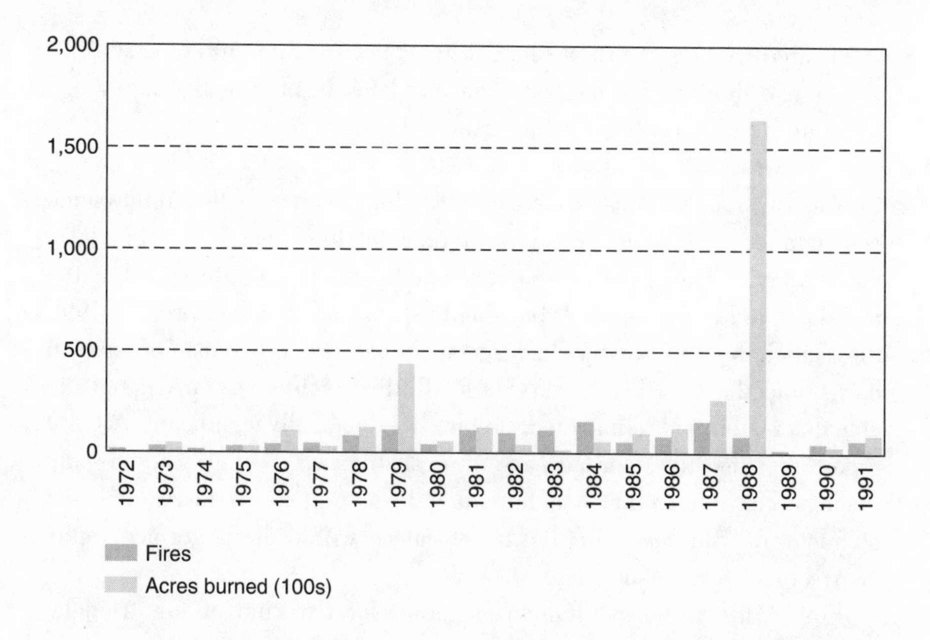

Prescribing an idea: natural fires in the national forests. The Forest Service changed its policy only after the 1977 fire season, itself the disastrous culmination of several years of western drought, particularly in California. Thereafter the service allowed prescribed natural fires that more or less followed climatic cycles, climaxing in the 1988 season, better known for the eruption of fires outside its jurisdiction in Yellowstone National Park. That effectively ended the experiment. The reconstruction is actually worse than the graph implies because, unlike in the previous decade, the prospects for growth are poor.

In the climax to Flannery O'Connor's novel *The Violent Bear It Away,* a boy raised according to a grotesque fundamentalism kindles a fire.

> There, rising and spreading in the night, a red-gold tree of fire ascended as if it would consume the darkness in one tremendous burst of flame. The boy's breath went out to meet it. He knew that this was the fire that had encircled Daniel, that had raised Elijah from the earth, that had spoken to Moses and would in the instant speak to him . . .
>
> When finally he raised himself, the burning bush had disap-

peared. A line of fire ate languidly at the treeline and here and there a thin crest of flame rose farther back in the woods where a dull red cloud of smoke had gathered.

In a sense that is a description of what the violence of the Yellowstone conflagrations, nurtured on an environmental fundamentalism, had also borne away. What began as a kind of oracle and apocalypse ended as the mundane and the languid. A burning bush became a brush fire. By 1992 the National Park Service had reconstituted fire programs for sixteen parks, and the U.S. Forest Service for thirteen wilderness areas; but the actual acres burned plummeted, no longer ecologically significant. A GAO report lamented the retarded pace of reconstruction. What remained—the unglamorous intellectual and bureaucratic mop-up—was to redefine the character of wilderness fire and to relocate it within the larger geography of American fire management.

Fire, wilderness, and humans together form a kind of fire triangle. Clearly fire is indispensable to natural areas. Most American ecosystems in fact suffer from a fire famine. Unburned fuels stockpile in alarming quantities, the combustion equivalent of a toxic dump. But what kind of fire regime is most appropriate is not obvious, nor what kind of Superfund might pay for restoration. Likewise, fire is inextricably bound to humanity, part of our heritage as a species, perhaps our one unique biological niche. The synaptic charge among neurons of the human brain is as vital to the earth's fire circuitry as the electrical discharge between sky and ground that makes lightning. It can thus be argued that humans had a moral duty to exploit this biological imperative, that along with the power to use fire came the obligation to do so—to employ it wisely but to use it. Without fire neither nature nor humanity could be what they are.

So if fire is a point of discord, it is also a means of integration. It remains a *focus,* literally—its Latin roots meaning "hearth," and also altar, home, and family—for any human engagement with our surroundings. What we have lost is wilderness fire as an informing principle and a moral crusade. What we have gained is the continued presence of fire in the wilderness as a unique expression of earth, sky, fire, and humanity.

The practical problems of wilderness fire management are solvable. For the most part, except where fanaticism and metaphysical euphoria overwhelm field operations, they exist already. They are neither perfect in conception, nor flawless in execution. They fail from time to time, though less often as experience mounts. But then every fire practice fails. Prevention fails. Fuel treatments fail. Prescribed burning fails. Suppression fails. What makes natural fire programs distinctive is that they are based on a calculated ambiguity. They assume that we can have a fire that is both natural and anthropogenic, a fire that is equally and simultaneously wild and regulated. They proclaim that it is possible to control a free-burning lightning fire through the application of knowledge, encoded in prescriptions and forecasts, and that it is possible to deliberately ignite fires that, by virtue of their wilderness surroundings, assume the qualities of wild nature.

The techniques for managing such fires can work only if there is some social consensus regarding the philosophy of values behind them. This debate has revolved around two issues, one concerning the source of the fire and the other the nature of control over a fire once it has begun. The first questions whether lightning is the only legitimate origin for the ignition of wilderness fire, or whether anthropogenic sources can also serve. (Can you have an "impure" fire on "virgin" lands?) The second asks whether, if lightning is accepted, under what conditions it can be "managed" and still remain a "wilderness" fire. (Can mortals create a vestal flame?) Or more broadly, can humans manage a wilderness from which in principle they are excluded?

These dilemmas disappear, however, if their premises are reexamined. The land was not virgin, and it has been an ecological and moral duty of hominids since the days of *Homo erectus* to actively manage fire. Because anthropogenic fire is a symbiosis between humans and the earth, the decision not to burn can be as ecologically fatal as promiscuous burning. Removing anthropogenic fire from many environments may be less an act of humility than of vandalism, no different than damming rivers or hunting moas to extinction. The contradictions embedded in such concepts as the prescribed natural fire are only paradoxes. They are no different logically than the dense mass of paradoxes around which orbit most of the great advances in modern science and philosophy.

The revolution in twentieth-century thought—call it modernism—has grappled with just such questions of self-reference and self-inclusion and of the complementarity of observed opposites. Modernist physics, philosophy, and mathematics couldn't resolve those issues; they had to accept them as paradoxes. Is an electron a particle or a wave? Can a mathematical system be both complete and consistent? Can we know precisely the position and velocity of a particle, or does the act of observing interfere in such ways that it limits the information available? Does the set of all sets include itself? The register is long, and it has expanded to other realms—the Antarctic Treaty, for example, which artfully grants and denies sovereignty in the same breath.

Once incorporated as paradoxes—which impose limits on what we know and how we know it—those fields of inquiry expanded enormously. On Gödel's example, systems of thought could be complete or consistent but not both. They can be consistent but at the cost of ignoring important pieces, or they can be complete but at the price of logical closure. This is exactly the philosophical situation with wilderness fire. We can have a wilderness that is consistent yet missing parts, or one that is complete yet ideologically flawed. We can have a vestal fire that is pure but inconstant, or one that persists but only through human tending. Wilderness fire is modernism as land manager.

The question has no technical, scientific, or even logical solution. The paradox simply is. It may take many forms but it has no further derivation. Deny the paradox and you preclude the possibility of practical management. Accept it and, despite its irreducible ambiguities, you are liberated to act. If mathematicians can live with Gödel's proof and logicians with Russell's paradox, if physicists can function with the principles of Heisenberg and Bohr, of indeterminacy and complementarity, then the fire community can cope with the theoretical paradoxes of wilderness fire management without lapsing into scholasticism or lethargy. Certainly those intellectual scruples offer no excuse to withhold fire or to deny anthropogenic burning. Fire belongs in wilderness, and not only because sites are small or oddly shaped or have suffered disruptions from European contact but because it was always there. Together lightning and people made the elastic matrix that defined the fire regime.

These intellectual concerns are nonetheless hardly trivial. There is a passage in Dostoyevsky's *The Possessed* in which a fire breaks out on the thatched roofs of a village, the scene of revolutionary unrest. A visiting noble asks what one man, scrambling on a ladder, is doing. "He is putting the fire out, your Excellency." "Not likely," comes the reply. "The fire is in the minds of men and not on the roofs of houses."

So, too, was wilderness fire a fire in the minds of men and women. It would not be addressed solely on the thatched roofs of the Sierra Nevada, the Selways, or the Mogollon Rim. It had to speak also to the mind and heart. It had a symbolic role for its nurturing culture as powerful as its ecological role in reserved biotas. The purpose of the vestal flame has never been simply to preserve fire but to preserve a society, to keep alight its national identity and destiny, to instruct its people in who they are and how they should behave. Wilderness fire spoke to just such aspirations, that American civilization had origins more noble than land speculation and strip-mining, and a future that transcended shopping malls and the Cold War.

Such visions are an inextinguishable part of the complex character of anthropogenic fire, the reconciliation of nature with culture, belief with practice, hope with history. And if the task seems Promethean, wilderness fire managers might well echo the words that Aeschylus attributed to the original Prometheus more than two millennia ago. In *Prometheus Bound* he has the tortured Titan proclaim that he "caused mortals to cease foreseeing doom." To this the chorus asks: "What cure did you provide them with against this sickness?" Prometheus answers, "I placed in them blind hopes . . . I also gave them fire."

The Summer We Let Wild Fire Loose

Let's all but bring to life this old volcano,
If that is what the mountain ever was—
And scare ourselves. Let wild fire loose we will—
—Robert Frost, "The Bonfire"

Everything about the fires seemed exaggerated. Groves of old-growth lodgepole pine and aging spruce-fir exploded into flame like toothpicks before a blowtorch. Towering convective clouds rained down a hailstorm of ash, and firebrands even spanned the Grand Canyon of the Yellowstone. Crown fires propagated at rates up to two miles per hour, velocities unheard-of for forest fuels. A smoke pall spread over the region like the prototype of a nuclear winter. Everything burned. It was as though the ancient caldera had, as geologists have warned it one day will, rumbled to life, spewing forth geysers of flame and sending out streams of fiery lava across the plateau, obliterating everything in their path. Approximately 45 percent of Yellowstone National Park burned in some fashion, almost a million acres in all. At its height, the Greater Yellowstone Area spent $3 million a day and converted the park into a black hole that sucked in firefighting resources from everywhere in the country—this while officials adamantly insisted that only winter would extinguish the fires. As a display of natural power the fires were a staggering, compelling, humbling, costly, inflating spectacle. Everything seemed exaggerated by an order of magnitude.

But then Yellowstone is an environment that encourages hyperbole. It is a huge place, populated by big animals, and endowed with an immense symbolism. Like a celestial body so massive it warps space and time, Yellowstone exaggerates everything it touches and distorts perspectives around it. The fires were, to some, an ecological holocaust, the ground zero of the national parks, nature's *Götterdämmerung.* To others, they were an eruption long predicted and much overdue, the invisible hand of nature working to correct biotic imbalances. Yellowstone's immensity—geographic and symbolic—allowed no middle ground.

When Americans reserved wildlands as forests and especially parks, it was not obvious which fire practices were appropriate. To early administrators it seemed patently clear that they had to eliminate wanton anthropogenic burning. They removed American Indians; barred frontiersmen, loggers, sheep and cattle pastoralists, slash-and-burn farmers; and regulated tourists. But there were still rashes of fires started by accident or lightning. It was difficult to justify suppressing anthropogenic fires, often set for traditional purposes, while allowing other fires to range freely. The apparent solution was to exclude all fire. The most important trials came during 1885–86, when New York experimented with fire protection on its Adirondacks Preserve and the U.S. Cavalry assumed the administration of Yellowstone National Park. Fires greeted M Troop as it rode into the park, and the troopers extinguished sixty other blazes that summer. Their example inaugurated fire protection by the federal government.

But during the 1960s, new thinking rubbed against old practices. The outcome was a rekindling of ideas about appropriate fire practices. For the National Park Service the critical document was the Leopold Report (1963), which urged the restoration of fire to wildlands; for other federal agencies, the Wilderness Act (1964), which compelled them to reassess their acquired pyrophobia. Ecological arguments for preserving fire in natural areas merged with powerful ideological currents that swirled around the concept of wilderness. Policy reforms followed. In 1967–68 the Park

Service released a new set of administrative guidelines that encouraged controlled burning and by implication the accommodation of natural fires. The Forest Service adjusted its policies in 1972, then reformulated them in 1978. "Prescribed" fire—controlled burning conducted under a specified set of conditions to meet specified goals—became the rage.

But which fire practices were suitable? It was increasingly evident that suppression was not a neutral act; it did not quick-freeze an ecosystem, which changed by having fire withheld as surely as it changed by being burned. There were problems with prescribed burning, too—escape fires were common, outcomes were not always predictable, and prescribed fire was, after all, a form of calculated intervention that left some people with the uneasy feeling that it was out of character with an untrammeled wilderness. What to do with the heritage of aboriginal fire was never resolved. Besides, wildland fire is far from a precision instrument. It is not some kind of Bunsen burner that can be turned on and off or regulated at will. Starting in the mid-1970s, most of America's disastrous wildland fires resulted from breakdowns in prescribed burning.

There were choices to be made, no escape from the imperative to make them, and no clear codes for guidance. There were good fires and bad fires: good fires that were wrongly suppressed, bad fires that were wrongly allowed to burn; good fires that went wild and became bad fires, and bad fires that, through reclassification, were declared good. It was possible to exhaust oneself in extinguishing a fire and be told that it would have been better to have let the fire burn; to stand by and "monitor" a natural fire that subsequently ravaged the landscape; or to ignite a prescribed fire to enhance the habitat of an endangered species such as Kirtland's warbler, only to have the fire escape, kill a firefighter, and burn down a nearby village. At the end of their shovels firefighters found a smoking existentialism.

Gradually a consensus emerged that fire of some sort belonged in the landscape, and there was slow, costly, grudging agreement on how, where, and when it could be restored. The most spectacular fire practice of the revolution, however, involved the co-opting of naturally ignited fires into overall fire management plans. In the absence of hard knowledge and techniques acquired by trial and error over millennia, there was ample room for experimentation and ideology. Initially a naive philosophy prevailed that held that wilderness fire management required only the restoration of a natural process to a natural landscape. This proved adequate tinder to start new programs but not to control them. But enthusiasm evaded intellectual ambiguities; field practitioners devised practical techniques; and ecologists confirmed the necessity for some kind of fire. It seemed sufficient to remove fire control and leave to nature the establishment of a natural order. Natural fire, or the artful ambiguity of the "prescribed natural fire," thus became the symbolic epicenter of the new thinking. And Yellowstone quickly declared itself the paragon of natural fire management.

Yellowstone's great asset was its size. Fire management in northern forests must grapple with large, often huge fires. Those big fires do most of the biological work; they perform in ways that many smaller fires cannot replicate even in aggregate. Those big fires are, for wilderness purists, the perfect symmetry of spectacle and symbol. A fire program that excludes big fires is a sham. But big fires require long times and large areas. In the northern Rockies high-intensity fires that engulf whole stands are relatively infrequent; they may require several centuries before the biota arrange themselves into a suitable fuel array to support such a fire, and then drought, winds, and ignition have to appear in the proper sequence. Once initiated, they are by nature crown fires—racing through the canopy, catapulting firebrands by the thousands, some over many miles before they descend and kindle spot fires. They become uncontrollable by any human technology. Ultimately they are contained only by a geography generous enough to transcend the normal spans of fuels and weather.

More than geography alone is involved, however. Managing natural fires involves a complex, often maddening choreography between fire control and fire use. In essence a naturally ignited fire is accepted if it burns within a specified geographic area and according to a prescribed range of behaviors and effects. If the fire threatens to leave that prescription, then fire organizations must take some control actions either to push it back into prescription or to extinguish it. Thus the ability to control fires is fundamental to the concept of natural fire management. One of the precepts of the new era, however, was the belief that it was possible to substitute information for intervention, that a sophisticated knowledge of fire behavior could replace traditional fire-control technology. It was the role of science-based models to advise a fire organization as to which fires to accept and which to suppress, and to make that evaluation early enough in the cycle to give fire suppression the chance to be effective. This introduced still greater flexibility into the decision process and to accommodate the big fires that a natural fire program requires there had to be a healthy margin for error.

The best insurance is land. The larger the land base, the less robust the state of knowledge needs to be and the less gargantuan the resident fire-control machinery. An immense land base—more than two million acres—is precisely what Yellowstone had. It was almost inconceivable that any fire complex could exceed the park's capacity to absorb it. Yet since it first proposed a natural fire program in 1972, Yellowstone has had everything a natural fire program needs except, ironically, large fires. Through 1987, 235 fires had burned 34,157 acres, the largest fire covering 7,400 acres. For a program that had loudly, if hyperbolically, proclaimed its special status, this was a pittance. The absence of large fires was an embarrassment. Without those fires Yellowstone's claims to primacy were bombast.

At the same time other parks and forests acquired considerable knowledge about prescribed and natural fires. Increasingly, Yellowstone fell behind national fire practices. Its once prominent suppression organization gradually disintegrated into vestigial bits and pieces, not exactly sure of its role and shunted into other park concerns such as running a helicopter. Its elaborate on-site monitoring of early fires gave way to a onetime evaluation of smoke reports, usually from aerial photos. The fire committee, which made decisions, became clubbish, haughty over its highly personalized knowledge. Yellowstone did not field a prescribed fire program: it had a let-burn program. Instead of prescriptions it had a philosophy. Instead of control technologies it had size. Its decision process regarding lightning fires had more in common with Roman augurers examining chicken entrails than with science-based technologies, however flawed, exploited elsewhere. The park no longer staffed a major fire-suppression organization yet was unwilling to commit to a genuine program of natural fire monitoring. Instead Yellowstone's immensity—symbolic and political no less than geographic—insulated it from the pressures that elsewhere led to internal reform.

National Park Service review panels urged Yellowstone to upgrade its plans and programs, particularly after the 1981 season. The park replied with the old story of former superintendent Jack Anderson, who once declined the post of national director because he regarded anything less than Yellowstone as a demotion. In 1985 the Rocky Mountain regional office hired an outside fire specialist to rewrite Yellowstone's fire plan. After ten weeks he left behind a compromise document that specified a regular procedure for making decisions about lightning fires but one that, at the park's insistence, failed to include any written guidelines for the criteria on which those decisions should be made. In a parting memo to the regional office and national Branch of Fire Management he recommended the park recharter its fire program, a proposal quickly discarded. The park never submitted the draft plan for public review. When the fires came it did not even follow the flawed decision charts of that still-unapproved plan.

Then came the fires of '88. If its size had lulled Yellowstone into complacency, it is also true that its size saved it from devastation. The 1988 fires went beyond the known scale of Yellowstone fire history. They were larger than any event in the history of the region's oldest trees. The impact of fires on this order is simply not known, and in the absence of knowledge exaggeration comes easily. It has been simple, in particular, to anthropomorphize the conflagrations, to make the elk into the homeless of Yellowstone, helpless victims in desperate need of emergency shelter and breadlines. The fires were not an environmental holocaust. They burned under heterogeneous conditions and will yield heterogeneous results. The wilderness will lose no species or vital processes. But the fires will reconstitute the ecosystem in fundamental ways. They have given the biotic kaleidoscope that is Yellowstone another twist, revealing broad new patterns. The Greater Yellowstone Area was, as its theorists believed, vast enough to absorb what may be a five-hundred-year, perhaps a thousand-year, event.

It is not at all obvious that, had Yellowstone adopted the best practices,

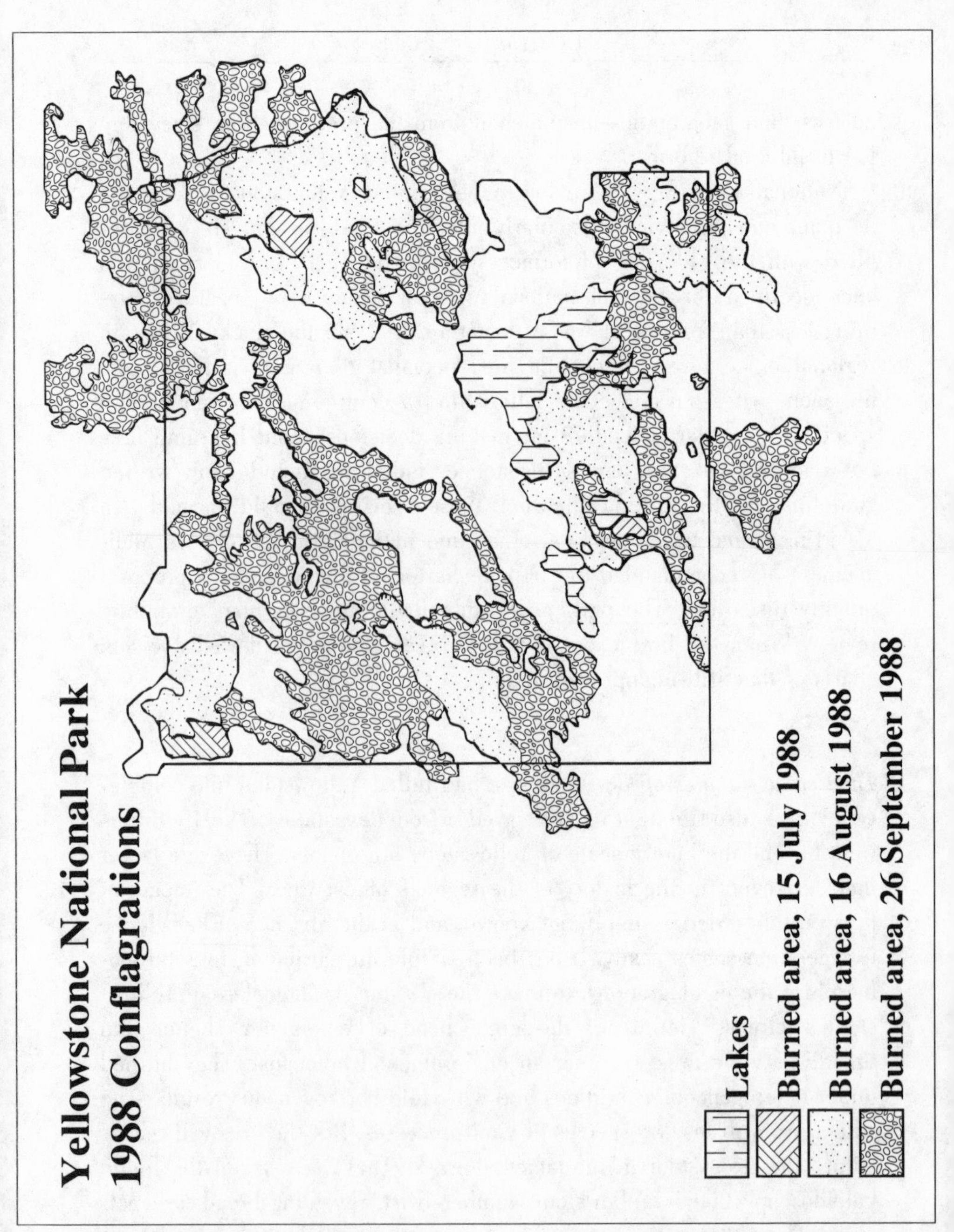

The fires grow. So do costs: the end tally is still not complete but will exceed $130 million. Almost 45 percent of the park burned, although the intensity of the burning varied tremendously.

it could have altered in any substantial way the outcome of the 1988 fire season. Its choices were limited. If suppression was to be effective, it had to be implemented very early. Yellowstone wanted big fires—it deserved them—and no one anticipated in the early weeks how relentless the conflagrations would become. Once that initial stage passed, there was little to do but throw suppression resources at the fire and spend money. Yellowstone got fires commensurate with its ego.

Equally, it is not clear that Yellowstone understood then, or now, the fires' larger context. Even as the summer worsened, the park continued to insist that the issue was not the fires but pyrophobia, that it was really battling to educate the public and the fire community about the ecological value of fire, that its enemy was not a smoke pall the size of Wyoming but Smokey the Bear. It repeated mantralike lines from its old fire plan to the effect that "nature is amoral" and that it was inappropriate to apply human values to wilderness events. In fact much of the public accepted fire, and prominent critics of the park wanted more fire, not less; but they wanted fire according to a schedule of prescribed burning conducted by humans, not according to the capricious and barely controllable ignitions kindled by lightning. The park bristled. The invisible hand of nature, not human calculation, would dictate the outcome. It was illegitimate to intervene. It cheapened the wilderness to predetermine its composition and dynamics. It was unfair to judge the park for a wholly natural, amoral event.

This position is both flawed and cynical. Once the fires became large, they were a power no different from hurricanes or earthquakes over which humans had little control. But the park could have altered the fuel structure in advance of the outbreak, and it could have intervened when the fires were small. The Forest Service held 218 fires in the surrounding forests to less than ten acres. It suffered only one major failure, the Canyon Creek fire that roared out of the Bob Marshall Wilderness when chinook winds sent an avalanche of flame down the east slopes of the Rockies. But of the thirty-one prescribed natural fires declared in the Greater Yellowstone Area, twenty-eight were in Yellowstone National Park. This reflects decisions by humans, not nonnegotiable demands from nature.

Park apologists have tried to deflect the failure onto fire suppression.

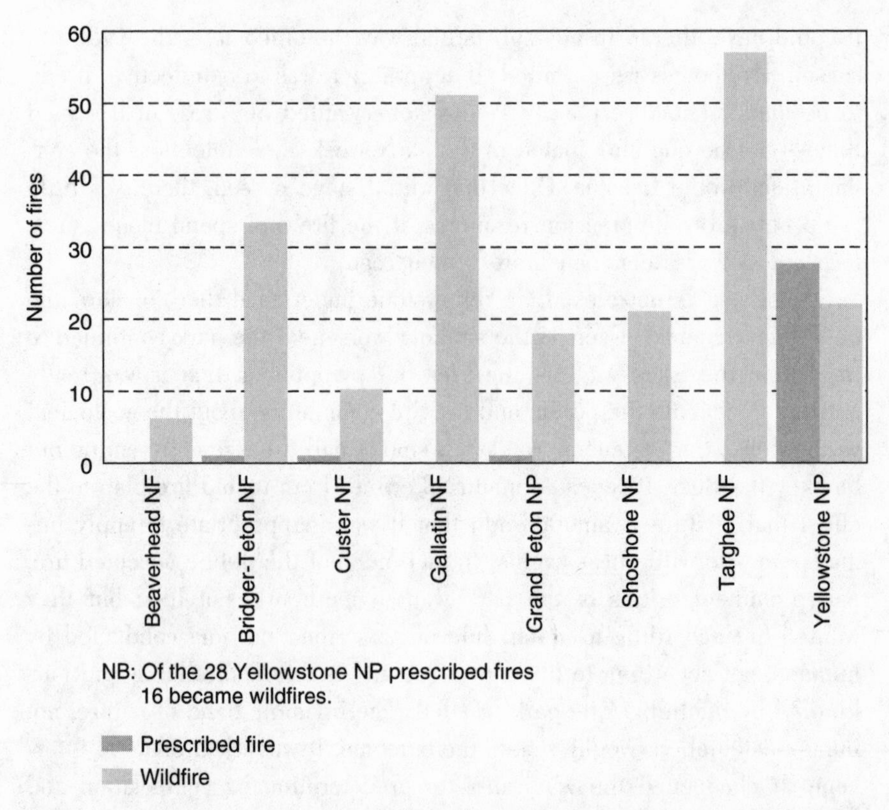

Yellowstone exceptionalism: the profile of fires, wild and prescribed, in the Greater Yellowstone Area. The preponderance of "prescribed natural fires" in Yellowstone Park is the product of agency decisions, not natural inevitability. The prescribed fires on the Bridger-Teton and Custer national forests were eventually declared wildfires.

They point out, correctly, that in the long spectrum of human and natural history fire control in the contemporary mode is an aberration. But so is the more recent concept of using only natural fires to advance land-management objectives. The norm is the human use of fire everywhere and for every conceivable purpose. Both natural fire and modern fire suppression are anomalies because they propose alternatives to anthropogenic fire. Yet it was this middle terrain—deliberate fire usage—that Yellowstone purged as interventionist.

The Yellowstone ideology insisted that the park should reflect nature's will, not American ideas about what nature should be. But this is a fallacy. All reserved lands already reflect human ideas about nature and what kind of relationships should exist between such lands and society. Nature offers no external standard. If nature is amoral, then it provides no basis for human judgment. Not to intervene is as much a moral act—and has as little sanction from nature—as intervention. The appeal to an "amoral" nature did not absolve Yellowstone from its decisions but only disguised the real nature of that ethos. Or it did until the fires forcibly deconstructed that text.

The mantra also avoids fundamental questions about how "natural" the fires of 1988 were. They had burned after a century of unprecedented human activity at Yellowstone—not only the installation of a new set of human practices but the abolition of ancient ones, including anthropogenic fire. Almost certainly a century of fire suppression influenced the character of the 1988 fires, but it is exceedingly difficult to spell out how. Fire exclusion involved more than direct firefighting; it meant the virtual elimination of anthropogenic ignition. Probably this accounted for a lot of the "lost" burning. It would not, after all, take many abandoned campfires or escaped hunting fires during critical years to rack up very large acreages.

Besides, climate, not fuels, dictated the overall fire complex. Everything burned. The one fascinating exception was the near immunity of the northern winter range, which fire-scar evidence had revealed to be the most frequently burned site in the prepark era. How the range managed to escape the fires (in fact, succeeded in retarding fire spread) during the largest fire complex in at least two centuries was never explained. The most likely reason is that the range, prime hunting grounds, had been subjected to regular anthropogenic burning very early or late in the season.

But the confusion worsens. Some 60 percent of the fires that swept through Yellowstone began outside the park or started from anthropogenic sources. Perhaps the most absurd episode involved the Mink fire that began in the Teton National Forest and was fought vigorously (at a cost of more than $1 million) before the prevailing winds drove it across Yellowstone's southern border. Park officials declared that they would accept the burn as a prescribed natural fire and withdrew control forces. Eventually the fire spilled back over the border and the Forest Service had to fight it again.

If these fires are natural, as park officials suggest, then why is prescribed fire unnatural? If it does not matter to a lodgepole pine whether the fire that burned it came from lightning or a campfire, then why does it matter to an elk whether it dies at the paws of a grizzly or from a hunter's bullet? If anthropogenic fires can enter the park because once within the wilderness they become "amoral," then why can't hunters, loggers, or geothermal drilling companies? Probably it does not matter to the lodgepole pine or the elk what kills it, but it does matter to humans. The appeal to an amoral nature does not remove park officials from their own moral universe. If the park does not express cultural values, then it has no reason to exist. Nature may be amoral; Yellowstone National Park is not.

Behind this appeal to moral agnosticism lies, once again, the hugeness of Yellowstone. In a smaller or more fragmented ecosystem intervention would be mandatory, and administrators would have to confess openly to the ends and means of their manipulation. Yellowstone, however, could cultivate its peculiar ideology in almost self-reflexive isolation—granted a bold margin for error and plentiful room for speculation; confident of its enormous importance to the National Park Service and the American public. It could, that is, until wildfire was really let loose. The great fires liberated the core issues much as crown fires seared away the waxes that held lodgepole pine seeds within their serotinous cones. In the ash bed of a million vivid acres those questions could take root. The fires magnified almost to the point of caricature not only fire management but also the whole spectrum of management concerns that infest Yellowstone.

The fires of '88 have revealed the limitations of all attempts to impose ideas on an indifferent nature. Knowledge is inevitably imperfect, technology always flawed, and the need to act ever present. What the public should demand is not complete control, which is a chimera, but a good-faith effort. It can expect the park to exploit the best practices available and to reconcile its ambitions with public opinion. The charge against Yellowstone is ultimately not ignorance, indifference, or incompetence but bad faith. It did not do what it said it was doing, it disguised its real beliefs behind a screen of agnosticism, and it refused to accept the humbling state of its ignorance.

Incredibly the park evaded the worst criticism, much as it had for decades evaded the philosophical hypocrisy and practical contradictions of its programs. When the fires flared onto public television, commentary was often misinformed. That allowed for a public relations counterattack. The park and its collaborators hammered at the early media mistakes, and made them, not the park's fire program, the object of future discussion. It diverted criticism into a discussion of national policy and away from Yellowstone's execution under that policy. It convened a panel of scientists to review the ecological consequences of the fires, then buried their report. When an interagency board reviewed federal fire policy, all references to particular programs (including Yellowstone's) were removed in the final editing. The result was to distribute Yellowstone's failings everywhere. Ironically, it took this purist park to demonstrate the full range of anthropogenic fire ecology. The ecological consequences of the Yellowstone fires were felt all over the United States, and even in Canada and Australia. Everywhere, it would seem, except at Yellowstone.

Yet in the end those who held that Yellowstone's size was its salvation were correct. It is apparent that the Yellowstone ecosystem is a lot tougher than our knowledge of it. Yellowstone will absorb the fires. Some of the ecosystem will recover slowly, other portions will erupt in biotic profusion; all the burns will serve as a fascinating, world-class monument to fire ecology. The fires of '88 brought the old volcano that is Yellowstone back to life.

If the fires were exaggerated, that is, after all, the Yellowstone way. Its size took small errors that elsewhere might have been overlooked and leveraged them into colossal breakdowns. What might have been buried in the backcountry—as the Alaska and Bob Marshall Wilderness fires of 1988 were—roared across television screens. That same hugeness also meant that fires that elsewhere could have consumed an entire park involved less than half of the Greater Yellowstone Area. The fires proved bigger than our understanding and vaster than our technologies; they also proved larger than our egos. The biota will survive with fewer scars than will our presumptions. The real ecological tragedy was the loss of marginal programs and the damage to prescribed burning that was experienced outside the Greater Yellowstone Area.

As a complex the fires burned in two swaths clustered along the northern and southern tiers of the park, sparing a broad middle corridor. In a symbolic sense, that may be what the fires do to the management of the Yellowstone environment. They may burn away the extremes and leave a less exaggerated but still robust middle. Between the burned and unburned landscapes, one can imagine a kind of oscillating biotic current, and between Yellowstone before and after the fires, one can envision a dialectic that restructures philosophy and practice around the great middle landscape of anthropogenic fire. If Yellowstone can build on that corridor, its fire program may follow its biota into recovery, and the park may in some future summer again let wildfire loose.

INTERMIX FIRE

The Fire
This Time

In 1943 Roy Headley, recently retired as chief of the Division of Fire Control for the U.S. Forest Service, completed his exhaustive treatise, "Re-thinking Forest Fire Control." He noted, pointedly, that there is "still much confusion and conflict as to the objectives" of fire control; that in the old days this was not a serious difficulty because there was "no danger of overdoing protection"; that there had to be some kind of "ultimate objective" if fire protection was to make sense. He then proceeded to list sixteen commonly held purposes.

Roy Headley was a practical man who had spent a career in fire control. He had been a member of the California cohort that invented the concept of systematic fire protection. He had formulated the "economic theory" of forest protection as an objective of fire control; had worried about the managerial impact of the emergency firefighting fund and bristled over the implications of the 10:00 A.M. policy; had even experimented with "let burning" in remote sites as a fiscally responsible alternative to all-out fire control. The conservative Headley, a career administrator of fire programs, even protested against the use of the term *smokechaser* because it conjured up the image of per diem firefighters wandering after intangibles in the woods. Thus, when Roy Headley proposed as one objective of fire control that it was essential "to maintain our civilization," he noted wryly that, even during World War II, to say this "was to take ourselves pretty seriously." Yet, he quickly conceded, this was an "attempt to formulate an ultimate objective"; and alone among the register it proposed a "final end," not a means to an end.

Headley understood that fire control could not survive as a task unto itself. It had to serve a larger social purpose, and in the fifty years since he labored through his never-published magnum opus, it has. What Headley did not appreciate was the way that society would change, and the way in which those changes would redefine the purposes of fire protection and redirect its firepower. In the postwar era fire protection shared in the nation's affluence and assumed a role in the national security state. When wilderness values proposed another vision of America, fire control, now rechristened fire management, retooled to serve it.

But still a restless American society moved on ("I move; therefore I am," observed one French critic of its auto culture). Although powerful as a cultural icon, wilderness was also remote, geographically distant, and intellectually abstract. The thrust of the wilderness movement had been to segregate people from wilds, or to restrict their contact to certain prescribed protocols. Wilderness fire was spectacular but abstract, like an Ansel Adams photo of Yosemite, an opera of the wild—its theater grandiloquent but far removed from the vernacular of American life. Now demographics put people and fire together again, not through a haze of wilderness smoke but by the incineration of houses.

Outside wilderness, land use became an omelette of scrambled jurisdictions. By the time Yellowstone immolated the wilderness fire era, a new principle—call it the intermix fire—was restructuring American geography. Its essence was that it resisted segregation, or packed an ever tighter mosaic of land use. Houses and wildland fuels thronged together as in a medieval bazaar. Not only the mix but the process of mixing made the compound volatile. The damaging fires of American history had typically erupted during periods of rapid transition: the introduction of farmers into the prairies, the spread of herders across the southern pineries, the arrival of slash and burners in the north woods, the construction of suburbs in California chaparral. The emergence of the intermix fire followed this black legend.

This time, however, the process came almost everywhere, and it was the mix itself—the dramatic kneading of wild and urban—that made the land attractive. The wild was made habitable by its association with the urban; the urban became amenable through its proximity with the wild. The best of both was in reach, one world to each hand. But that proximity also made the two abrade like a match on wood. America was being resettled anew, or more newly unsettled, into a transitional state that promised to remain permanently suspended. Unlike past precedents these fires would intensify, not vanish, as settlement progressed.

Here was, in brief, a cameo of American civilization, unwilling to give up either its cities or its wildlands, coveting both, and watching as its swelling population and economy crammed that polarity closer and closer together toward a critical mass. Wildland fire reconnected to American society, though in ways no one had anticipated. And fire management found itself pursuing the spot fires and flaming fronts of social change with as much vigor as it had once chased smoke in the backwoods.

There were plenty of examples of the new fire, some spectacular but none dominant. Prototypical fires swept through mixed landscapes in Florida, North Carolina, Michigan, Nevada, California, Arizona, Idaho, Colorado—any place that mingled houses with natural fuels and ignitions. Not until the latter 1980s, however, was the process recognized as a national problem, or accepted as a major issue for the fire community. A confusion of names matched the profusion of examples, an indirect index of how robust and varied the new fire had become. For a while "wildland/urban interface fire problem" claimed something like an official imprimatur. The "intermix fire," however, is equally apt and more succinct.

By 1989, after Yellowstone had shattered the dominance of wilderness fire, the United States hosted an international conference that proclaimed the phenomenon as a global concern, or at least one that afflicted an industrialized world still struggling to redefine its rural heritage. Australia fashioned an urban bush; Canada compressed forests and towns in places such as Jasper and Banff, and watched the boreal forest march south through former prairie and farms; Europe saw its rural landscape, particularly in the Mediterranean, unraveled into fire fuses, its coasts reclaimed by the villas of urbanites and tourists. The pressures of a global economy busily redefined the relative power of urban metropolises and rural peripheries, and recalibrated the landscape values of an affluent elite. The intermix fire is one result.

In the United States the story begins with demographics. Between 1950 and 1990 America added 100 million people to its population. Most poured into metropolitan regions, which sprawled outward in the often eccentric patterns defined by commuter corridors. Metropolitan statistical areas had occupied 5.9 percent of U.S. lands in 1950; by 1987, that proportion increased to 16.2 percent. Meanwhile the farm population shrank to less than 3 percent even as agricultural land remained, at a national level, more or less constant. The repopulation of rural lands—first those accessible to metropolitan areas, then those in "wilderness" counties—came as a result of a secondary outmigration from urban centers. Between 1970 and 1980 the population of counties adjacent to or containing federal wilderness grew by 13.4 percent, and from 1980 to 1987, 24.3 percent, compared with 6.9 percent for all nonmetropolitan counties. The so-called rural renaissance was not, however, a revival of rural economies so much as repopulation of formerly rural landscapes by an exurban population.

But this, the outmigration of the city, is only part of the equation. If the city went into wildlands, it was also true that wildlands came to the city. And there was that vast murky landscape, once tended by agriculture, that was abandoned to a scrub of houses and woody weeds under the blows of a global commodities market. The burning suburbs of Spokane, Santa Barbara, the California Sierras, and the Front Ranges of Colorado were classic examples of the first. The incineration of Oakland and Los Angeles were species of the second. The resurgence of fire in Florida, North Carolina, and Michigan were examples of the last. Add them together and they sum up a new problem fire. If they lack the conceptual purity of wilderness fire, that, after all, is the essence of the problem. The era is a mixture of fire types as well as houses and wildlands. It is to the American landscape what multiculturalism is to American society.

Clearly the new settlers are not ruralites in the sense that they, like their predecessors, live off the land. Instead they live on it. Their income comes from elsewhere, as do their values and expectations. Typically they want urban amenities but without an urban setting. They expect urban services such as sanitation, police, education, and fire protection but not urban bureaucracies, taxes, and hassles. They want a rural setting without having to rely on a rural economy. They want the best of both worlds, and are willing to fall through institutional cracks to get them. They assume that fires occur elsewhere. And when fires do strike, often they expect that someone else will fight them.

From the perspective of fire protection the intermix environment is often the worst of all worlds. It slips between the jurisdictions of urban and wildland fire agencies, and outside effective county or rural fire districts—the point of many of these communities is, after all, to live beyond the range of service taxes. There is little zoning for fire control. There are few building codes to reduce hazards such as wooden roofs. There is scant pressure to reduce wildland fuels around dwellings. Open spaces that serve as buffer zones shrink as houses and woodlands expand. True rural residents engaged the land, and controlled fuels by gathering wood, clearing, grazing, plowing, and so on. Exurbanites do none of this, which leaves fuels unimpaired. Narrow roads to sheltered homesites, rustic wooden houses with shake-shingle roofs, lush vegetation dripping over walls and roofs, distance from prying officials and taxes—all this is why the exurban communities were created. To render them fireproof is to begin to re-create the environments from which the residents fled in the first place.

To leave them alone, however, brings considerable social costs. Inevitably fires occur and must be fought, and this usually means action by outside agencies such as the federal land or state forestry bureaus, followed by emergency aid of one kind or another. These costs are distributed over all of American society. Not only is this ineffective, it is unfair. It repeats the classic inequity by which profits are privatized and losses are socialized. Moreover, as the process progresses, the isolation of the earliest groups disappears. New developments threaten the older ones. The ability of public land officials to manage land or fire becomes more and more problematic, a complication of ends and a confusion over possible means.

The deadly irony of all this is well conveyed in the cover photo to the "Report of the National Wildland/Urban Fire Protection Conference" issued in 1986, which shows a crown fire, which began as a prescribed burn to improve the habitat of Kirtland's warbler (an endangered species) but then escaped, bearing down on the village of Mack Lake, Michigan. The town burned to its rock foundations and one firefighter, a tractor-plow operator, died in the attempt at control. There it all was, up in smoke, an allegory of the American fire scene: the wrong fire at the wrong time in the wrong place.

A reordering is under way. One of the unexpected consequences of the Yellowstone conflagration is that it liberated the fire community from the tyranny of wilderness fire, now left to fallow, and has allowed it to pursue a new agenda. Pushed by fatal fires and pulled by the prospects of political payoffs, it is engaging the new colonization of America. The best minds, the greatest investments in infrastructure, the most strenuous politicking—all are focused on the intermix fire, which has rightly assumed the status of a legitimate problem fire.

One outcome is a new level of integration in national fire protection. This goes beyond the experience of interagency fire coordination among the federal land agencies. And although it builds on the tradition of cooperative fire that the Forest Service has promoted since the 1911 Weeks Act, it greatly expands its range of participants to include alliances with rural volunteer departments, urban fire services, and even the National Fire Protection Association. Much as the intermix fire has blurred the boundaries between wild and urban landscapes, so the American fire establishment has had to reconsider the boundaries that have traditionally divided the various fire services and the kinds of jobs they do.

Until recently wildland and urban fire services evolved along segregated tracks, and neither had much to do with rural fire brigades. Urban and wildland fire agencies fought their own types of fires in their own ways with their own kind of firefighters. They served different clienteles, have staffed their ranks with firefighters of very different motivations, and have interpreted their jobs in almost incommensurable ways. Urban firefighters save people; their cherished image is that of lifesaver; their nightmare is a high-rise fire. Wildland firefighters encounter forests, grasslands, and brush; they more resemble an army engaged in a kind of moral equivalent of war against nature; they measure losses in acres, not lives or property. Rarely—Southern California excepted—have the two cooperated in routine operations. Almost never has anyone examined them together either as a social problem or a public service. No histories pair them. Little art joins them.

The intermix crisis, however, has stirred the lot into a common cauldron. Beginning with California and spearheaded by wildland agencies, these unexpected allies have hammered out a mutual language to describe fires and equipment, have evolved joint tactics so that they can work together on a given fire, have trained crews equally adept at fighting fires in houses and woods, and have groped toward a mutual understanding of the burning commons they share. Collectively they have promoted a second-order fire-protection system, one that relies on a further integration of existing resources rather than on the invention of new ones. It is a process of resynthesis and rethinking that, if successful, will begin to plaster in the cracks through which small communities fall and to plow under or pave the vacant lots of the intermix landscape that breed fire.

What the scenario has lacked, however, is a fire of sufficiently dramatic power to announce the intermix fire in the way that the Yellowstone conflagration advertised the bold death of the wilderness fire era. The succession of large fires that punctuated the 1980s, even those that consumed hundreds of houses, was overshadowed by the drama of even bigger burns in the still wild West. The story remained fire and wilderness, not fire and city. Critics and journalists had no formulas, no set pieces by which to interpret a forest fire that burned down cities or urban fires that burned like forests. That is, they had none until fire blasted the Berkeley Hills in Oakland, California, and then visited Southern California in the fall of 1993 like a wandering plague.

. . .

The Oakland holocaust shows both the power of the intermix fire and its liabilities. For all the attention lavished on the episode, it seems to have produced no lasting public discourse. It came and went, like the 1989 earthquake. Almost no one moved to establish a congruence between what happened there and what was occurring in Grayling, Michigan, and Palm Coast, Florida. Certainly the 1991 fire burned more or less as the 1923 Berkeley fire had, the difference in damages reflecting seventy additional years of urban development. In that sense California had learned nothing and forgotten nothing. Yet the fire had the potential to proclaim something more than California exceptionalism. It could have stood—might still stand—as a visitation from an inchoate intermix fire.

The fit is not as exact as logic alone might demand. No one would confuse the Berkeley Hills with Pattee Canyon outside Missoula, Montana, or Dude Creek some miles up the Mogollon Rim from Payson, Arizona, also stocked with houses and swept by wildfires. But the same social values that flung trailer parks and shake-shingled houses into the woods also encouraged the woods to grow around a city. Instead of exporting the city into wildlands, affluent urbanites imported wildlands into the metropolis. The critical fuels were eucalypts, brush, and wooden roofs. Build a city out of forest materials and it will burn like a forest. The fire swept through a city remade in an exurban image.

This was an intermix fire. The question remains whether it was fought like one. Almost certainly lawsuits will ask whether the Oakland fire department followed standard practices, or whether it failed to recognize that this fire was a hybrid that called for different procedures and for which urban fire services had to seek different allies. But then it may be that no one wants to claim this fire. Urban fire services resemble baronies, each secure in its fiefdom, reluctant to sally far outside its walls, despite the California emergency plan that allows engines to go wherever they are needed throughout the state. Wildland fire agencies, although accustomed to fires that race across jurisdictions, view cities as almost extraterrestrial landscapes, their fire departments closer to police and ambulance services than to anything that should fight fire in the wilds. Some may wish the Oakland fire to remain as only another sign of decaying urban life, not unlike the nearby Nimitz freeway still unrepaired from the shock waves of

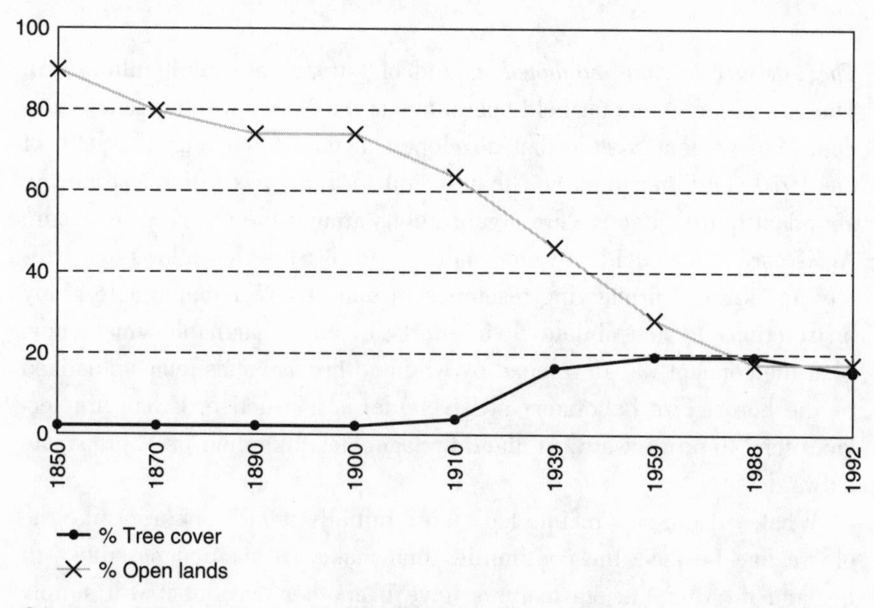

Growing a forest in the city. Holocaust followed soon after shrinking undeveloped lands met expanding urban afforestation.

the 1989 earthquake. Others may squabble over the symbolism like lawyers over liabilities. The 1993 fire bust relocated the spectacle to Hollywood's back lots. But fires in Malibu were familiar, like reruns of *I Love Lucy* or *M*A*S*H* (whose set a wildfire had once incinerated).

Both those who would dismiss the fire and those who would seize it for synecdoche are right. The intermix fire defies clean categories—but so does the society that sustains it. What had been apart is now squeezed together. Fires that had raged voyeuristically in the remote hinterlands of Alaska and Wyoming have now come to city and suburb. The confusion over responsibility for such fires only reflects the cross-purposes of American society, ultimately an expression of the ambivalent character of American life. The truism is that wherever you have wildlands, regardless of whether they originate from natural or social causes, you will have wildland fire. This holds even where houses sprout among the trees like mushrooms pushing through forest litter, or where wildlands intergrow with a city like ivy through a lattice. That, in fact, is precisely the lesson of the intermix fire.

. . .

The intermix fire has promoted a kind of bureaucratic multiculturalism. The common grounds for field operations is the National Interagency Incident Management System that developed in the fifteen-year aftermath of the 1970 California fires and that sought to find ways to orchestrate the abundant but disjunctive fire organizations around the Los Angeles basin. At its core is the Incident Command System (ICS), which allows for virtually any kind of firefighting resources in almost any amounts across any jurisdictions to consolidate their efforts in any imaginable emergency. That the concept was developed by wildland fire agencies (conceptualized by the Forest Fire Laboratory at Riverside) is instructive. Urban fire services tend to turn inward; wildland fire agencies, like their fires, propagate outward.

What will the new regime look like? Initially it will look a lot like the old regime because the continuities that make a transition possible will demand it. After the era matures (give it another decade) it will simply look different. It is probably pointless to speculate in much detail about its future appearance. There are too many players; too many permutations possible; too many chance events waiting to intervene. As William James said regarding the pragmatic definition of truth, "By their fruits ye shall know them, not by their roots." We won't really understand the era of intermix fire until it expires.

But continued outbreaks help define its dimensions the way seismic tremors map the contours of a magma chamber. The geography of American fire is following the geography of American society, metropolitan, bicoastal, and mobile. Fire loads are shifting from remote sites, valued for their absence of people, to once-rural sites crowding with restless immigrants. Likewise institutional firepower is moving from agencies that specialize in wilderness fire to those that grapple with intermixed landscapes. Even regions such as the northern Rockies that a decade ago pointed to the Selway-Bitterroot wilderness as the epitome of fire management now seize on fires around Boise, Spokane, and Missoula as a model of their ruling concerns. The change, moreover, serves as a reminder that the natural fire load on public as well as private lands is an artifact of decisions about how that land will be used.

All this is rapidly remaking the institutional landscape of American fire. However inevitable and politically shrewd the conversion to the intermix fire, the move has its costs. It makes more difficult the reconstruction of wilderness fire management after the Yellowstone catastrophe; it downsizes prescribed burning from a treatment of choice (endowed with almost metaphysical power) to one tool among many, and a practice not easily exploited within the intermix landscape; it threatens to divorce from field operations a fire research program that like a wobbling top was seeking to stabilize its funding and define its purposes around the drama of global climate change; it plunges fire management into the feisty politics of small towns, counties, unincorporated subdivisions, zoning boards, planning commissions, building codes, and real estate developers. This is a landscape as far removed from the tradition of wildland fire as hydrants are from stockponds and self-contained breathing apparatuses from bandannas. This is not the landscape that most members of the fire community had grown up in, or the kind of fire they expected to fight when they joined up.

Any reconnection of fire with American civilization must involve individuals as well as society. The intermix fire, however, promises to alienate a good fraction (perhaps a majority) of those fire specialists who had grown up with a wildland fire establishment that dealt with *wildland* fires. They may see the alliance with urban fire as a forced merger, a hostile takeover of purposes, that threatens not only their jobs and how they do them but also their self-esteem, their identity. It is clear that a turn away from wilderness fire is not a turn back to previous eras. The world will look stranger, not more familiar. When the fire community goes home again, it is more likely to a trailer park than to a lookout tower or a fireguard cabin.

The new era must inspire as well as serve. It is not enough that the intermix fire problem be understood; it must also be felt. Wilderness fire and its predecessors did not occupy their informing roles because of logic alone; they flourished because they mobilized emotional commitments and moral energies from many individuals. That experience became vital to the identity of many persons, at least a generation or more, in the fire community. Similar conversions are an old attribute of wildland fire protection in America. It has to do with the fact that wildland firefighting has relied on young men (and more recently young women); it has to do with the capacity of wildland firefighting to function as a rite of passage; it has to do with the impact that even a season or two can make on the lives of its participants. I know that is how it affected me.

I began my association with fire, as a recent eighteen-year-old, in the summer of 1967 on the North Rim of Grand Canyon. This was the last year the Park Service operated under the 10:00 A.M. policy. For me fire management will always recall that first summer. Wildland fire will always mean two smokechasers somewhere deep in the woods, a flaming snag, a broken saw, the smell of smoldering duff, wind throwing sparks across an evening sky, C rations and a P-38, paper sleeping bags, the awesome crack of a felled ponderosa, headlights through silhouettes of fir and aspen, sunset across the canyon, moonlight on smoke, and youth. The special youth of an eighteen-year-old growing up. The youth to which Norman MacLean referred when he wrote of "watching the universe's four elements at work—sky, earth, fire, and young men." Without that experience I would never have stayed at the rim, would never have pondered the meaning of fire and humanity, would not have tried to organize life on earth around fire in the way fire had organized my life at the North Rim.

Others in the fire community have their own (I suspect, similar) stories. For many of the younger generation that story will be intimately connected with the great controversy—one that embraced surprisingly large constituencies in our culture—about wilderness and what to do with fire in the wild. It is a transition I narrowly missed. So while my head tells me that prescribed burning is fire, too, that a confined fire is really a managed fire, that exurban landscapes, Scot air-packs, and dual-purpose engines are a form of firefighting, that fire management must engage not only nature but also society, my heart remains with that flaming snag. Anything else is something else.

So I return once more to Roy Headley. One of the most vital connections between wildland fire and our civilization, I suspect, may be the former's special role as a rite of passage. If the intermix fire is to succeed as a defining problem, if it is to rally the kind of fierce conviction that previous problem fires have tapped, it will have to discover similar rituals, experiences, sentiments. That will not be easy. To those initiated into purer wildland fire, intermixed fires won't *look* right, and they won't *feel* right. It will be a strange fire, in the biblical sense, vaguely sacrilegious, impure by virtue of being contaminated with alien fires.

It was possible for many members of the fire community to adapt to wilderness fire because, if the practices were different, at least the landscapes looked the same. That will not be true for intermix fires. To new recruits, perhaps older and drawn from volunteer fire departments, already absorbed into adulthood, there may not be that shared experience of initiation in the woods so common to wildland firefighters. For the new era it will not be enough to count burned houses, legislate building codes, speak abstractly about future scenarios, or stage palace coups in Washington bureaus. The intermix fire will have to establish psychological continuity with wildland fire communities of the past. The issue goes beyond professional dedication and bureaucratic gamesmanship. It goes literally to the heart of wildland fire management. I believe that this continuity can be found because it has to be found.

I left the North Rim many times over fifteen seasons, but I remember best the autumn when the snow was falling and an October night was fast approaching. At the large meadows by the park entrance, winds whipped the snow into whirls and clouds. For an instant—a blink of the mind—I blanked into reverie. The snow turned to smoke, the car headlights to headlamps. There was a clamor of shouts; the rasp of a chain saw; the thunk and scraping of pulaskis and shovels; the snap of falling branches; a rush of flames. Through the smoke and noise I could see a fire. It was a fire I will always carry within me.

The fire next time—whatever it is—will have to be a fire that we carry within us all.

Nouvelle Southwest

The fire bust of June 1990 followed a classic formula, but one intensified in puzzling ways.

Fire seasons in the Southwest obey a natural rhythm of wet and dry, a two-cycle engine for which lightning provides the spark. This interplay takes several forms. Part is topographic, in which the Southwest's fabulous terrain creates differentials between moisture and aridity. Slopes that face south or north betray different levels of moisture. So do plateaus that range from high to low. As a result some place is nearly always dry and somewhere nearly always wet. The greater cause, however, is climatic, the summer monsoon. Surges of moisture from the south strike mesa and mountain, and thunderstorms tower up like spumes of surf on a boulder; rain descends in drying veils; lightning kindles fires that flash from the peaks like beacons.

The process is spotty, like a handful of popcorn scattered on a skillet.

One moment there is a deluge, the next a flood of desert sun. Ideally, there is enough storm to hurl lightning and wind, not enough to quench burning snags. Add to this winds that splash out from thunderheads like water from an overturned bucket, spillage that makes for dust storms in the desert, firestorms in the mountains. Altogether it is one of the great ecological rituals of the region, and it accounts for the fact that the Southwest has the highest concentration of lightning fires in the United States.

The figures are astonishing. Between 1960 and 1974 there were twelve days in Arizona and New Mexico when more than 100 lightning fires started; on June 28, 1960, lightning kindled 143 fires. In 1970 lightning ignited 100 fires on July 18, and the next day brought 100 more. On June 24, 1971, 103 lightning fires burned 75,713 acres. The Southwest's national forests average more fires per year than any other region; they have the second-highest rate of burned acreage, from both wild and controlled fires; and critical fire weather occurs here with greater frequency and persistence than anywhere else in the nation. Yet despite prodigious numbers of fires, there are few truly devastating burns. The sheer number of ignitions, plus the exquisite minuet between rain and fire, assure a certain equilibrium that balances large numbers of fires with smaller sizes of individual fires. As the monsoon persists, the wet triumphs over the dry. The number of fires rises steadily from May to August but the average size of those fires diminishes.

Among the fire provinces of the United States, the Southwest stands, as Clarence Dutton once wrote of the Grand Canyon, as a grand ensemble. It combines elements of all the other regions, of all the problem fires that have motivated and baffled the nation's evolving fire establishment. They are all here in a mélange that makes this, year in and year out, the most interesting fire province in America. In the realm of wildland fire, the Southwest is always close to the center, always the partner of whatever region may lead the dance. Appropriately the national training facility (the fatuously named National Advanced Resources Technology Center) resides at Pinal Air Park outside Marana, Arizona.

But while fire busts are frequent, they are rarely immense and even more rarely fatal. The Dude Creek fire outside Payson, Arizona, on June 26, 1990, shocks because it was both. It took familiar elements—some ancient, some modern—and compounded them into an event that looked grotesque, alien, out of character. Perhaps it was. The fire savaged forests, houses, and a fire-suppression organization. When a microburst of wind drove the fire through a squad of firefighting prison inmates, killing six and hospitalizing four others, it prodded debate about how fire and American society could coexist there, which is to say, about what the character of each had become.

The fire blew up on the hottest day ever recorded in Phoenix (122.5 degrees Fahrenheit), which argued that it was an old story intensified. But it burned through a forest vastly different than in presettlement times, through summer homes and trailer parks that had little historic precedent, through a society that had sought to eliminate fire of all kinds. Old and new had come together with as much force and fury as wet and dry. In that sense the Dude Creek fire says something vital about life in the contemporary Southwest.

That extraordinary fire load is not simply a product of natural processes. For millennia humans have busily restructured the geography and seasonality of southwestern fire, sometimes complementing and sometimes countering the old order. Lightning had to compete not only with rain but also with aboriginal firesticks. Human inhabitants added other sources of ignition in the service of hunting, raiding, foraging, and horticulture, and as an inadvertent byproduct of a seasonal nomadism whose routes became trails of smoke from campfires, signal fires, and escaped fires of diverse origins.

"The most potent and powerful weapon in the hands of these aborigines," concluded S. J. Holsinger, of the General Land Office, at the turn of the century, "was the firebrand. It was used alike to capture the deer, the elk, and the antelope, and to vanquish the enemy. It cleared the mountain trail and destroyed the cover in which their quarry took refuge." Obviously, burning on this scale "must have exerted a marked influence upon the vegetation of the country. Their fires, and those of the historic races, unquestionably account for the open condition of the forest. . . . The high pine forests were their hunting grounds, and the vast areas of foothills and plateau, covered with nut-bearing pines, their harvest fields." The aboriginal fire regime was itself in transition as peoples migrated into or departed from the region.

There is an old adage in firefighting that says the fine fuels drive the fire. Fine fuels include grasses, conifer needles, low shrubs, the portion of the fuelbed that reacts most quickly to changes in moisture and heat, that most readily combusts. It determines the ease of ignition and the rapidity of fire spread. Under aboriginal rule, fine fuels blossomed, and the Southwest burned easily and often. Lightning and firestick competed to see which would burn a particular site or in what season. The density of that competition fashioned an intricate honeycomb of burned and unburned sites. In dry years fires simmered for weeks, smoldering and flaring as the opportunity permitted. The principal check against conflagrations was simply the magnitude of low-intensity burning on all sides.

There are eyewitness accounts to the burning, but the most compelling evidence was recorded in the land itself, the golden grasslands, hillside montages of brush and prairie, and most spectacularly oak and pine savannas. Early explorers spoke enthusiastically about the great natural parklands of the region in which mature ponderosa pines marched in majestic columns through grassy glades. In 1882 Captain Clarence Dutton, exploring the Kaibab Plateau for the U.S. Geological Survey, exulted that

> the trees are large and noble in aspect and stand widely apart, except in the highest parts of the plateau where the spruces predominate. Instead of dense thickets where we are shut in by impenetrable foliage, we can look far beyond and see the tree trunks vanishing away like an infinite colonnade. The ground is unobstructed and inviting. There is a constant succession of parks and glades—dreamy avenues of grass and flowers winding between sylvan walls, or spreading out in broad open meadows . . . The way here is as pleasing as before, for it is beneath the pines standing at intervals varying from 50 to 100 feet, and upon a soil that is smooth, firm, and free from undergrowth. All is open, and we may look far into the depths of the forest on either hand.

For his report on a prospective wagon road through northern Arizona, army surveyor E. F. Beale wrote in 1858 that

> We came to a glorious forest of lofty pines, through which we have travelled ten miles. The country was beautifully undulating, and although we usually associate the idea of barrenness with the pine regions, it was not so in this instance; every foot being covered with the finest grass, and beautiful broad grassy vales extending in every direction. The forest was perfectly open and unencumbered with brush wood, so that the travelling was excellent.

The explanation for the character of these semitended fields is that only a tiny fraction of ponderosa pine seedlings survived the near-annual onslaught of fire flashing through bunchgrass. Great trees that toppled over ripped up the ground at their roots, creating pockets of grass-free soil for a few years; so did the fallen trunks when, after a period of decomposition, they burned to white ash. In the critical years that followed, seedlings thrived, and reached a state in which they could survive routine fires. Around Flagstaff, for example, fire-scarred pines testify to fires that burned an average of every 1.5 years. They are mature trees, oddly grouped, clustered in ways that betray their origin in the churned-up soil of old root holes, or aligned along the trend of fallen boles. The macrogeography of such forests depended on the microgeography of fire refugia.

The paradox that the land was both burned and forested baffled some observers, such as the Norwegian naturalist and explorer Carl Lumholtz, who witnessed an astonishing profligacy of aboriginal burning across the border with Mexico. "These Indians, the pagans as well as the Christians, keep up the custom of burning off the grass all over the sierras during the driest season of the year . . . [so that] fires are seen continually burning day and night all over the mountains up to the highest crests, leaving the stony ground, blackened and barren, but the forests stand green." That was the rub. Despite this fantastic amount of firing Lumholtz became convinced that "the continuous, immense forests here could never be destroyed by the Indians" because, paradoxically, all this chronic burning inoculated the forests against wildfires. They ensured the forests' "indestructibility."

. . .

All this changed with the advent of European colonization. Settlers introduced some new ignitions and removed several old ones, but it was by utterly restructuring the regional fuel complex that they remade the fire regimes of the Southwest. Generally colonization made itself felt in the New World primarily through mixed agriculture organized around plowed fields. This impact was muted in the Southwest, however. Regional aridity, hostile tribes, distance from major markets, the slow movement of westering Americans, the retarded admission of Arizona and New Mexico into statehood—which meant that the vast majority of lands remained public—all militated in the Southwest against the kind of pervasive agricultural settlement that typified most of the American frontier. Logging and landclearing remained relatively local; farming concentrated on irrigation rather than lands fire-flushed for nutrients. Indians, sequestered onto special reservations, became inconsequential as a source of fire. When Gifford Pinchot visited Arizona in 1900, he watched a distant Apache setting "the woods on fire" to improve his hunting, trailing fire like a broken lance in the dirt. But the battle of opposing firearms was over. By the end of World War I public land agencies aggressively attacked all fires, regardless of their source.

Instead, settlement followed hoof, not ax. Pastoralism prevailed—first through Hispanics, then through partial adoption by select tribes of indigenes (such as the Navajo), and then with mounting force through Americans. Livestock came to the Southwest in immense numbers. Cattle and sheep—the sheep were "ten times worse," Pinchot insisted—hit the region like a shock wave, disassembling fire regime after fire regime in ways that may prove irreversible. Flocks roamed in the hundreds of thousands, pounding forests to dust. (Pinchot was wrong: the cattle were worse because sheep browsed on tree seedlings and cattle did not, allowing them to blossom unchecked.) When the big ranches collapsed, hundreds of smaller homestead ranches took up the slack. The epidemic of herds continued. What had fed the flames of fast combustion now stoked the slow combustion of metabolizing livestock. Well before systematic fire control, cattle and sheep cropped fire from the land, and they did so with a thoroughness that engine companies, smokejumpers, and helitack crews have never equaled.

In the old Southwest grass had infiltrated every landscape, and had dominated some environments. Now herbivores seized every blade and pursued the succulent grass into every niche. No place was spared. Rolling hills of oak, high desert grasslands, mountain meadows, slopes dappled with chaparral, open pine forests like the columns of the great cathedral of Seville—the relentless hoofs and hungry teeth found them all. Thanks to his herds the reach of the rancher far exceeded the grasp of his own numbers. He became the biotic conquistador of the Southwest, the ecological equivalent of Cortés, and his herds the analogue of the Tlascalan confederates who reduced Tenochtitlán to rubble and shipped its plundered wealth across the seas. There was no sanctuary, no refuge from the conquest. The indigenous fires went the way of the grizzly bear and the mountain lion. Only on Indian reservations could fire survive in anything like its former state, and then only if tribes did not take up herding with the same ruthlessness apparent elsewhere in the region.

The evidence is written widely if complexly in the land. The finely bounded mosaic that had constituted the Southwest scene smeared; desert succulents, mesquite, juniper, and chaparral replaced grasses in desert basins and across high plateaus; brush congealed into jungles; open forests, once dappled with glades of sun and shadow, snarled with downed logs, dense tangles of understory, and young groves of pine and fir reproduction "thick as the hair on a dog's back." In 1902 S. J. Holsinger observed that "in Arizona you will find no young forests of any considerable extent antedating a period of forty years, and almost all of the regrowth has sprung up within the last quarter of a century." Surveying southern Arizona in the early 1920s, Aldo Leopold reasoned that "one is forced to the conclusion that there have been no widespread fires during the past 40 years." Forty years later Charles Cooper mapped the peculiar age structure of Arizona pine and determined that the forest derived from a small number of cohorts, all of which became established during the favorable climatic periods of the early twentieth century but which survived because they were spared fires. Others disputed the climatic argument for broadcast regeneration—1919 as an *annus mirabilis,* for example—but agreed with the outcome. What had been restricted to microsites and select times now thrived everywhere, year after year. Trees and brush multiplied like fruit flies in a jar of bananas. They spread, a scabby reaction to a vast ecological infection.

The fine fuels—the grasses and forbs—that had carpeted aboriginal Arizona now massed into three-dimensional jungles that readily transformed surface fires into crown fires. Loggers aggravated the scene by culling whatever old growth they could reach. Bark beetles, fungi, and dwarf mistletoe infested the thickets that, in the absence of grass and fire, sprang up in unnatural profusion. In places young growth—two meters high and seventy years old—existed in a comatose state, unable to grow and unwilling to die, waiting until fire could shock it back to life. A century after Beale rejoiced in the open pinelands of the region, Wallace Covington estimated that tree density had exploded from 23 per acre to 851, tree basal area from 23 to 315 square feet per acre, and crown closure from 8 to 93 percent. Where Dutton had praised the wooden colonnades of the Kaibab, tree density had ratcheted upward from 55.9 to 276.3 per acre, tree basal area from 44 to 245 square feet per acre, and crown closure from 16 to over 70 percent. Herbage had correspondingly plummeted, from 1,000 to 112 tons per acre in northern Arizona, and from 589 to 117 on the Kaibab. Near Flagstaff, a site that had once been covered 83 percent by herbaceous plants had only 4 percent in 1990. More ominously, fuel loads rocketed from 2 and 0.2 tons per acre, respectively, to 44 and 28 tons per acre. The land metamorphosed from a pine savanna to a forest tangled in dog-hair thickets. In recent years the woody invasion has also included houses. While the ultimate reasons for this biotic drama reside in the character of settlement, the immediate cause has been the elimination of fire.

If it is not everything, timing accounts for much of this condition. An interesting study has compared the forest structure on the Chuska Mountains in the Navajo Reservation to that elsewhere in northern Arizona. The Navajos acquired livestock from the Spaniards, and their herds soon swelled. Huge numbers of sheep and goats came to the Chuskas by the 1820s. As the flocks advanced, fires receded. But the rapid reforestation that followed American grazing later in the century did not immediately occur here; browsing by sheep and goats pruned back reproduction, and regeneration apparently had to wait for a favorable climate, which oc-

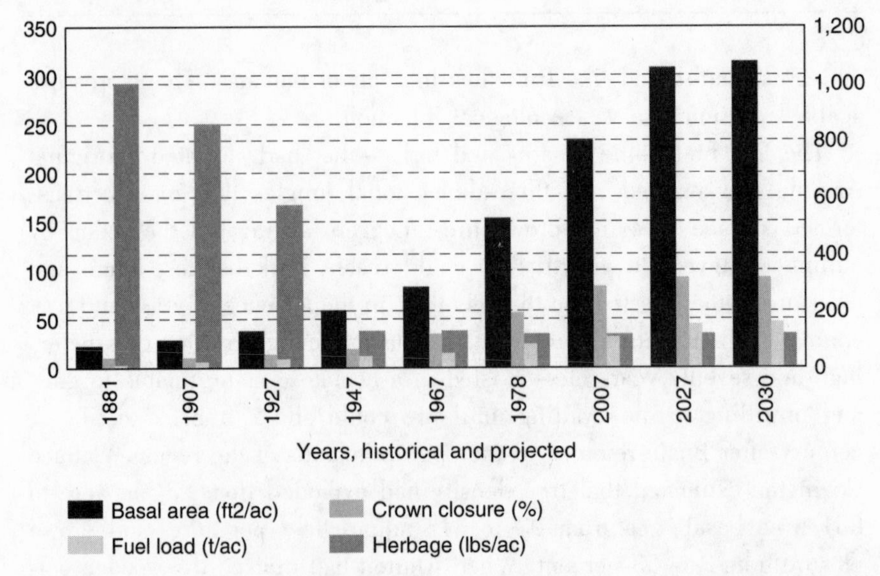

Southwest Passage: changing forests and fire regimes, Flagstaff, Arizona, historical and projected. Note, in particular, the dramatic reversal of grass and wood.

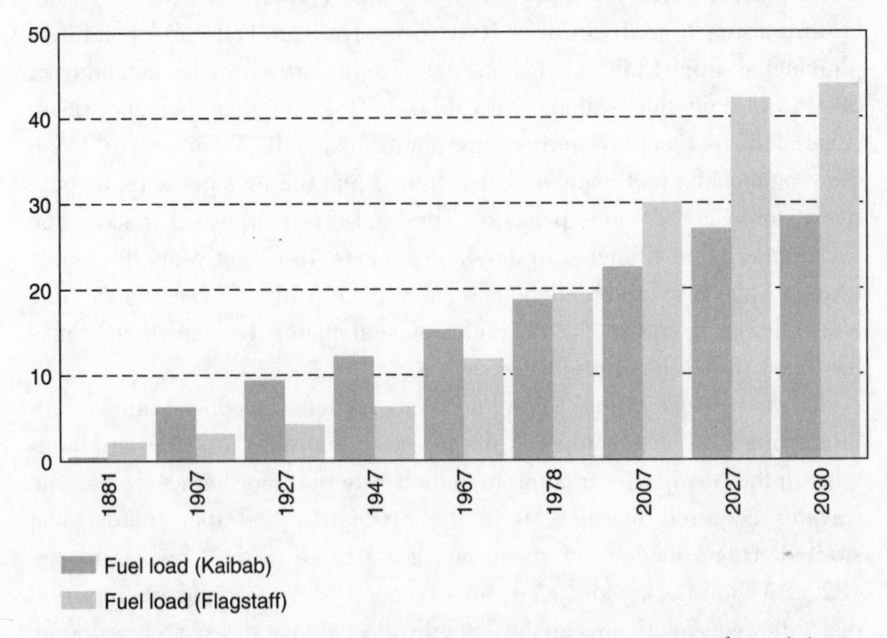

Priming the charge: increasing fuel loads on two pine sites in northern Arizona, historical and projected. The total biomass is less relevant than its forms; wood added to boles contributes little to increasing fire hazard, but understory shrubs and reproduction add greatly.

curred throughout the region in the early decades of the twentieth century. The congestion of the Chuska forests thus came synchronously with that elsewhere in the Southwest, the product of a beneficent climate that promoted regeneration and the absence of fire thinning made possible by intensive grazing.

The march of woody weeds was only the beginning, however. Biodiversity declined, and those creatures inimical to ranching or dependent on the old fire regime melted away like Arawaks before smallpox. Worse, the land started to erode. In southern Arizona a spectacular cycle of arroyo cutting began in eerie lockstep to the cattle invasion. Elsewhere there were slides, debris flows, and garden-variety siltation on a bigger scale than ever known before. The cumulative outcome was a colossal degradation of the landscape for which the tree, elsewhere a talisman of land health, was ironically an emblem of decay. By the early 1920s the declination had reached its nadir. Even cattle and sheep could no longer thrive on the land and had to be shipped to feed pens for fattening. Economically, ranching depended for the most part on public subsidies, even as it remained a political power, that relationship not being entirely coincidental. The debate over the relative contributions of climate and anthropogenic activities continues, but clear-eyed observers of the day had no doubt about the chain of causality. In 1933 Aldo Leopold wrote epigrammatically that "when the cattle came the grass went, the fires diminished, and erosion began."

By the time organized fire protection arrived it had only to confirm the fire ban announced by overgrazing. Fire suppression was, at first, an exercise in regional mop-up. Aboriginal fires were sequestered onto reservations. Livestock had perverted fuels and quelled the impact of lightning fires; fires had little to burn, or they burned amid open forests, easily extinguished by pine boughs and blankets. Even ranchers who sought to "green up" spring pasture by burning found it difficult to do so.

But, of course, fire could never be abolished. Tremendous extents of the Southwest were committed to public reservations for Indians, forests, parks, the military, wildlife refuges, and other purposes. These lands remained quasi-natural, persisting in forms that would not yield to farm or city. Something like the native fuels endured, though often leavened with pyrophytic weeds. Lightning, too, continued, and it interacted with whatever fuels flourished on the rugged landscapes it struck. Fire endured.

As decades rolled by, however, the fires that escaped now burned with unprecedented intensity and magnitude. Increasingly the saga of settlement moved from irony to tragedy. Fire had enhanced biodiversity; fire exclusion, through hoof and later shovel, destroyed it. Anthropogenic burning had improved fire control; fire suppression worsened it. Ranchers had sought to replace the wild with the domestic, and they had done so by exchanging sheep and cattle for grizzlies and cougars; but in the process they had also replaced domesticated fire with feral fire. If the land became less suitable for wild fauna, it became progressively more prone to wild fire.

That trend continued despite the New Deal's investment in conservation measures, including the CCC. Crown fires—large by virtue of their intensity as well as their size—increased from 10,127 acres per year in the 1940s to 15,117 acres per year in the 1980s, despite a progressively increasing investment in high-tech firefighting. Fires that had rarely exceeded 3,000 acres in presettlement times now routinely reached 10,000 to 20,000 acres. It became apparent that to remove fire was as powerful an ecological act as to introduce it. In 1972 the Tall Timbers Research Station mustered a task force that singled out the pine forests of the Southwest as a case study in the consequences of fire exclusion. Slowly, grudgingly, the fire establishment admitted that its successes, ever more costly, were self-defeating.

These concerns were not restricted to the Southwest, but when, some twenty-five years ago, the clamor for reform shook the national fire establishment, the Southwest quickly responded. For the new era this meant that somehow, in some form, fire had to be retained in the landscape. Where it had disappeared, it had to be restored. Prescribed fire for fuel reduction, for conversion of woodlands to pasture, for wilderness ecology, and for improved wildlife habitat became commonplace if not extensive. Excepting the South, the Southwest practiced more broadcast burning than any other region. But it was not enough.

Nowhere has anyone reintroduced fire as fully as it has been removed. Restoration failed to keep pace with even annual requirements, much less to make inroads in reducing a century's backlog of burning; those fires did little more than make the minimum payment on an environmental credit card charged to its maximum limits. Often the fuel situation has worsened, particularly as logging exploded on the region's national forests in the 1980s and slash proliferated. It is easy to fund a dramatic firefight; tough to justify the quiet burning that, if done properly, does not become a public spectacle. It was difficult to restore fire without restoring the other conditions that had helped sustain it. The end could follow only from the means. Somehow the accumulated fuels had to be disposed of.

The new Southwest is a product of the old. Those fuel loads on public lands are a kind of environmental debt, like toxic dumps, that will take decades of determined action to clean up; it is not clear that either the resolve or the money is there to do it; the backlog is too great, and the requisite social consensus too elusive. The logging of large trees inflames environmentalists; the removal of small trees does not suit the economics of sawmill logging; the proposed conversion of saw mills to pulp factories worries tourists and summer-home residents. Besides, air quality considerations regulate the pattern of open burning. Woodsmoke must compete with industrial sources—smelters, coal-fired power plants, automobiles—for its share of the regional airshed. In 1975 and 1979 air-pollution alerts in metropolitan Phoenix resulted, in part, from an overload of broadcast burning on the Mogollon Rim.

And now exurban sprawl plasters the private lands within and around the public domain with houses, adding all the problems of a suburban environment but with few of its correctives. It increases the complexity (and expense) of fire protection without improving the prospects for fuel treatment, broadcast burning, or fire services. The developer is replacing the rancher, and summer homes and trailer parks and tourists, the throngs of sheep and cattle that previously penetrated every meadow and forest paddock. The outcome promises to be similar. The fuel situation worsens, without a corresponding change in ignition.

Like shots fired in the dark, nature's analogue to drive-by shootings, sooner or later lightning will hit the right combination of fuel, wind, and terrain with perhaps fatal effects, particularly when lightning discharges, as it does here, with the scatter of a shotgun blast. Increasingly it appears that wildfire is the only legally and politically acceptable form of burning.

As drought seized the region in the late 1980s, wildfires increased. The Dude Creek fire was its climax, and its prophecy.

The character of southwestern fire reflects the changing character of its human occupation. The classic Southwest blaze is a trying fire that exposes and assays, sometimes in dramatic fashion, the relationship between the natural landscape and the humans who live on it. It is difficult to reconstruct the impact of early humans, whose firesticks coincided with the colossal climatic fluctuations that ended the last ice age. Probably the Southwest featured a diffuse geography of burning in which firestick and grass created a regime that always simmered but rarely boiled, the intensity of fires following the tidal surges of rain and drought, the human firebrands smoothing out the jerky rhythms of lightning and wind, the very ubiquity of the burning helping to dampen catastrophic eruptions.

It is easier to document the dramatic alterations that have accompanied European settlement. The landscape mosaic became coarser, less able to absorb chronic disturbances. Fire vanished from some sites—flooded into irrigable fields, paved into cityscapes, eaten away by livestock, or swatted out by determined fire crews. In other sites, fire receded temporarily, only to return in altered but reinvigorated form. Elsewhere it was kept in check only through ever-increasing expenditures in fire suppression. The slow growth rates in the semiarid Southwest bought time; decades might pass before the consequences of fire exclusion became apparent or irreversible.

Now, that interdependence is again shifting. Whatever climatic changes may occur, human-inspired change is outstripping them. Overall, human activity is increasing the total number of fires even as it shrinks to a razor's edge the border between a controlled fire and a wildfire. Smaller numbers of fires break free, but these rage over larger areas and with greater ferocity. Like the interest on a compounding debt, wildfire threatens to claim an ever larger proportion of the region's fire economy. The rural landscape that had once helped buffer the urban and the wild continues to shrivel; in its place, houses insinuate into every nook and cranny of private land. There is little slack, small margin for error. The gradient between the wild and the urban steepens, building like an electrical charge. Eventually it will arc.

In Southern California, the intermix fire is a familiar morality play, almost a distinctive art form. Certainly the environment is built to burn. Clearly, the construction of expensive wood-shingled houses in mountains bristling with decadent chaparral and pummeled by Santa Ana winds is an act of hubris. But the worst fires are identified, rightly or wrongly, with arson. This transforms an environmental dilemma into a simple parable of human madness or malice.

It is more difficult to interpret fire in the Southwest, where lightning, not humans, normally supplies ignition. The relationships are more complex and balanced, the ironies more subtle, not readily decoded into simple dialectics. It is not clear, for example, whether the Dude Creek fire was a freak event, the fiery manifestation of a record-shattering heat wave, or the calling card for a new era in which, regardless of technological investments, the pressure of human population and the legacy of suppressed fires will combine to make a truly ungovernable fire regime. Was it the old amalgam, merely intensified? Or is it a new compound, volatile as nitroglycerin, ready to explode at the first stumble? It seems to be both.

Fire belongs in the mountainous Southwest, and unless the peaks flatten, the monsoon evaporates, the seasons homogenize, or the biota vanish, those fires will continue. They ought to. The issue is how to relate to fire—how to keep it from destroying people and how to keep people from transforming flame into a destroyer. Otherwise the border between the human and the natural will grind with greater and greater force, and out of that friction will come fire that no one wants and no one can control. The summer beacon will become a pyre on the mountain.

MOP-UP

Consumed by Either Fire or Fire

Yes! I know from whence I spring,
Never sated, like a fire.
Glowing, I myself consume.
All I seize and touch makes light,
All I leave behind me ashes,
Surely, flame is what I am.
—FRIEDRICH NIETZSCHE, *ECCE HOMO*

We only live, only suspire
Consumed by either fire or fire.
—T. S. ELIOT, *FOUR QUARTETS*

In a universe informed by fire, fire becomes a universal tool. To the prehominid fires of the earth, humans have added, subtracted, redistributed, and rearranged. Human societies have inserted fire into every conceivable place for every conceivable purpose, and they have done so for so long and so pervasively that it is impossible to disentangle fire from either human life or the biosphere. The alliance between hominids and fire is ancient, apparently dating from the time of *Homo erectus,* a part of our biological inheritance.

The era of anthropogenic fire, however, was neither inevitable nor inextinguishable. It began, and it may well end. One useful way to imagine that history is through a series of competitions. While combustion in various forms is reasonably well understood, the specific character of these competitions is not.

There was, initially, the competition with lightning. Burned areas differ from unburned, but not all burning yields identical effects. Anthropogenic fire practices, both ignition and suppression, alter the timing, frequency, size, intensity, and seasonality of fire regimes; and it is to these regimes, not to generic fire, that ecosystems adapt. Lightning and humans compete over the source materials, the biomass that serves as fuels. What one burns the other cannot. The classic index of environmental health in this combustion calculus is the soil, especially its organic component.

At the other end of this historical continuum, there is the growing competition between anthropogenic fire and fossil-fuel combustion. This affects fuels indirectly, resulting in changing agricultural and land-use practices, fuelwood gathering, a redefinition of natural resources, a global economy driven by trade, combustion-powered machinery, demographic revolutions, and so on. But the real competition between fire and furnace is over the available sink, the amount of emissions that the atmosphere can absorb without major dislocations. The atmosphere stirs local burning into a common cauldron of global combustion in ways that were not true before. A burned forest in Indonesia finds itself linked with automobiles in California, a firestick-kindled savanna in West Africa with coal-fired steel mills in the Ruhr. Airshed has replaced soil as the ultimate index of environmental health.

These two borders are porous, of course. In only a few places and for limited times have humans utterly replaced lightning (or other natural ignition sources), and in recent decades lightning has reasserted itself as a significant source of wildland burning. Likewise, fossil-fuel combustion has not abolished biomass burning, and in fact cannot do so. Again, changes in land use catalyzed by industrialization have in many places encouraged a proliferation of biomass, which in turn has led to renewed burning. And inevitably there is accidental fire. Even so, the advent of fire-wielding hominids does mark a boundary that did not exist before, and so does the spread of fossil-fuel combustion. That those boundaries are subject to incessant negotiation does not change the fact that they exist. The sea advances and retreats twice daily and rises and falls with the tides of climates but no one would deny the existence of a shoreline.

Between these mobile borders the history of fire can be understood as a competition between anthropogenic fire of various kinds. In some instances this involves literally a change in fire technologies, or the outright dislocation of one set of fire practices by another through the migration

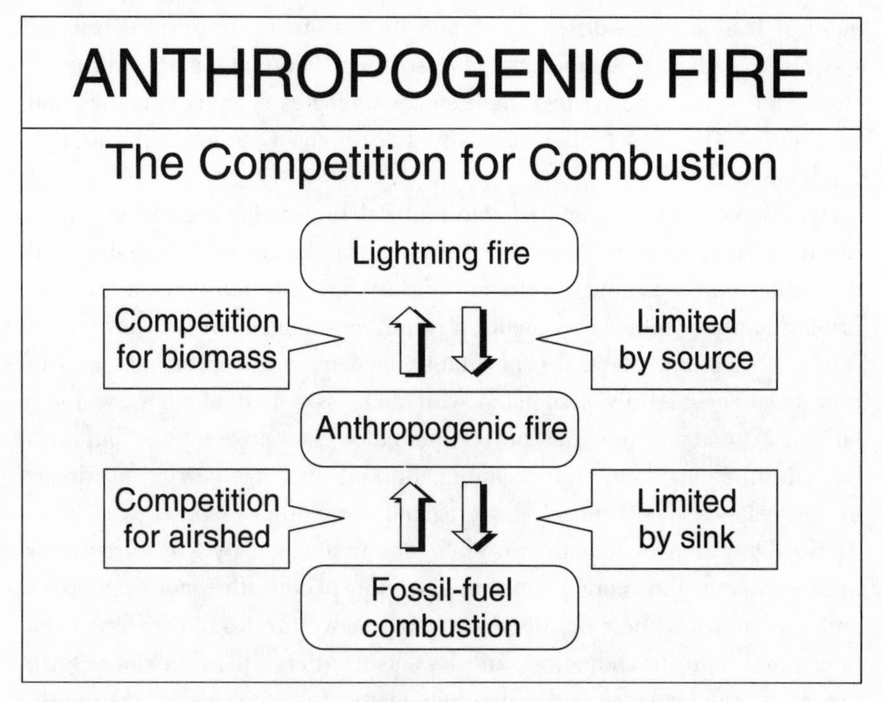

The competition for combustion.

and dislocation of peoples. In most cases, however, fire serves as an enabling device for a variety of technologies, and the competition is not between the torch and the match directly but among fire practices made possible by the interaction of fire with other practices—the competition between fire hunting and fire-assisted herding, for example. The fire regimes that result reflect the nature of these interactions in the form of exchanges, mergers, or extinctions.

Fire restructured the relationship between humans and their world. It furnished light and heat. It made possible a social life after dark; redefined social roles; warmed against the cold; demanded shelter and sustenance; served as communication media; supplemented ax, knife, and drill; and allowed cooking, which revolutionized diets and food gathering. Fire assisted almost all branches of technology. The tools for re-creating fire more or less at will—drills, saws, and flints—obviously derived from or coevolved with the technologies for striking, scraping, and drilling in stone and bone. Cooking inspired new technologies such as ceramics and metallurgy. The preservation of fire was an intensely practical as well as symbolic act.

Of course, fire also entered into cultural life, reshaping the cognitive world of humans as fully as their physical landscape. Fire worship and divination were a primitive religion; trial by fire, a primitive legal system; fire-based philosophies and myths, a primitive science and literature. As a source of heat and light, fire inevitably accompanied ceremonies, and in time became indelibly associated with them as a part of their symbolic milieu. Attendant fires not only illuminated the proceedings but also helped enlighten them with meaning. Burned offerings carried sacrifices to the heavens. The trying fire segregated dross from essence.

No dimension of human existence was untouched by fire, directly or indirectly, and the more remote in time the people, the more pervasive and apparent was their fire dependence. Remove fire from a society, even today, and both its technology and its social order will lie in ruins. Strip fire away from language and you reduce many of its vital metaphors to ash. It comes as no surprise to learn that—from the Aztecs to the Stoics, from the Christian Apocalypse to the Nordic Ragnarok—the myth of a world-beginning or world-consuming fire is nearly universal. The dominion of humans did begin with fire. And it may well end with it.

Fire's influence on the environment, however, extended beyond its valued service to humans, an aide-de-camp to wandering hominids. It was applied directly to the landscape. And it was this capacity that defined humanity's special ecological niche, that made fire something more than a surrogate for talons, fangs, fleetness, or muscles. Anthropogenic fire endowed whole ecosystems, not merely a species.

Of course, anthropogenic fire built upon preexisting associations. The human torch activated adaptations already resident in the biota; humanity's fire practices mimicked nature's own fire drives and slash-and-burn cycling by windstorm and lightning; and human-kindled fire rarely produced results through its own, isolated actions. Instead anthropogenic fire reshaped the structure and composition of landscapes, recalibrated their dynamics, reset their timings of growth and decay; fire accelerated, catalyzed, animated, leveraged; it combined with other processes to multiply their compounded effects beyond the sum of their individual impacts, to choreograph new rhythms and steps for the partners in this biotic ballet. Fire was a remarkably intricate and pervasive enabling device, without which other technologies or practices were often incompetent; there could be no slashing, for example, without the capacity for burning, no hunting without the means by which to maintain the habitat. Similarly, humans' ability to manipulate fuels redesigned the environment within which fires (either theirs or nature's) had to operate.

Thus fire interacted with the new flora and fauna that migrating humans introduced, with livestock in search of browse, with ax and sword as weapons of conquest, with plow and seed, and with humans' understanding of what they saw and how they should behave. It passed into the natural environment by means of the cognitive and moral world of humans. Knowledge (or error) acquired in one place could be transferred to another. Fire ecology had to incorporate the pathways of human institutions and knowledge as fully as biogeochemical cycles of carbon and sulfur. Patterns of trade and systems of belief became as fully a part of fire ecology as nitrogen, soot, and seed banks.

Aboriginal Fire

Anthropogenic fire made the world more habitable for those who held the torch. Aboriginal fires kept corridors of travel open; they assisted hunting, both by drives and by controlling the extent and timing of browse; they helped cultivate, after a fashion, many indigenous plants and simplified their harvest; not least, anthropogenic burning shielded human societies from wildfire by laying down controlled fields of fire around habitations in regions prone to natural ignitions. Fire was both cause and effect for the fact that humans preferentially lived where burning was possible and shunned unburned regions as uninhabitable, that the nomadism of hunting-and-gathering societies was intimately interdependent with a cycle of growth and regrowth that was itself contingent on a cycle of burning.

Fire remade the land into usable forms, much as cooking helped re-

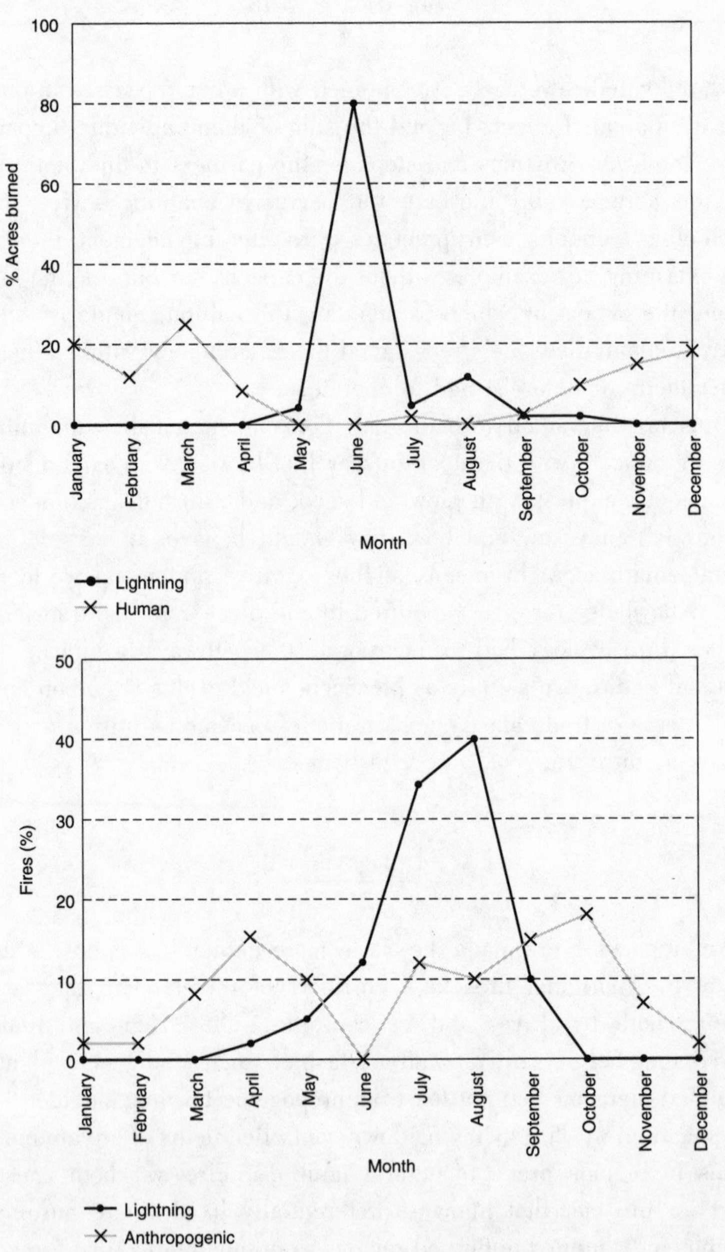

Anthropogenic fire regimes: adding fire. Two examples of controlled burning competing with lightning to define fire seasons. (top) Everglades National Park. Anthropogenic fire does not include arson or accident. (bottom) Northern Great Plains. Anthropogenic fire includes historically recorded fires by explorers and travelers.

work an intractable environment into food and as the forge refashioned rock into metals. When John Smith asked a Manahoac Indian what lay beyond the mountains, he was told "the Sunne"; his informant could say nothing more because "the woods were not burnt," and without fire both travel and knowledge were impossible. John DeBrahm noted also the purgative effect of burning, the purifying passage of humanity's ring of fire. "The fire of the burning old Grass, Leaves, and Underwoods consumes a Number of Serpents, Lizards, Scorpions, Spiders and their Eggs, as also Bucks, Ticks, Petiles, and Muskotoes, and other Vermins, and Insects in General very offensive, and some very poisonous, whose Increase would, without this Expedient, cover the Land, and make America disinhabitable." To a remarkable extent humans were able, through fire, to shape wholesale the environments in which they lived, to render that land accessible.

As an instrument of biotic forcing, fire was both subtle and pervasive. It promoted favored grasses, forbs, tubers, and fruits such as wild rice, sunflowers, camas, bracken, cassava, and blueberries; it helped harden the sticks that dug them up; it stimulated the reeds that, woven into baskets, carried the harvest; it cooked the gatherings, leached them of toxins, boiled them down into oils or sap, dried them for storage; it yielded the light by which the crop was discussed and celebrated. Fire helped gather chestnuts in the Appalachians, acorns in California, and mesquite beans in the Southwest. It drove off insects; in certain seasons humans lived in smoke, as reluctant to leave its sheltering cloud as to walk away from a campfire into a moonless night.

Perhaps the most spectacular practices involved hunting. It may be said that any creature that *could* be hunted by fire *was*. Torches assisted evening hunts, and made fishing at night notoriously productive. Smoke flushed out bears from dens, sables from hollowed firs, possums from termite-cored eucalypts. Fire drives were practiced for springboks in southern Africa, elephants in Sudan, turtles in Venezuela, rheas in Patagonia, kangaroos and maalas in Australia, boars in Transbaikalia. In North America fire hunting targeted bison, deer, and antelope; aboriginal Alaskans used it against moose and muskrat, Yuman Indians for wood rats, Californians for rabbits, Great Basin tribes for grasshoppers, Texans for lizards. Scandinavians used torches and bonfires to hunt bear, and smoke to attract reindeer; Siberians, for moose; Guam hunters, for that island's imported ungulates; Bahamians, for the gathering of land crabs.

But apart from the hunt itself, fire controlled grass and browse, recalibrating the calendar of renewal and the rich flush of nutrients that the springtime brought. Applied correctly, fire could inaugurate that first growth, or stimulate a second, late flush in the autumn. Regardless, the pattern of burning dictated the pattern of feeding, which is to say the annual migration of grazers and browsers and, of course, the hunters that followed them. Fire hunting and fire herding involved enticing as well as driving. Select sites could even be baited with the smoke that meant relief from flies or with the fresh grasses recharged by burning. Apaches set such traps for mule deer; Lapps, for reindeer; contemporary African poachers, for rhinos. Boreal tribes burned along traplines to keep them open and encourage foods attractive to the rodents upon which furbearers thrived. Fishing tribes in East Borneo burned the drifting sods of *kumpai* grasslands in dry years, transforming shady forests without fish into open lakes with fish.

This fire-mediated relationship was profoundly reciprocal. If humans used flame to promote their foodstocks, so wildlife often depended on anthropogenic burning to fertilize and ready the landscape and stimulate the fodder that they, the indigenous fauna, required in order to thrive. The land reflected this symbiosis, reshaped into fire-sculpted hunting grounds both large and small. Where fire could burn freely—where winds, terrain, and biota favored routine, expansive fires—great steppes, campos, llanos, prairies, savannas, cerrados, and grassy veld could result.

What Surveyor-General Major Thomas Mitchell said of 1830s Australia could stand as an epigraph for all those environments aboriginal peoples have fashioned through flame:

> Fire, grass, and kangaroos, and human inhabitants, seem all dependent on each other for existence in Australia; for any one of these being wanting, the others could no longer continue. Fire is necessary to burn the grass, and form those open forests, in which we find the large forest-kangaroo; the native applies that fire to the grass at certain seasons, in order that a young green crop may

> subsequently spring up, and so attract and enable him to kill or take the kangaroo with nets. In summer, the burning of long grass also discloses vermin, birds' nests, etc., on which the females and children, who chiefly burn the grass, feed.

The suppression of Aboriginal burning, coupled with the introduction of cattle, set in motion "such extensive changes in Australia as never entered into the contemplation of the local authorities."

And that—the realization of what the removal of aboriginal burning had done—is perhaps the most compelling testimony to its lost power. With fire removed, grasslands everywhere have receded (with the problematic exception of Brazil). What prairies have not been plowed under have suffered from spontaneous reforestation. Woodlands have marched south across the Canadian prairie at the rate of almost one mile a year in this century. Open forests have thickened, closed canopies, and often shielded a tangle of understory. Biodiversity has diminished.

Perhaps the finest expression of what fire can mean to aboriginal peoples comes from Alfred Radcliff-Brown's study of the Andaman Islanders. Fire, he writes,

> may be said to be the one object on which the society most of all depends for its well-being. It provides warmth on cold nights; it is the means whereby they prepare their food, for they eat nothing raw save a few fruits; it is a possession that has to be constantly guarded, for they have no means of producing it [not true], and must therefore take care to keep it always alight; it is the first thing they think of carrying with them when they go on a journey by land or sea; it is the centre around which the social life moves, the family hearth being the centre of the family life, while the communal cooking place is the centre round which the men often gather after the day's hunting is over. To the mind of the Andaman Islander, therefore, the social life of which his own life is a fragment, the social well-being which is the source of his own happiness, depend upon the possession of fire, without which the society could not exist. In this way it comes about that his depen-

dence on the society appears in his consciousness as a sense of dependence upon fire and a belief that it possesses power to protect him from dangers of all kinds.

The belief in the protective power of fire is very strong. A man would never move even a few yards out of camp at night without a fire-stick. More than any other object fire is believed to keep away the spirits that cause disease and death.

Agricultural Fire

The Neolithic revolution adumbrated these practices with others. Of course, fire alone could not create agriculture, but it is no accident that agriculture originated in fire-susceptible environments and reworked fire-adapted grasses into cereals at its biotic forges. Outside of flood-recharged alluvial plains, early agriculture was impossible without burning. At a minimum fire was mandatory for the expansion of farming and herding beyond their special environments of origin. Agriculturists needed fire to convert and catalyze, to impose an alien flora and fauna, to wrest a stubborn biome into new shapes. The metamorphosis was sometimes easy, for small adjustments could reconfigure fire practices suitable for hunting into those to serve herding, or to shift from the harvest of a fire-cultivated native flora to a fire-catalyzed alien one. But often the transformation was messy and complex.

The really revolutionary changes occurred not from fire per se but fire in conjunction with other practices. The violence of farming was expressed with ax and torch, the fire and sword of biological imperialism. Virtually all farming involved rotation and fire. The differences between shifting and sedentary cultivation blur, much as the border between gathering and gardening does. In one, a farmed site is rotated through the biome; in the other, a contrived biome is, in effect, rotated through a fixed site. Both require fire, or did until recent developments began to manufacture fertilizer and power from fossil fuels. Where fallow was lacking, farmers would transport wood, manure, peat, or other debris to the site and burn it as a surrogate.

The scope is breathtaking. Wandering farmers carried slash-and-burn agriculture into the most remote landscapes—Maoris in New Zealand; Melanesians in Fiji; American indigenes where maize or cassava were cultivated; Bantus across tropical and subtropical Africa; jhum-practicing tribes throughout the Indian subcontinent and Southeast Asia; and, of course, Europeans who practiced it across their varied frontiers, from the Finns in Karelia to the Norse in Iceland and Greenland, from Russians in the taiga to overseas colonists in the Americas, Africa, and Australia. Swidden extended to grasslands in Africa, heath and moor in northern Europe, boreal and tropical forests—any landscape that demanded a long fallow. Part of the agricultural revolution that preceded Europe's industrial revolution involved the conversion of "wasteland," largely organic soils, through a regimen of cutting and burning. Even in northern Europe swidden persisted into recent times—in Germany in the late 1890s; Belgium in 1908; France in the mid-1920s; Sweden, into the 1930s; and Finland, who, burdened with war refugees, revived it into the 1950s.

Meanwhile flocks of domesticated livestock advanced like shock troops. They reclaimed range previously fired for the hunting of wildlife; they frequently forced herders to keep open fields, first cleared for farming, as pasture, again through routine burning; they forced the creation of new grazing lands or degraded the old ones. In humid environments, from Australia's northern tropics to Siberia's taiga, the introduction of domesticated animals typically inspired more fire, with fire used as a flaming ax necessary to slash back the encroaching woods. In more arid grasslands, from Morocco to the American West, it commonly brought a reduction in burning as livestock cropped fuels that would otherwise feed flames. In those complex landscapes subject to transhumance, seasonal burning was normal in Iberia, Italy, France, Greece, the Balkans, Britain, and Algeria. In mixed economies fields once cleared for crops might, through subsequent burning, be kept in grass for pasture or hunting grounds. Inedible crop residues, such as wheat stubble, were routinely burned.

Eventually fire was itself domesticated no less than land, flora, and fauna. Wildland fire became agricultural fire. Field burning obeyed a new calendar, operated at reduced intensities, and altered the frequency of broadcast fire. For the most part, humans dictated these parameters. Where fire had once reflected the subtlety and complexity of the natural world, it increasingly assumed the regimen and personality of human society, responsive to the dynamics of the human mind, or what a flawed human will could impose of itself on nature.

Encountered Fire

Within the last half millennium two events have rewritten the history and geography of fire. The expansion of Europe begun with the Renaissance voyages of discovery set in motion a colossal mixing of the world's flora, fauna, and peoples. Old orders disintegrated, and new ones continue to emerge that have yielded a global economy, a global ecology, and a global scholarship. But even while these events were proceeding, imperial Europe became industrial Europe, and the pressures for global change accelerated. Anthropogenic fire was, as always, catalyst, cause, and consequence of these processes.

Some of these outcomes were universal to any export of hominid burning. The coming of Europeans to uninhabited islands such as Madeira or St. Helena was little different from the first-contact burning induced by the Maori in New Zealand. Some were specific to European exploration. Through its ships Europe began to join together what the earth had long ago sundered. The greatest divide was that between north and south, Laurasia and Gondwana. The transfer of Finns to North America, for example, proceeded naturally. The transfer of northern peoples, plants and animals, and fire practices to South America, Africa, Asia, and Australia proved more perplexing. In the Americas they largely replaced the indigenous, or created a vigorous hybrid. In Africa they stalemated; in Asia they were largely repelled. From the perspective of fire history, most of these encounters had ample precedents. But beyond the geographic linkages it forged, the expansion of Europe cascaded into an industrial revolution that made what followed profoundly different.

Access to a global market exposed large hinterlands to revolutionary reforms. Inevitably, economic exchange brought ecological change. Both by accident and deliberation, the world's flora and fauna became hopelessly intermingled and selectively exterminated, usually with the help of or at least in the presence of fire. Livestock from the Old World went to the New; cultigens from the New World went to the Old; plants and animals from both invaded Australia; Africans were forcibly shipped to the Americas, British convicts to Australia, Russian exiles to Siberia, and an unstable alloy of hope and despair inspired a century of emigration that reduced Europe's population in roughly the same proportion as the Black Death. Weeds, vermin, diseases, and insects, no less than cereals, cattle, citrus, and medicines, crossed over long-separated biotas with results that were sometimes productive, usually unexpected, and often catastrophic.

The new geography of global fire followed the evolving geography of European expansion. In some places burning increased to epidemic proportions as immigrant fires mingled with indigenous ones, all now cut loose from traditional moorings. In others, the fires flared, like a Bunsen burner speeding a critical reaction, then ceased, the experiment completed. In still others, they vanished into field and ceremony, little more than vestigial symbols. The interactions—the permutations in how fire, flora, and fauna could interact—were infinite, and no single formula can encompass all the outcomes. But in general the effect has been to reduce the amount of open burning. The earth has far less free-burning fire now than when Columbus sailed.

This observation is counterintuitive, or more properly it runs against the grain of Western colonial mythology, populated by Noble Savages and Virgin Forests, which requires that a discovered world of innocence be ravished by a decadent and cynical Europe. Thus the fire policy of America's National Park Service seeks to restore the state of nature that existed prior to First Contact—a definition that includes the native peoples who burned sequoia groves annually but not the German immigrants who swatted those fires out with pine boughs. It seeks to preserve lightning as an ignition source but to eliminate the anthropogenic fires with which these ecosystems have coevolved for thousands of years. It seeks to fashion a cultural ideal out of natural materials, and is every bit as much an artifact of twentieth-century American society as Michelangelo's statues are of Renaissance Italy or the Gothic arches of Mont Saint Michel are of medieval France.

Certainly European contact led to environmental change, very often degradation of one kind or another; in locales where landclearing was involved, contact often inspired horrific fires. The transitional period when fire practices and fire regimes mixed was metastable, and prone to violent reactions. But this scenario fails to recognize the geographic extent and longevity of anthropogenic fire. It sees the violent flash of European gunpowder but not the already smoldering land behind it. But when the smoke cleared—as it soon did—there was typically less fire after settlement or colonial rule than before. This appears to be universally true, not least of all for Europe. No one knows with the least confidence what the quantitative history of fire looks like, but it is not unreasonable to suggest that contemporary levels of burning are 20–25 percent of pre-Columbian levels.

As often as not the expulsion of fire was itself frequently a cause of environmental degradation and social disintegration. Fire removal attacked the habitat of fire-reliant humans as it did that of such fire-adapted organisms as longleaf pine, eucalypts, fynbos, kangaroos, geotrophic orchids, and rhinos. The more universal fire was as a tool—the fewer the alternative technologies that were available to a people—the more devastating was its removal. With the environment changed, at least partly through a new regimen of fire, a recovery of habitat was difficult or impossible. Bison could not return to tallgrass prairies that had reverted to woods. Springbok fled veld overrun with scrub. Grouse sickened on moors decadent with unburned ling.

The reasons for Europe's pyrophobia are several. For one, heartland Europe had slowly but steadily strangled fire from its lands. Peasant agriculture restricted, then practically expelled fire, as usage of every blade and stick intensified. Europe itself came to resemble a kind of fire, burned out in the center, flaming only along its perimeter. When it sought to export its agriculture, this model was the one it promulgated, such that after a burst of fiery landclearing agriculture became sedentary and land usage more obsessive. The process might take decades, if not a century or two, but that meant only that the neo-Europes were recapitulating the evolutionary history of Old Europe, which had demanded several millennia to tame its feral lands. What happened with wheat and cows happened also to fire as it joined other technologies on the European arks that rafted around the world ocean. But having transported the vestal fire to new lands, officials hastened to put it back in its ceremonial hearth as quickly as possible.

For another, Enlightenment Europe had created a species of engineering, forestry, that assumed special responsibility for the management of free-burning fire. Silviculture was a graft on to the great rootstock of European agriculture, and it correspondingly condemned burning of every sort. According to European agronomists, fire not only killed trees, it also destroyed the humus, upon which all life depended. Without that spongy cover soil erosion was inevitable, and environmental decay irreversible. Far better to exploit biological agents such as sheep or cattle, or to fertilize with compost and dung rather than with ash, or to sponsor labor-intensive surrogates such as weeding—anything but fire. Almost uniquely fire defined the border between "primitive" and "rational" agriculture.

Besides, fire threatened fixed property, and often the social relationships of rigidly ordered societies. Broadcast fire encouraged varieties of nomadism—the seasonal cycling of pastoralists, the long-fallow hegiras of swidden farmers, a mobility of population that made political control and taxation difficult. The ideal of the garden that prevailed in central Europe demanded that every person, like every plant, have its assigned place. Finnish and Russian swiddeners fled king and tsar by burning into the boreal forest; American pioneers and Boer pastoralists avoided tax-collector and company edict by carrying their fire-powered economies into continental interiors; Greek, Corsican, Sardinian, and other Mediterranean pastoralists ranged as freely as their flocks, defiant of political authority and confident in the power of fire to intimidate settled communities and harass agents of the state. Control over fire meant control over how people lived.

Forestry absorbed the practice doctrines, and in its shock encounter with other, oft-fired biotas, it hammered these precepts into a catechism of fire exclusion. Some wildfires would inevitably occur from the striking of native flint and European steel, from lightning, from arson, from accident, as, for example, railroad-powered logging sent its iron tentacles everywhere; some broadcast burning might be unavoidable in the early years of institution-building, an expedient compromise until surer control was possible; but the ultimate ambition was a forest without fire, an orchard of saw wood and pulp. As soon as it was politically and technically feasible, foresters instigated fire-control measures. As often as not, fire suppression was one of the most powerful means of controlling indigenes.

This exercise in technology transfer was easiest in North America, and messiest in the southern hemisphere where biotas were dramatically different and where, in places, indigenous peoples and livestock outnumbered European imports. There traditional burning persisted in defiance of European desires; often, in such circumstances, colonists adopted native fire practices, or hybridized with them in mutual hostility to the edicts of colonial administrators and the theories of European intellectuals. Fire policy became an expression of colonial rule, and firefighters a species of frontiersman, especially where land was reserved from folk access in the form of forests and parks. In response, natives burned illicitly, at once a protest and an attempt to restore traditional lands to customary purposes. Regardless of whether they retained political access to these reserved sites, unless they had fire they lacked biological access to the potential resources these lands held. The character of the land changed, often irrevocably.

All this did not pass unnoticed. Typically a debate ensued, often formal and even published, that pitted European standards against local practices, fire control against fire use. The most dramatic confrontations flared in British and French colonies (or former colonies) where foresters nurtured in Franco-German traditions were most aggressive at imposing policies that aspired to fire exclusion. In North America fire control triumphed; in India, Australia, and South Africa, awkward compromises resulted that first denounced controlled burning, only to recant and ultimately absorb the practice into official doctrine. In time, controlled fire returned to North America as well, although in a language ("prescribed burning") that denied the legitimacy of its earlier incarnation ("light burning"). The same debate is currently under way in Brazil as that country seeks to keep fire where it enhances and to remove it from where it damages.

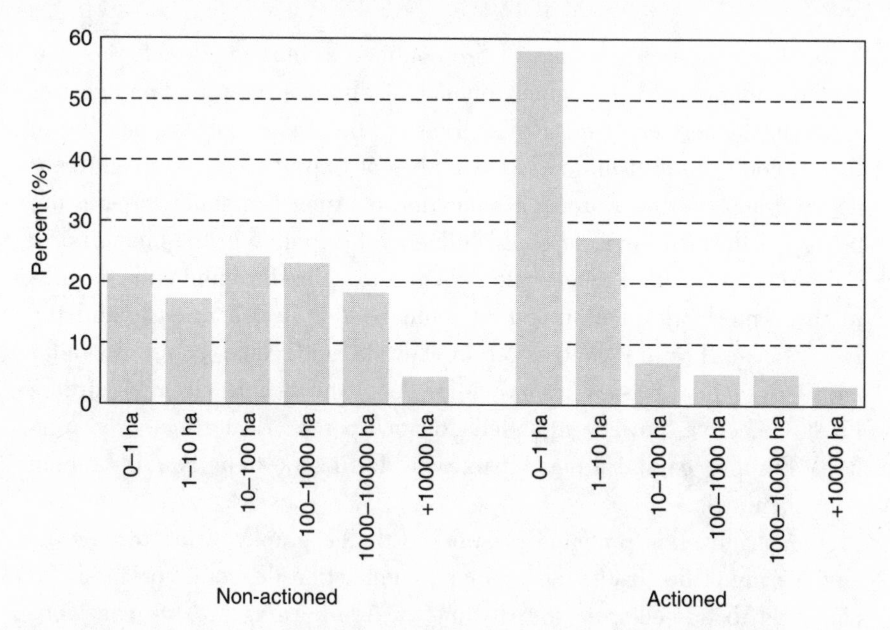

Anthropogenic fire regimes: removing fire. Effects of acting and not acting on fires in northern Ontario, 1976–1988. Fewer large fires occur, not only because firefighting suppresses first ignitions but because fires no longer linger on the landscape over lengthy seasons where they are subject to repeated runs as conditions permit.

Industrial Fire

Not least of all, industrial Europe began to sublimate the power of fire into machines. Controlled combustion steadily replaced controlled burning. Technology invented new devices to illuminate rooms, warm houses, bake bread, harden ceramics, shape metals, and the other myriad tasks fire had once performed. New sources of energy led to new sources of mechanical power and transportation, the pathways by which nutrients would be cycled in this revolutionary new model of nature's economy. Guano could be shipped from Pacific islands to fertilize European fields; clover could replace indigenous grasses unpalatable to Eurasian livestock; chemical herbicides and tractors could substitute for fallow burning to control weeds.

No longer was the ecology of fire confined to burned sites. It was now cycled and ramified throughout human institutions as well. Through foresters in the service of the British Empire, the impact of teak burning in Burma could be transmitted to the fynbos of Cape Colony. Through technology transfer—the European education of American students—the fire history of the Schwarzwald could influence fire policy in the chaparral of California. Through French colonialism and scientific publications, fires in the hinterlands of Montpellier could be felt in Madagascar and the Ivory Coast. Postfire succession in Nevada and Natal could be influenced by exotic grasses from Central Asia and acacias from Australia. Fire's selective power expanded to incorporate domesticated grazers from Europe, capital from Japan, and fertilizers from Saudi Arabian petrochemicals.

Industrialization profoundly reworked the geography of fire through its engineering of fire itself. Increasingly, combustion depended on fossil fuels, long abstracted from the rhythms of free-burning fire, its emissions outside the mechanics of ecological scavenging. Always in the past fire had depended on life, on the interaction with the oxygen and fuels that life generated, on the ecological fugue between fire and fuel played over eons of evolutionary trials. While this symbiosis had made fire possible (even necessary), it had also restricted its dominion. But industrial combustion burned without regard to the living environment. It literally stood outside the ancient ecology of fire—and outside the traditional social mores that had guided anthropogenic fire use.

But industrialization's impact extended further: its reach exceeded its grasp. It redefined what in nature was a resource and what was not, which land uses were appropriate and which were not, which regions were accessible and which weren't. It compelled a full-blown redefinition of nature, rank with new values and new perceptions, many derived from the rowdy efflorescence of a global scholarship, modern science, that accompanied and sought to explain European expansion. It set into motion a counter-reclamation that redefined the basis of European land use away from traditional agriculture. The revolution demanded not only new combustion technologies for furnace and forge but also new fire practices for field and forest.

Like planets orbiting binary suns, ecosystems now had to revolve

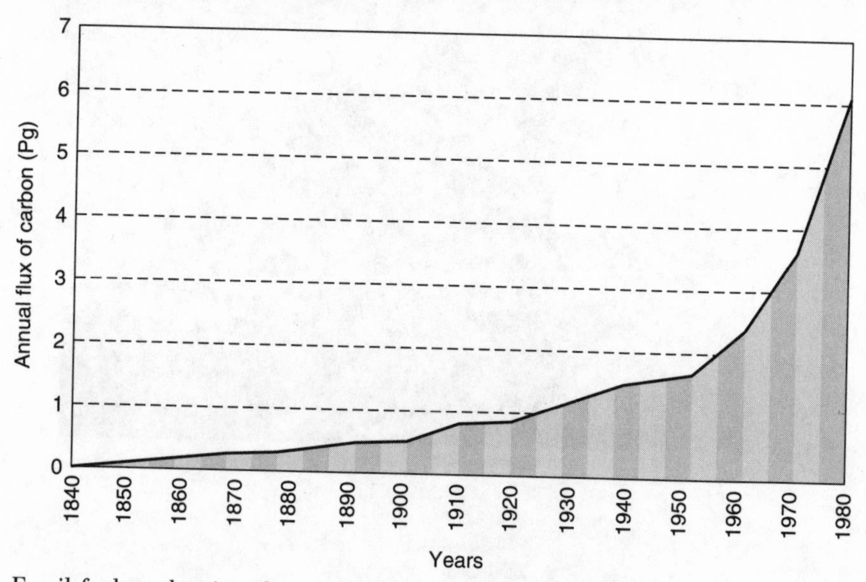

Fossil-fuel combustion: direct effects. Most of this annual flux of carbon is in the form of carbon dioxide, thus competing in the atmosphere with emissions from burning biomass.

around the gravitational field of industrial combustion as well as biomass burning. The interplay between the two became increasingly dense, inextricable, unpredictable, even contradictory, the ecological equivalent of the three-body problem in celestial physics. Industrialization could strip forests through logging and, equally, restore woodlands by abolishing the need for fuelwood. Industrial societies could subject the most remote site to exploitation, yet simultaneously propose special categories of wildland that were, by law, to remain immune to human consumption. For these new landscapes new fire practices had to be invented, behavior sanctioned by neither nature nor the precedent of human history. The upshot has been the most fundamental restructuring of anthropogenic fire and global fire regimes since the Neolithic.

One outcome of this extraordinary expansion was that Europe established itself as a standard and censor of planetary fire. The fire practices of its industrial and agricultural heartland—Germany, northern France, the Low Countries, England—were accepted either as the

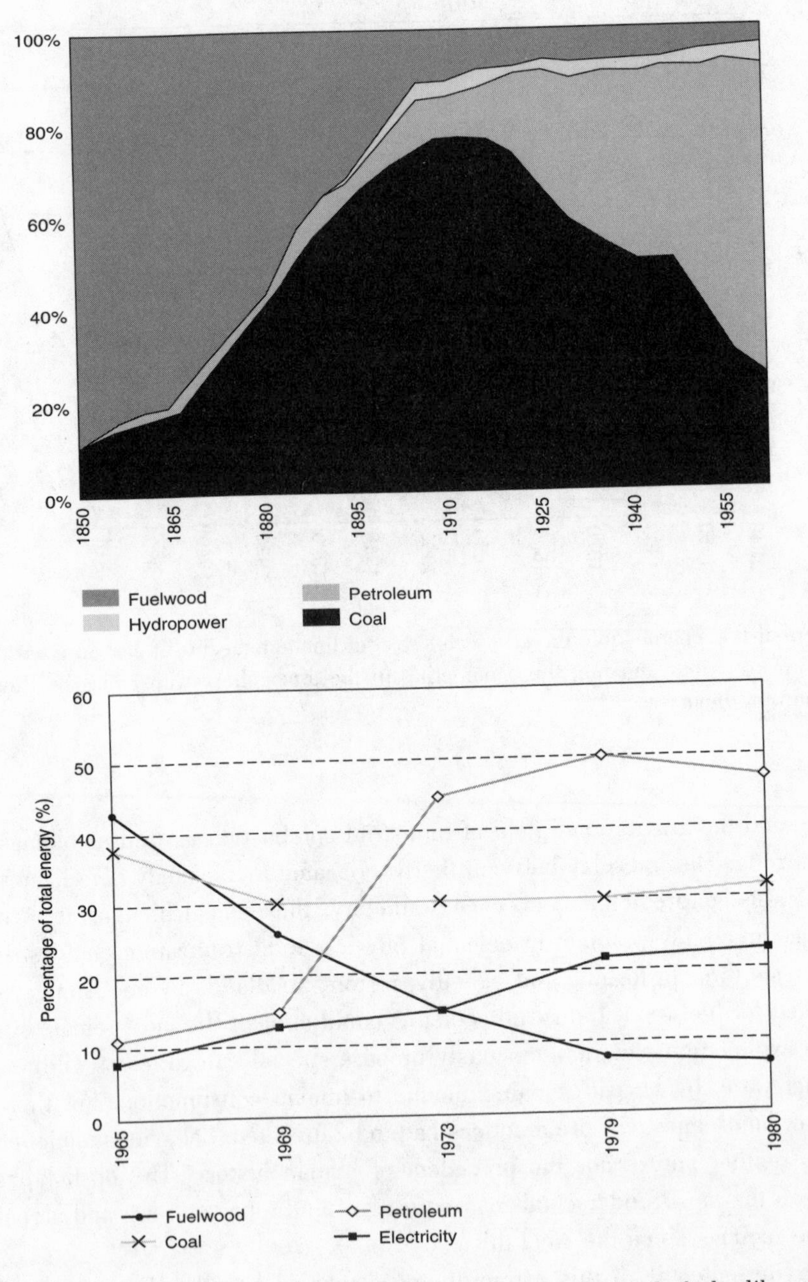

Fossil-fuel combustion: indirect effects. The shift to fossil fuels liberates fuelwood, which if not consumed or contained in some way becomes available for wildland burning. The examples here are from the United States (top) and the Republic of Korea (bottom).

norm or as the goal to which developing nations should aspire. Not incidentally these nations were themselves among the important colonial and scientific powers, and they, unlike Spain and Portugal, distrusted free-burning fire.

To a remarkable extent fire practices and norms have flowed one way—from Europe outward. The U.N.'s Food and Agriculture Organization (FAO), for example, has dispatched Germans, Finns, and French to Mauritania, Nepal, and Patagonia to advise on forestry and fire, but has not sponsored Senegalese and Filipinos to help reintroduce fire to Sweden and Austria. The trend continues, qualified by the emergence of North America as a global fire power, although until recent decades North Americans also accepted the anomalous European fire scene as normative. It is perhaps no accident that current alarms over fire as an instrument of global change—with its implicit condemnation of biomass burning—have emanated from Europe.

Global Fire

The contemporary geography of earth's fire is far from stable. Perhaps it never has been. Even where the rhythms of returning fires show some regularity, two idiographic tendencies have worked to upset any putative cycles. One is climate, the other humanity. It is among the alarming trends of contemporary times that, through fire, climate and humanity have begun to interact in new ways.

Under the impress of industrial combustion, the ancient dialectic between hearth and holocaust has skewed. In the past decade large wildfires have ravaged landscapes as diverse as Borneo and Canada, Australia and Manchuria. Fire has been implicated in world-ending scenarios, the fast burn of nuclear winter, the slow burn of a greenhouse summer, the epoch-ending extinctions of the fires along the Cretaceous-Tertiary border. The public imagery of fire has become both vivid and confused, with conflagrations applauded at Yellowstone National Park and denounced in Amazonia. Nearly everywhere, free-burning fire is identified with global environmental havoc, a torch ready to kindle the accumulated fuels of the Apocalypse. Again, fire will be both cause and consequence. Uncontrolled combustion may not only provoke global warming but also result from it, in the form of wildfire, as boreal and temperate forests become colossal tinderboxes.

The transition from flame to furnace has demanded novel indices by which to measure the new world order of fire. In particular, air has replaced humus as the ultimate yardstick of fire effects. The role soil served as an indicator of environmental health for agricultural societies, the atmosphere has assumed for industrial states. Combustion will be held to be bad or good depending on whether it aggravates or ameliorates the world's airsheds, and through them, planetary weather and climate. Fires for foraging and hunting must compare not only with fires for farming and herding but also with those that power turbines and automobiles; the byproducts of field and forest burning with those from the burning of coal and oil. Even so, Earth remains a fire planet, and humans remain fire creatures. Neither can forsake fire and be what they are.

It is clear that the dominion of fire is changing with more than usual speed, and that it is pursuing pathways not sanctioned by evolutionary trial. This new pyrogeography takes many forms, and is rewriting the registry of fire excesses and deficits. Where change is sudden, where new land use or misplaced fire practices introduce ignitions, upset fuels, or scramble the wild with the rural or urban, wildfires rage. While the particulars vary, the outcome has been a global surge in uncontrolled burning—in Siberia, Indonesia, Amazonia, the Mediterranean littoral, and Oakland, California. Elsewhere anthropogenic fire, once rare, has more or less permanently established itself like a naturalized weed. Still other landscapes—grasslands, brushfields, and forests—suffer from a deficiency of burning, a fire famine. After all, biodiversity can be lost as surely through fire exclusion as through fire excess. In the United States the best minds and most aggressive programs over the past two decades have sought to *restore* fire to wildlands.

But industrialization has gone further. Much as early hominids sought to replace the flame with the torch, and as early agriculturalists sought to substitute domesticated burning for wildfire, so modernized societies have striven to replace wildland combustion with industrial combustion. The furnace supersedes the hearth; the power of fire engines, the power of torches. The critical environments are mechanical, literally within machines, and those portions of the atmosphere and biosphere that directly exchange gases with them. Combustion is no longer necessarily even associated with flame. This has rendered the status of anthropogenic burning unclear.

Whether anthropogenic fire is increasing or not is indeterminable. But its relative contribution appears doomed to diminish. Lightning fire will reassert itself in the boreal forests, in the abandoned fields of the Mediterranean, and in nature preserves. Industrial combustion through fossil fuels continues to accelerate, with major escalations promised in the near future from India, China, and other rapidly developing nations. The consequences for anthropogenic fire are both direct and indirect. In some places fossil fuels can substitute for biomass burning or animal energy. Everywhere industrial combustion integrates once-unconnected landscapes to a global market which in turn profoundly influences the distribution of surface biomass and the fire practices that are applied to it.

These substitutions, however, have been incomplete and unsynchronized, and will likely remain so. Industrial combustion burns with savage indifference to fire ecology, without regard to time of day, season, climate, or biota. Because it depends on fossil fuels, there is no ecological feedback between fuel and fire, no biological linkage among combustion, nutrient cycles, and pathways of succession; the "fires" burn outside the parameters of natural ecosystems, often beyond the capabilities of organic scavenging and biochemical recapture. Instead of liberating nutrients, their byproducts may overload or poison the environment. For industrial combustion, unlike wildland varieties, ecosystems must be more than reshaped; they must be wholly reconstituted, complete with new pathways of energy, new cycles for biogeochemical compounds, new "mechanical" creatures, and, of course, new fire practices.

Those practices are far from established. Worse, they are challenging more traditional expressions—re-forming the domain of anthropogenic fire, subjecting earth, particularly its atmosphere, to a combustion load in excess of what it can likely absorb, all this while operating outside the folkways that have traditionally guided anthropogenic fire. They are rewriting the history of fire on earth, and redefining the special relationship between humans and fire. Increasingly humans are less keepers of the flame than custodians of the combustion chamber. That change, if it continues, has enormous ramifications for the natural world. The transition from torch to furnace has unsettled the fire geography of the planet, and this, in turn, has upset the moral geography of anthropogenic fire.

Over the millennia anthropogenic fire has mediated between human society and its natural environment. If it broke one regime, it joined another. Through a kind of ecological pyrolysis, it had dissolved certain relationships among species, yet through an equally compelling process of biotic welding it had fused new ones. If fire segregated humanity from the natural world, it also bonded humans to the living world through the fuels they shared and the fire-powered dynamics it made possible. What it granted in power, anthropogenic fire demanded in responsibility. But the new world order is breaking this legacy in fundamental ways. Industrial fire threatens to choke out agricultural fire, much as agricultural fire practices once weeded out aboriginal fire.

Anthropogenic fire may be lost altogether. Contemporary primitivism typically denies anthropogenic fire a role in the management of parks, wilderness, or natural reserves. As a projection of human agency, such fires, so it is argued, represent an unwarranted and unbalancing intervention into the natural order that these sites seek to preserve. Equally, the magnitude of industrial combustion is forcing biomass burning to compete with fossil-fuel burning for limited airsheds. Taken together these processes suggest that fire is unnecessary and undesirable. In fact, it is essential. Biodiversity will fail without an appropriate fire regime. Carbon sequestration and other ecological methods for modulating industrial emissions must also rely on anthropogenic fire. What must change is the context and its expressions.

The competition between furnace and flame has thrown into confusion the combustion calculus of the earth. How much combustion can the earth absorb? What is the relationship between fire and combustion? How much industrial combustion can be added, and how much wildland burning withdrawn, without ecological damage? It is not obvious that there is more fire on the earth now than in the past. In fact, a good case can be made that there is not enough fire, or rather that there is a maldistribution of fire—too much of the wrong kind of combustion in the wrong kind of places, too little of the right kind of burning in the right places or at the right times. So what is an appropriate standard for anthropogenic fire? Can guidelines be found in the earth's fire history, or must humanity confront an existential earth, silent as a sphinx—its ecosystems, climates, and fire creatures all intrinsically unstable? The longevity and pervasiveness of anthropogenic fire not only complicate the quest for a natural norm but may condemn it as metaphysically meaningless.

But norms of some kind are needed. Which fires are good and which are bad? What is the right proportion of fires? How much fire should be tolerated, or even promoted? Increasingly intellectuals are looking to the atmosphere to supply these standards, and to the threat of anthropogenic climate change as a means to furnish the moral and political arguments for reform. Air has replaced humus as the European measure of environmental health; the atmosphere is integrating at a global scale effects that had previously been considered local or regional.

It is not just that fire may be changing world climate, but that the climate of world opinion is compelling a change in our ancient relationship to fire. The case for anthropogenic burning has not found an advocacy equal to that against it, which is to say, for fossil-fuel combustion. Yet fire can countermand certain greenhouse effects, and may, in the future, be an essential thermostat for global climate. If free-burning fires release greenhouse gases, for example, they also deposit elemental carbon as a residue. The biota recapture the gases, and the carbon persists, virtually indestructible, in partial compensation for the exhumation of ancient hydrocarbons burned as fossil fuels. Aerosols and sulfur emissions reflect incident radiation, thus enhancing cooling; fire effluents often serve as nuclei for clouds, further altering the radiation balance. It could be argued, plausibly, that anthropogenic fire has retarded—or may be necessary to retard—the advent of a new ice age, which is surely closer to the climatic norm in which *Homo sapiens* has had to live than the climate of the past century. (Since the Pleistocene 80 percent of the earth's climate has been glacial, and much of the last millennium was dominated by the Little Ice Age.) The world may need more fire and less combustion. More grassland burning can lead to greater elemental carbon storage. Proper

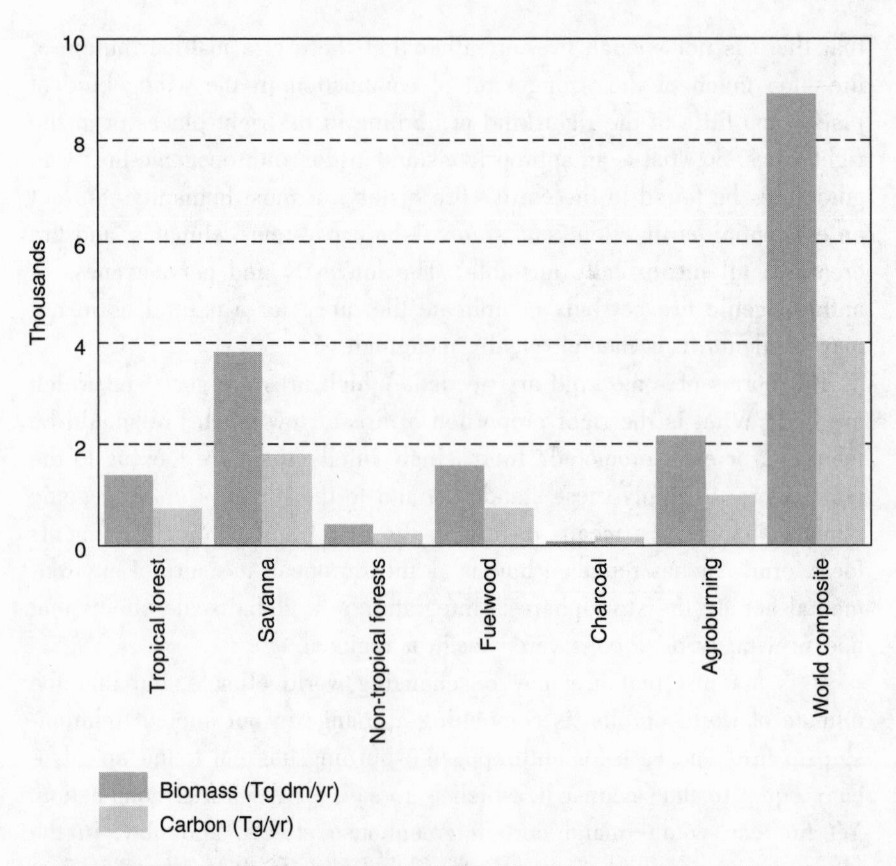

Global biomass burning, estimated for 1990. The African savannas dominate the planetary numbers, with agriculture, fuelwood, and tropical forest burning a strong complement. The low values for temperate and boreal forests reflect industrialization and fire suppression.

burning in forest and shrubland can improve long-term productivity and net carbon storage. Much of the natural world that preservationists seek to protect coevolved with anthropogenic fire and to remove that fire regime can be catastrophic; to replace anthropogenic fire with lightning fire alone does not restore a former, prelapsarian state but more likely fashions an ecosystem that has never before existed. But no iconography has emerged to argue the case for the legitimacy, utility, or necessity of anthropogenic fire; there is no counterimagery of fire as a global hearth, no Smokey the

Torch to advance the case for anthropogenic burning as benign stewardship of natural areas, no vision of fire as shield and filter.

Instead climate, like cancer, has become a universal touchstone to condemn unwanted environmental practices. In particular, Europe's air pollution has become so serious that the continent can now project that local deterioration across the globe, just as in centuries past it projected the reckless destruction of its soils. There are good causes for alarm over atmospheric abuse, but it would be tragic if global powers once again misapplied their ignorance and mistreated fire. The attempted suppression of fire by Europe's colonial powers was an ecological disaster, though one often camouflaged because it accompanied other, equally damaging and more visible practices. The extinction of indigenous burning was often a critical prelude to or catalyst for other extinctions. European fire is an anomaly, not a norm. However humans try to manipulate the earth's climate, fire management, not fire extinguishment, will be vital.

Today European-dominated policy stands to repeat these errors if, once again, it categorically seeks to expunge fire from the landscape. Carbon cannot be sequestered like bullion; biologic preserves are not a kind of Fort Knox for carbon. Living systems store that carbon, and those terrestrial biotas demand a fire tithe. That tithe can be given voluntarily or it will be extracted by force. There can be net changes in the earth's fire load, but to speak of eliminating burning is not only quixotic but dangerous. Eliminate fire and you can build up, for a while, carbon stocks, but at probable damage to the ecosystem upon the health of which the future regulation of carbon in the biosphere depends. Stockpile biomass carbon and you also stockpile fuel, the combustion equivalent of burying toxic waste. Cease controlled burning and, paradoxically, you may stoke ever larger conflagrations. Refuse to tend the domestic fire and the feral fire will return.

For millennia—on much of the earth, for hundreds of millennia—humans have directed those obligatory fires. They have paid the fire tithe on behalf of their resident ecosystems. But now the situation has blurred; the liberation of fossil fuels has changed the identity of *Homo sapiens* as a fire creature; new fire technologies have broken down the old order without yet installing a replacement. But the issues involve more than just atmospheric pollutants and human-governed combustion. With or without

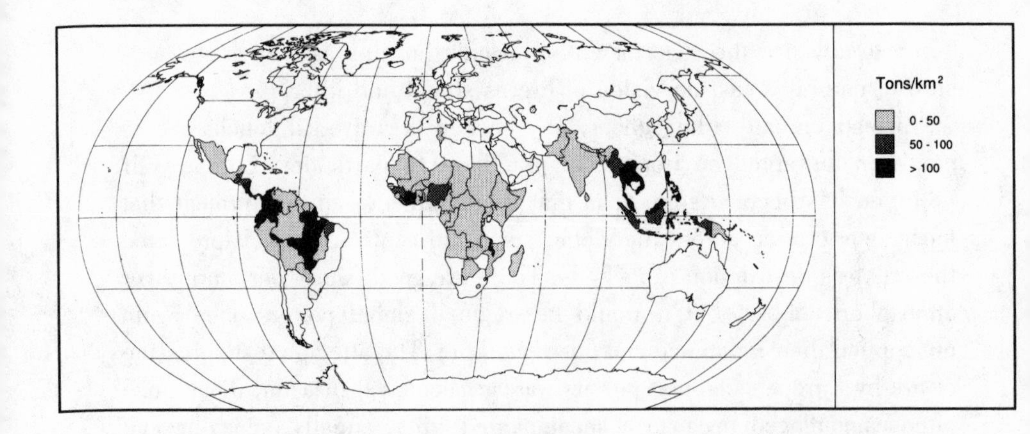

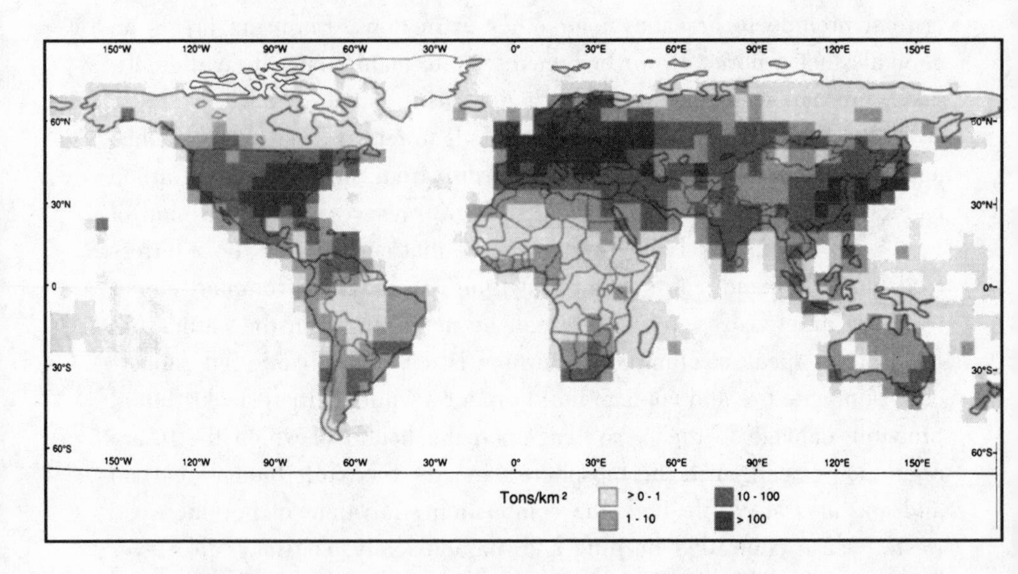

The global geography of combustion (1980): Top map shows the net flux of carbon from terrestrial ecosystems, thus compounding both deforestation and biomass burning. Bottom map displays carbon emissions from the combustion of fossil fuels.

humans, with or without industrial combustion, the earth will burn: fire is not exclusive to humans. Rather it joins us to a complicated biosphere for which we, as unique fire keepers, enjoy a special niche and obligation. There is a legitimate place for anthropogenic fire, and it must be defended against both those who seek to replace the flame with the furnace and those who wish to abolish the torch for the lightning bolt.

Natural reserves have a special role in all this. While their practical dimensions may be small, their symbolism is important. They testify that humans were intended to be the keepers of the flame, not its extinguishers. They shock us to the realization that fire forces decisions, that to apply and withhold fire may be equally powerful acts. They remind us that it is through fire that we still know the world, and best relate to that great Other, Nature. They testify that to preserve and use fire is the oldest of humanity's ecological duties, its most distinctive trait as a biological organism, the first of its quests for power and knowledge, the genesis of its environmental ethics.

We hold the flame for the planet. Earth's trial by fire is our own.

AFTER THE LAST SMOKE

Flame and Fortune

Sourdough mountain called a fire in:
Up Thunder Creek, high on a ridge.
Hiked eighteen hours, finally found
A snag and a hundred feet around on fire:
All afternoon and into night
Digging the fire line
Falling the burning snag
It fanned sparks down like shooting stars
Over the dry woods, starting spot-fires
Flaring in wind up Skagit valley
From the Sound.
Toward morning it rained.
We slept in mud and ashes,
Woke at dawn, the fire was out.
The sky was clear, we saw
The last glimmer of the morning star.

—Gary Snyder, "Myths and Texts: Burning, Section 17"

The Area* • June 26

Another day in the cache. SWFFs sent with Stieg to repair fence. *T'oo bah'ih,* they call it: "no good." Dry, hot, windy. Fire weather great; fires lousy.

Fires everywhere but on the Rim. Took coffee break at the Lodge and watched the smokes on the Hualapai Reservation. Joe says they burn to clear off the chained juniper and the hotter the fire the better. Everything smoked in to the southwest. Kent sighted what looked like a smoke to the south of Kendricks Peak, on the Coconino; Forest Service action. When he took the situation report on Affirms this morning, Joe says, half the national forests in Arizona have a project fire. Kent and The Kid drove to North Rim Tower for lunch and a look and found nothing.

Ran heliport—routine shuttle flights between rims. Mostly just moved in and out of the fire cache, painting and sharpening tools and wrestling with the saws; the Big Mac is the only one that works, which Joe says is about normal. Later The Kid attacked Joe with a fedco and we nearly had an all-out water brawl. If Kent hadn't sabotaged the pumper, Joe would have retaliated by flooding The Kid out of the cache.

Listened to Uncle Jimmy reminisce about the Old Days—lots of fires, no jerry jobs, no idiot bosses. BS, no doubt. Says we have to wait for the monsoon. Says that all the fires we see around the Rim have been set by people, but almost all our fires are set by lightning. Says we have to wait for the monsoon lightning.

After everyone had left for Lodge, I walked to the helispot and watched stars over the Canyon. Strong, crazy winds along the Rim. There are always winds along the Rim. You can hear them from the cabin. The Rim is a strange, violent place. It simply, directly joins plateau and canyon, and there is nothing to grade between them. Kent says he heard a visiting geologist call it a geographic rite of passage. But the Rim is where we live

*"The Area" refers to the Park Service–developed area around Bright Angel Point. "Cache" (or "fire cache") is the standard term for wildland fire's equivalent to a fire station. "SWFF" stands for Southwest Forest Firefighters, in this instance a squad of Navajos. "Big Mac" is slang for the larger 895 model McCulloch chain saw. "Affirms" is the national computer network for storing and calculating fire-danger rating. "Fedco" is the trade name for a five-gallon backpack pump.

and work. That's where our cabin is. Joe says most of our fires occur along the Rim, which figures. Uncle Jimmy says the Rim is a time as much as a place.

Read, and wrote in journal. Lots of time to write, not much to write about. Have entered nothing for several days. Glad I didn't join the rest of crew at the Lodge. Things got pretty rowdy in the saloon and the rangers had to shut it down. Uncle Jimmy says that's what happens when there aren't enough fires.

The Area • June 29

It seems crazy to live on the North Rim and not see the Grand Canyon the tourists see, so yesterday—my lieu day—I left the Area for the scenic drive. When I got back Stieg told me about the fire at Ribbons Falls, along the Kaibab Trail.

Everyone went save him and me. Had I been here, he says, I would have gone too. I may be happy I missed it. Joe says that it was a typical Canyon fire—most of it burned out before anyone arrived (a bitch of a hike, too: everyone had to jog down with packs in noon heat). Joe and The Kid followed the flames up some talus slopes and chased them into pack-rat middens. Winds were too squirrelly to fly in or out so they had to hike back up. Waited for the shadows. Too bad they didn't have any fishing tackle. I watched some of the fire from Bright Angel Point. Joe says I can go by myself next time.

This afternoon the Forest Service had a big fire break out in Moquitch Canyon. We hung around the cache, waiting for a call as backup. But nothing came through. The Kid climbed the tree tower by the office and watched the smoke billow skyward. He says he could see clouds building up on the Peaks, too. Uncle Jimmy said, yes, he thinks the monsoon is coming. Then tonight we all went to the Lodge (I was more or less told I was going) and sat on the veranda and watched the storms plaster he San Francisco Peaks with lightning. Everything was pitch black—cloudy overhead, no moon, just lightning silhouetting the Peaks in orange and yellow. Joe said it's converging on us, it'll be our turn soon. As we left I thought I could hear thunder rumble through Bright Angel Canyon.

Walked home; the others drove.

Cape Final • June 30

Fire at Cape Final. Everyone went—SWFFs too. Fire along the Rim. Joe and The Kid had to go into the Canyon to catch part of the fire that slopped over into Canyon brush; the rest of us stayed on the Rim, hotspotting; still don't have a complete line around the fire. Got the blue pumper in. McLaren (Park fire officer) called for slurry and a plane (B-17?) from South Rim airport made two passes. Hard time finding fire until it flared up. Lots of confusion. Uncle Jimmy disgusted. Joe says it's about time; says we ought to name it the Finally fire.

Cape Final: farthest point east on the Rim. Sticks out like a sore thumb, literally. Near nothing else. Kent says we may as well be in Chuar Amphitheater. No source tree, and may not be lightning-caused. A tourist might have walked out on Fireroad E-7, not far from here. No one knows.

Rations at 2100 hrs. Mopped up until midnight. Kent and The Kid sent back to the Area for sleeping bags and headlamps and more rations.

Cape Final • July 1

Mopped up all day. Worked in early dawn, then had breakfast. Kent found a swath of white ash and thinks fire started there. Two crews working pumpers, and we have lots of water, though Uncle Jimmy and a SWFF (Henry John) stayed with shovels and fedcos; says too much water is bad for your technique. Says we still don't know how to fight fire. We had some snag fires early in the season, but not everyone had arrived and the rookies like me hadn't even gone through fire school and we haven't really worked together since then. Joe told me to put down my pulaski and take care of the real mop-up. He meant my times.

Keep track of every hour, he said. Nothing fancy, just the fires and times. Don't trust them to do it for you. They never make mistakes in your favor. Better yet, he said, fill out the report yourself. No report, no pay. I have my pocket notebook in my firepack. Put it in your fireshirt, he told me.

Broke for dinner around sunset and sat on the Rim. Not a great view—the Kaibab puts the Canyon into shadow, but sky colors wonderful. Took half hour dinner so we could claim another hour of OT. Left some handtools and a couple of canteens and a fedco in case more smokes show up after we leave. The Kid says he'll return tomorrow to check and pull up the flagging. Always check the fire twenty-four hours after the last smoke, he reminds me.

The Area • July 2

Spent day in cache. The Kid and Uncle Jimmy checked the Finally fire and declared it out. Fixing tools and re-outfitting pumpers. Hot, cloudless.

Filled out a time report for the first time. This would normally be done by the fire boss, but I figured it would be good practice and that way Uncle Jimmy could make sure my times agreed with his. He spent day in Fire Pit* completing the report. Every so often he would emerge and shout obscenities about the idiotic codes that you have to use to record everything and how he couldn't find the right code and what the hell difference did it make if the fire burned in fuel model C or U because it burned on the Rim which meant it burned in every kind of fuel. Then he would get a cup of coffee or fuss with the saws or pumpers before going back to the Pit.

Heard an incredible rumor. The rangers have coveted some Pivetta boots for their uniforms but haven't had any way to pay for them, so Kent says they told Uncle Jimmy to give them some overtime—charge it to the fire—enough to cover the costs. They say the fire was so disorganized, so many folks came and went, that no one would know. Uncle Jimmy refused. He says if they had any sense they would connive to get to the *fires* rather than to the *money*. But Kent thinks the reports were doctored after Uncle Jimmy sent them on.

A couple of the SWFFs came by looking for rides to the Lodge. Everyone else had gone, so I said I would drive them. Stayed for a couple of hours.

Copied jottings from notebook into my log pretty much as is. Can't write and fight both.

* A term used to describe the fire organization's office.

Swamp Ridge • July 4

Monsoon lightning. Fantastic storm throughout afternoon—bolts every few seconds, lights in cache and Pit blinking with each flash, rain and hail piling up on asphalt. Dispatched in pairs, one fire crew regular and one SWFF. Five fires so far. Sent out with Tommie Begaye to fire on Swamp Ridge.

Long drive, through national forest, then back into Park. Recon 1* gave us a bearing of 172°, corrected; snag fire; about half mile from Swamp Point road. Flagged our route in. Tommie spotted smoke, about forty yards from compass line. Gigantic ponderosa, base half rotten. Cleared out swath along lean, then dropped tree. Put in a line. Bucked up and split open bole and scraped embers out and cooled with dirt and water. Little surface spread: needles and duff still wet from rains. Both of us made an extra trip back to pumper for fedcos and rations (only brought in one pack initially: took saw and fedco and handtools instead). Dark when we left fire for vehicle. Mopped until now, around 2300. Tommie has gone back for sleeping bags. Should finish in the morning. Decided to name it the Independence fire. Listening to radio.

Half of crew on the Grail fire at Lancelot Point. Drier there and a patch of bug-killed snags, and fire got caught in strong winds along Rim and downdrafts from thunderheads. Uncle Jimmy says it's probably an acre, with maybe a dozen snags that need to be dropped. Lots of mop-up. Reinforcements came in by helo—found old helispot not far from fire. Other fires about like ours. No one can raise Joe, somewhere in Kanabownits Canyon, and Chuck may send out a search party. Plan to go to Grail fire and help mop tomorrow. Uncle Jimmy says more fires will probably show up in morning. Got Joe on radio: says his fire was too easy and there won't be enough overtime and he wants another.

Twilight Zone, North of Point Sublime Road • July 5

Six fires this morning. Everyone dispatched from their fire to another except for the Grail fire, where two were left to drop snags and baby-sit.

*The radio call name for the park's aerial observer.

Recon 1 discovered smoke not far from us along Swamp Ridge, and we not only arrived there early but got pumper in (only a couple hundred yards from W-4B fireroad). Green ponderosa, little spread by ground fire, fire mostly confined to lightning scar. Dropped and hosed down tree, then scratched a line. Over by noon. Sent to make initial attack on larger smoke in Twilight Zone—that's what Joe calls it. No fireroads here. Long hike on compass bearing (22°, corrected) from pumper. Joe and Henry John followed our flagging in. Tommie and I already had one small snag down and were ready to fell another. Joe said to finish it off. He said he and Henry would help clear a drop site, then dig line while Tommie and I bucked up the trunk. Took all afternoon. The Kid and two SWFFs joined us after dinner; brought in some fedcos and rations and canteens. All other fires out except Grail fire. Everyone else will go there tomorrow, and looks like they will call for Forest Service assistance—we're spread too thin. Uncle Jimmy wants to sling in some cubitainers of water and fedcos by helo to speed along the mop-up at Grail. No storms today, but Uncle Jimmy thinks there may be more fires—sleepers, he calls them. Monsoon only starting.

Need to get new pocket notebook. This one nearly filled. Joe was right. If I hadn't kept score, I wouldn't remember which times went with which fire.

The Area • July 6

Grail fire still being mopped up, but everyone else back in Area; one FS* tanker crew at Grail. Cache in uproar. Spent morning putting everything back in order—saws, pumpers, firepacks, times. Have to be ready for next bust. In afternoon, after Recon 1 gave us an all-clear, we returned to check old fires. Found a few smokes on Independence fire and dumped rest of canteens on them and worked the duff with shovels. Mop-up—it's like proofing, it never ends. Waited twenty minutes. No smokes, so left and pulled flags. The tough fires are on the Rim.

* "FS" is an abbreviation for "Forest Service."

The Area • July 7

Rechecked Twilight fire—one small smoke in duff. Spent afternoon extracting a fire report from my notebook. As fire boss I have to submit reports (that is, times) for both Independence and Twilight fires. Much more fun to fight fires than report them. The Kid says it's indecent to make fires continue by having to write them up. Joe says it's like double mop-up. Affirms calls for drying spell over next few days, then new surge of monsoon moisture. Everyone back tonight, and I joined them at the Lodge, but too tired for more than a couple of beers.

The Dragon • July 10

Big smoke reported on the Dragon around 1600 two days ago. Chuck called for a helo, and Uncle Jimmy, Henry Goldtooth, and I flew for initial attack. Winds too gusty to land; backed off on full power. Returned to Harvey Meadow, dropped Henry off, and returned back to the Dragon and landed this time. Kent told to round up some reserves and drive them to the Dragon trailhead and hike in. Rest of regular crew and SWFFs to be flown in until dark. Uncle Jimmy and I flagged route from helispot to fire. Fire on the Rim, a few acres in size.

Winds blustery. We jogged around fire with shovels and canteens. Lots of surface fire, whipped into red whitecaps by winds. Some fire apparently spilled over the Rim into the Canyon. Rim fire burning in a shallow basin, so we began lining it along the flanking ridges. Joe and Henry arrived. Sounds of helo overhead, the scrape and thunk of handtools on rock and duff and root, the whoosh and snap of flame. Work slow but heart fast: adrenaline flashed like lightning. More arrivals. Kent reached trailhead, but still had a couple-hour hike ahead of him; would arrive after dark. Sunset painted sky on fire.

Uncle Jimmy organized us into two squads—regular fire crew and SWFFs. Almost lost line at one point (flame advancing like a surf), but Henry John rallied SWFFs and they cut a scratch line and burned out and held. Joe found a good fire burning over the Rim; but Uncle Jimmy wanted nothing to do with it at night. Kent and his mob arrived, and one fee collector crashed beneath a pine, and a fern feeler wandered around in a daze, and a thirsty ranger from the inner Canyon grabbed a quart canteen filled with chain oil, thinking it was water, and took a big swallow of 10W30; pretty well lubed for the night. Last helo flight brought in hot meal from Lodge. The Kid and Tommie Begaye sent back to the helispot to get it. Very dark: only light comes from fire, boiling like lava. Cold. Sweat congealed. Joints stiff. Hungry. Can hear winds over and around the Rim.

Ate late and by headlamp. Quiet; everyone tired. Joe and I sent on patrol. Kent told to take his mob and find a place to bed down; they'll begin mopping up in the morning. Uncle Jimmy sat down on a small rise, with his back to the Rim—probably trying to take notes, get our times straight. Joe says we'll have to write a narrative for the fire because it is big and will cost more. We hiked around the perimeter, knocked fire out of the base of one snag then lined it, and located a large flat rock that overlooked most of fire. Stars thick as embers.

The Dragon is a curious place, a peninsula that thrusts south into the Canyon, nearly segregated from the Kaibab by erosion. It is all Rim: Canyon and Plateau nearly perfectly balanced, so that you are never far from either. Lots of fires. Fire signs everywhere. Lightning scars on many trees. Charred bark, scabs of brush where ponderosa burned out, bare limestone paves ground. Eerie, yet compelling. Joe says the Dragon has the highest fire load on North Rim. He says no one can be on the crew long and not experience a fire on the Dragon, and no one can claim to have fought fire without coming here. No one else comes here; no rangers, no tourists, no one except the fire crew. Uncle Jimmy said in fire school that the Dragon was located at the apex of the Rim's fire triangle. Cute phrase, I thought. But the Dragon isn't cute: it's all fire and all Rim.

Listened to the radio. Stieg didn't come. Suddenly refused to get into helo, and Kent had already departed, so he was left behind. Another fire was located just before sunset, however—small snag fire on Crystal Ridge. Stieg took Bundy (garbage collector), and our helo, on a return flight, helped guide him into smoke. Now his radio was stuck on broadcast, and we listened as he explained to a skeptical Bundy how they would drop the tree. Bundy didn't want to spot for Stieg under the tree, wanted to stand back with a fedco and squirt water on him if any branches fell. Hilarious. Joe said that Stieg told him he planned to transfer to the rangers soon.

Said it was too bad he froze up about flying because he had never been to a Dragon fire before. I guess he had had enough without the Dragon. Joe said Stieg deserved what was happening to him. Finally whatever jammed Stieg's radio open broke free. Joe called him and kidded him about not coming to a real fire. Then I saw it.

It was just a glow at first, an orange specter over the Rim. I saw tongues of flame and then Joe and I both heard it and we stared as the whole Canyon flank seemed to erupt into fire and Joe grabbed his radio and warned Uncle Jimmy who ran to the Rim in time to watch a thin stream of flame rush up to the Rim, then Uncle Jimmy shouted for the rest of the crew to get their asses out of their sleeping bags and get some tools because the whole fucking Canyon was on fire and if we didn't hold it at the Rim we would lose everything. Then it got really interesting.

A wild, true Rim fire. For half an hour everything was a frenzy of men and noise; saws whined and coughed, trees fell in crashes, handtools scraped and chipped in atonal splendor, radios and voices shouted over the roar of flames. Not once but several times long strips of flame rushed out of the Canyon. We stood on the Rim, our faces flushed with firelight, our backs in darkness. Uncle Jimmy was everywhere, shouting and cutting. But the line held. By the time we crawled back into our bags, a false dawn edged into view.

The fire went on for two more days. We dropped retardant from a PB4Y2 for several hours that first morning to prevent another blowup. Uncle Jimmy reasoned that the fire had slopped over the Rim then crept along the surface down-slope, drying out the brush above it, then caught some updrafts and swept up through the desiccated crowns. He didn't want that to happen again. Then mop-up. All day. Sent Kent and the irregulars home; fire was lost on them. Joe and I donned fedcos and plunged down the slope into the scene of the night burn, the rocks slippery with slurry, careful to stay in the burned-out zones. Mopped up along slopes. Hot. Crummy night, slept only from exhaustion. Then another day of the same. Joe and I again had to go over the Rim and mop up. But that evening we were all flown home. Uncle Jimmy declared the fire out. He told me that Stieg would transfer soon and I could take over the blue pumper if I wanted.

Nothing left but the paperwork. Glad I took notes. Things were so confused that no one could agree on exactly what happened or to whom or when. Uncle Jimmy looked at my notes and gave me his and told me to write up a narrative. For a fire like this, he explained, you have to have a narrative as well as the coded report. You can't leave it in your pocket, he said. It has to be a public document. I hardly knew where to begin. First there is nothing to write about, then there is too much. It was as though a whole season had to be distilled into one bust, one fire. No report, no pay, Uncle Jimmy reminded me. All flame and no fortune. Write it.

Everyone planned to meet at the saloon and talk about the Dragon and fire and what a great time we had. The Kid suggested someone ought to write it up, and Joe said I was, and everyone laughed because they knew no one would read the narrative, only the times, and a real smokechaser would only want to go to another fire, not write up an old one. Sitting in the Pits, Joe calls it. Uncle Jimmy told everyone to go home and get ready for some serious drinking. Then he asked if I felt any different. Yeah, I said. I feel richer.

Walking to the Lodge tonight I saw Tommie Begaye on the other side of the road, mad that he had missed a ride and that I had chosen to walk rather than drive. This road *t'oo bah'ih,* he yelled. He ran over and then ahead of me. I ran past him. The pavement glistened like black scales. Then we both ran, shoulder-to-shoulder along the road, all the way to the rest of the guys, all the way to the Rim.

Notes and Additional Reading

My intention has been to bring to a larger audience something of the significance of fire on Earth. Detailed scholarship on the subject must depend on books already published, and on those yet to come. Much of my research for this particular book has derived from travels and conversations rather than from archives, yet even minimal standards of intellectual respectability require some documentation. And to the extent that this book aspires to introduce readers to a large (and perhaps new) dominion of knowledge, it ought to contain the rudiments of a basic library.

The literature on fire is vast. The International Association of Wildland Fire sponsors a computerized bibliography that incorporates more than forty thousand entries, a compendium that grows with quarterly installments and will quickly crash any personal computer purchased before 1992. The entries consist of largely fire-specific references in the natural sciences or land management; that sprawling literature on geography, history, environmentalism, to say nothing of cultural history, stands outside it. There is no point in recapitulating this marvelous encyclopedia or other published bibliographies.

What follows instead is a stripped-down register of some of the core books, articles, and conference proceedings that collectively define the field of fire scholarship. Each contains its own particular bibliography. In addition there exist three journals dedicated to wildland fire: *International Journal of Wildland Fire,* published under the auspices of the International Association of Wildland Fire; *Fire Management Notes,* published by the U.S. Forest Service; and *International Forest Fire News,* edited by Johann Goldammer and published through the FAO. For each of the separate essays that make up *World Fire,* moreover, I include specific references, some that come from the generic list, others new, along with citations for those places in the text in which I quote some source.

In the final analysis books such as this one deal with constructed worlds and are accepted or not according to the believability of what they describe. Something there is that doesn't love the wall between genres, but fences make for good neighbors and even limited scholarship can serve as a useful corrective, the intellectual equivalent of replacing the stones that topple down from the wary border that divides fiction from nonfiction.

General Literature on Fire

Alexander, M. E., and G. F. Bisgrove, eds. *The Art and Science of Fire Management,* Information Report NOR-X-309 (Forestry Canada, 1990).

Artsybashev, E. S. *Forest Fires and Their Control.* Russian Translation Series 15 (A. A. Balkema, 1984).

Bachelard, Gaston. *The Psychoanalysis of Fire.* Trans. A.C.M. Ross (Beacon Press, 1964).

Bartlett, Harley H. "Fire, Primitive Agriculture, and Grazing in the Tropics," pp. 692–720, in William Thomas, Jr., ed., *Man's Role in Changing the Face of the Earth,* vol. 1 (University of Chicago Press, 1956).

———. *Fire in Relation to Primitive Agriculture and Grazing in the Tropics: Annotated Bibliography,* 3 vols. (University of Michigan Botanical Gardens, 1955–61).

Batchelder, Robert, and Howard Hirt. *Fire in Tropical Forests and Grasslands.* Technical Report 67-41-ES (U.S. Army Natick Laboratories, 1966).

Baumgartner, David M., et al., eds. *Prescribed Fire in the Intermountain Region* (Washington State University, 1989).

Biswell, Harold H. *Prescribed Burning in California Wildlands Vegetation Management* (University of California Press, 1989).

Booysen, P. de V., and N. M. Tainton, eds. *Ecological Effects of Fire in South African Ecosystems* (Springer-Verlag, 1984).

Brown, Arthur A., and Kenneth P. Davis. *Forest Fire: Control and Use*, 2d ed. (McGraw-Hill, 1973).

Chandler, Craig C., et al. *Fire in Forestry*, 2 vols. (Wiley-Interscience, 1983).

Collins, Scott L., and Linda L. Wallace, eds. *Fire in North American Tallgrass Prairies* (University of Oklahoma Press, 1990).

Conklin, Harold C. *The Study of Shifting Cultivation*, Studies and Monographs VI (Panamerican Union, 1963).

Conrad, C. Eugene, and Walter C. Oechel, tech. coords. *Proceedings of the Symposium on Dynamics and Management of Mediterranean-type Ecosystems*, Gen. Tech. Report PSW-58 (U.S. Forest Service, 1982).

Crutzen, P. G., and J. G. Goldammer, eds. *Fire in the Environment: Its Ecological, Climatic, and Atmospheric Chemical Importance* (John Wiley and Sons, 1993).

Davis, James B., and Robert E. Martin, tech. coords. *Proceedings of the Symposium on Wildland Fire 2000*, Gen. Tech. Report PSW-101 (U.S. Forest Service, 1987).

diCastri, Francesco, and Harold Mooney, eds. *Mediterranean-type Ecosystems: Origin and Structure* (Springer-Verlag, 1973).

diCastri, Francesco, et al., eds. *Mediterranean-type Shrublands* (Springer-Verlag, 1981).

Drysdale, Dougal. *An Introduction to Fire Dynamics* (Wiley-Interscience, 1985).

Food and Agriculture Organization. *Wildland Fire Management Terminology*, FAO Forestry Paper 70 (Rome, 1986). In English, Spanish, German, French, and Italian.

Ford, Julian, ed. *Fire Ecology and Management in Western Australian Ecosystems* (Western Australian Institute of Technology, 1985).

Fraser, James. *Balder the Beautiful: The Fire-Festivals of Europe and the Doctrine of the External Soul*, 2 vols. (Macmillan, 1930).

———. *Myths of the Origin of Fire* (Macmillan, 1930).

Freiburger Waldschutz-Abhandlungen. *VW-Symposium Feuerökologie*, 2 vols. (Freiburg im Breisgau, 1978).

Freudenthal, Herbert. *Das Feuer im deutschen Glauben und Brauch* (de Gruyter, 1931).

Gill, A. M., ed. *Fire and the Australian Biota* (Australian Academy of Sciences, 1981).

Goldammer, Johann G. *Feuer in Walkökosystemen der Tropen und Subtropen* (Birkhäuser Verlag, 1993).

———, ed. *Fire and the Tropical Biota* (Springer-Verlag, 1990).

Goldammer, J. G., and M. J. Jenkins, eds. *Fire in Ecosystem Dynamics. Mediterranean and Northern Perspectives* (SPB Academic Publishing, 1990).

Goudsblom, Johan. *Fire and Civilization* (Allen Lane, Penguin Press, 1992).

Hallam, Sylvia. *Fire and Hearth: A Study of Aboriginal Usage and European Usurpation in South-western Australia* (Australian Institute of Aboriginal Studies, 1979).

Harris, David R., ed. *Human Ecology in Savanna Environments* (Academic Press, 1980).

Hazen, Margaret Hindle, and Robert M. Hazen. *Keepers of the Flame: The Role of Fire in American Culture 1775–1925* (Princeton University Press, 1992).

Hough, Walter. *Fire as an Agent in Human Culture,* U.S. National Museum, Bulletin 139 (Smithsonian Institution, 1926).

Huntley, B. J., and B. H. Walker, eds. *Ecology of Tropical Savannas* (Springer-Verlag, 1982).

Jones, Rhys. "Fire-stick Farmers," *Australian Natural History* 16 (1969), pp. 224–28.

———. "The Neolithic, Paleolithic and the Hunting Gardeners: Man and Land in the Antipodes," pp. 21–34, in R. P. Suggate and M. M. Cresswell, eds., *Quaternary Studies* (Royal Society of New Zealand, 1975).

Koonce, Andrea, et al., eds. *Prescribed Burning in the Midwest: State-of-the-Art* (Fire Science Center, University of Wisconsin–Stevens Point, 1986).

Korovin, G. H., and N. A. Andreev. *Aerial Forest Protection* (Moscow, 1988) [in Russian].

Kozlowski, T. T., and C. E. Ahlgren, eds. *Fire and Ecosystems* (Academic Press, 1974).

Kuhnholtz-Lordat, G. *La Terre Incendiée* (Editions de la Maison Carrée, 1938).

Levine, Joel S., ed. *Global Biomass Burning* (MIT Press, 1991).

Lewis, Henry T. *Patterns of Indian Burning in California: Ecology and Ethnohistory* (Ballena Press, 1974).

Lewis, Henry T., and Theresa Ferguson. "Yards, Corridors, and Mosaics: How to Burn a Boreal Forest," *Human Ecology* 16 (1) (1988), pp. 57–77.

Long, James N., ed. *Fire Management: The Challenge of Protection and Use* (Utah State University, 1985).

Lotan, James E., et al., tech. coords. *Proceedings—Symposium and Workshop on Wilderness Fire,* Gen. Tech. Report INT-182 (U.S. Forest Service, 1985).

Luke, R. H., and A. G. McArthur. *Bushfires in Australia* (Australian Government Printing Service, 1978).

Lyons, John W. *The Chemistry and Uses of Fire Retardants* (Wiley-Interscience, 1970).

———. *Fire* (Scientific American Books, 1985).

McPherson, Guy R., et al., compilers. *Glossary of Wildland Fire Management Terms Used in the United States* (Society of American Foresters, 1990).

Ministère de l'agriculture et de la forêt. *Revue Forestière Française* XLII, "Espaces Forestiers et Incendies," numero special, 1990.

Mooney, Harold A., and C. Eugene Conrad, tech. coords. *Proceedings of the Symposium on the Environmental Consequences of Fire and Fuel Management in Mediterranean Ecosystems,* Gen. Tech. Report WO-3 (U.S. Forest Service, 1977).

Mooney, Harold A., et al., tech. coords. *Proceedings of the Conference, Fire Regimes and Ecosystem Properties,* Gen. Tech. Report WO-26 (U.S. Forest Service, 1981).

Nodvin, Stephen C., and Thomas A. Waldrop, eds. *Fire in the Environment: Ecological and Cultural Perspectives* (U.S. Forest Service, 1991).

Perles, Catherine. *Préhistoire du Feu* (Masson, 1977).

Peters, William J., and Leon F. Neuenschwander. *Slash and Burn: Farming in the Third World Forest* (University of Idaho Press, 1988).

Pyne, Stephen. *Burning Bush: A Fire History of Australia* (Henry Holt, 1991).

———. *Fire in America: A Cultural History of Wildland and Rural Fire* (Princeton University Press, 1982, 1988 pb.).

———. "1990. Firestick History," *Journal of American History* 76 (4) (1990), pp. 1132–41.

———. *Introduction to Wildland Fire* (Wiley-Interscience, 1984).

Raumolin, Jussi, ed. "Special Issue on Swidden Cultivation," *Suomen Antropologi* 4 (1987).

Robinson, Jennifer. "The Role of Fire on Earth: A Review of the State of Knowledge and a Systems Framework for Satellite and Ground-based Observations," Ph.D. thesis (University of California, Santa Barbara, 1987).

Sauer, Carl. "Fire and Early Man," pp. 288–99, in John Leighly, ed., *Land and Life* (University of California Press, 1963).

Schroeder, Mark J., and Charles C. Buck. *Fire Weather,* Agriculture Handbook 360 (U.S. Forest Service, 1970).

Sigaut, François. *L'agriculture et le feu: Role et place du feu dans les techniques de preparation du champ de l'ancienne agriculture européene* (Mouton, 1975).

Society of American Foresters and American Meteorological Society. *Proceedings, Conference on Fire and Forest Meteorology* (alternate years).

Stewart, Omer C. "Fire As the First Great Force Employed by Man," pp. 115–33, in William Thomas, Jr., ed., *Man's Role in Changing the Face of the Earth,* vol. 1 (University of Chicago Press, 1956).

Stokes, Marvin A., and John H. Dieterich, tech. coords. *Proceedings of the Fire History Workshop,* Gen. Tech. Report RM-81 (U.S. Forest Service, 1981).

Tall Timbers Research Station. *Proceedings, Tall Timbers Fire Ecology Conference,* 16 vols. (Tall Timbers Research Station, 1962–1976, 1991).

United Nations Economic Commission for Europe, et al. *Seminar on Forest Fire Prevention, Land Use and People* (Greek Secretariat General for Forests and Natural Environments, 1992).

Vakurov, A. D. *Forest Fires in the North* (Nauka, 1975) [in Russian].

Valendik, E. N. *The Struggle with Big Forest Fires* (Nauka, 1990) [in Russian].

Van Nao, T., ed. *Forest Fire Prevention and Control* (Martinus Nijhoff/Dr. W. Junk Publishers, 1982).

Viegas, Domingos Xavier, ed. *Proceedings, International Conference on Forest Fire Research* (Coimbra, 1990).

Wade, Dale D., compiler. *Prescribed Fire and Smoke Management in the South: Conference Proceedings* (U.S. Forest Service, 1985).

Walstad, John D., et al., eds. *Natural and Prescribed Fire in Pacific Northwest Forests* (Oregon State University Press, 1990).

Weeden, Paul, et al., tech. coords. *Proceedings, International Wildland Fire Conference* (U.S. Forest Service, 1989).

Wein, Ross W., and David A. MacLean, eds. *The Role of Fire in Northern Circumpolar Ecosystems* (John Wiley and Sons, 1983).

Wright, Henry A., and Arthur W. Bailey. *Fire Ecology: United States and Southern Canada* (Wiley-Interscience, 1982).

Size-Up

There is less scholarship regarding the meaning of fire for humans than one might believe. Johan Goudsblom assembles most of it in *Fire and Civilization,* although his perspective is largely urban and sociological. Still useful is Walter Hough, *Fire as an Agent of Human Culture;* Loren Eiseley, "Man the Fire-Maker," *Scientific American* 191 (1954), p. 57; and John Pfeiffer, "When Homo Erectus Tamed Fires, He Tamed Himself," *The New York Times Magazine* (December 11, 1964), pp. 58, 61, 65–72. For the ecological implications, see Carl Sauer, "Fire and Early Man," and E. V. Komarek, "Fire—and the Ecology of Man," *Proceedings, Tall Timbers Fire Ecology Conference* 6, pp. 143–70. For a provocative analysis of how early hunters influenced the vegetation, see Wilhelm Schüle, "Vegetation, Megaherbivores, Man and Climate in the Quaternary and the Genesis of Closed Rain Forests," pp. 45–76, in J. G. Goldammer, ed., *Tropical Forests in Transition*

(Birkäuser, 1992), and "Landscapes and Climate in Prehistory: Interactions of Wildlife, Man, and Fire," in Goldammer, ed., *Fire in the Tropical Biota.*

Donald Worster's essay appears in a roundtable on environmental history published in the *Journal of American History* 76 (4) (March 1990). Quotes from Aldo Leopold are from *A Sand County Almanac* (Ballantine, 1970), passim; "Grass, Brush, Timber, and Fire in Southern Arizona," *Journal of Forestry* 22 (October 1924), p. 2; and Curt Meine, *Aldo Leopold* (University of Wisconsin Press, 1988).

The story of the Gotland fire is based on a tour of the site in August 1992 made possible by Olle Zackrisson and his colleagues in the Department of Forest Ecology, Sveriges lantbruksuniversitet, Umeå.

Hotspotting

Fire Flume

The source material for this essay comes from my book *Burning Bush: A Fire History of Australia.* Please consult it for the full citations. Anyone interested in Australian fire could begin equally well with Gill, ed., *Fire and the Australian Biota;* Luke and McArthur, *Bushfires in Australia;* Hallam, *Fire and Hearth;* or Ford, ed., *Fire Ecology and Management in Western Australia.* The Australian literature is unusually rich, and growing exponentially.

Veld Fire

The best distilled source of information is Brian van Wilgen, et al., "Fire Management in Southern Africa: Some Examples of Current Objectives, Practices, and Problems," pp. 179–215, in Goldammer, ed., *Fire in the Tropical Biota.* For more comprehensive surveys see Booysen and Tainton, eds., *Ecological Effects of Fire in South African Ecosystems* and *Proceedings, Tall Timbers Fire Ecology Conference* 11, "Fire in Africa" (1971). An excellent summary of fynbos fire is contained in Richard Cowling, ed., *The Ecology of Fynbos* (Oxford University Press, 1992). The other principal biomes are the subject of extensive reports issued by the South African National Scientific Programmes.

Other examples of fire practices, particularly hunting and shifting cultivation, can be found in Bartlett, *Fire in Relation to Primitive Agriculture and Grazing in the Tropics: Annotated Bibliography,* 3 volumes. For the story of marsh rose, see C. Boucher and G. McCann, "The Orothamnus Saga," *Veld and Flora* 61 (2), pp.

2–5 (1975). For a brief but effective historical survey, see D. P. Bands, "Prescribed Burning in Cape Fynbos Catchments," in Mooney and Conrad, tech. coords., *Proceedings of the Symposium on the Environmental Consequences of Fire and Fuel Management in Mediterranean Ecosystems,* pp. 245–56.

Queimada Para Limpeza

Probably the best summary of fire are the articles published in Goldammer, ed., *Fire in the Tropical Biota,* in particular, R. V. Soares, "Fire in Some Tropical and Subtropical South American Vegetation Types: An Overview," pp. 63–81; L. M. Coutinho, "Fire in the Ecology of the Brazilian Cerrado," pp. 82–105; P. M. Fearnside, "Fire in the Tropical Rain Forest of the Amazon Basin," pp. 106–16; J. B. Kauffman and C. Uhl, "Interactions of Anthropogenic Activities, Fire, and Rain Forests in the Amazon Basin," pp. 117–34; and for the impact of remote sensing, J.-P. Malingreau, "The Contribution of Remote Sensing to the Global Monitoring of Fires in Tropical and Subtropical Ecosystems," pp. 337–70, and Y. J. Kaufman, et al., "Remote Sensing of Biomass Burning in the Tropics," pp. 371–99. PREVFOGO has published an epitome of its fire seminar, *Conclusões do I Seminario Nacional sobre Incendios Florestais e Queimadas* (Brasília, 1993), and promises the full proceedings soon. For the region's biotic history, see Paul A. Colinvaux, "The Past and Future Amazon," *Scientific American* 260 (5) (1989), pp. 102–8, and Ghillean T. Prance, ed., *Biological Diversification in the Tropics* (Columbia University Press, 1982).

No one has assembled the human history of Brazilian fire yet. The story of buried charcoal is uncovered in Robert Sanford, et al., "Amazon Rain-Forest," *Science* 227 (1985), pp. 53–55, and Christopher Uhl, et al., "Disturbance and Regeneration in Amazonia," *The Ecologist* 19 (6) (1989), pp. 235–40. For a revisionist view of population and pre-European civilization, see Anna Roosevelt, *Parmana* (Academic Press, 1980). The *Handbook of South American Indians* holds scattered references to fire practices, particularly relative to farming. John Hemming's *Red Gold* (Harvard University Press, 1978) and *Amazon Frontier* (Macmillan, 1987) feature numerous references to fire in warfare. One is also reminded of Claude Levi-Strauss's inquiries into mythology, *The Raw and the Cooked* and *From Honey to Ashes,* both of which include fire in South America. Many explorer references to fire (but far from all of them) are contained within Bartlett's three-volume opus *Fire in Relation to Primitive Agriculture and Grazing in the Tropics.* Particularly revealing are the travels of August de Saint-Hilaire. These have been republished in Portuguese. See, for example, his observations in *Viagem as Nacentes do Rio São Francisco,* Reconquista do Brasil, vol. 7 (Ed. da

Universidade de São Paulo, 1975), pp. 51, 151–52, 164–65, 180–85. The citation from Netto comes from "Additions a la flore Bresilienne," *Ann. Sci. Nat.*, 5e Ser. V (1886), pp. 158–201.

Recent ethnographic studies with important meaning for the fire history of the region include Anthony B. Anderson and Darrell A. Posey, "Reflorestamento Indigena," *Ciencia Hoje,* volume especial Amazonia (December 1991), pp. 6–12, and William Balèe, "Indigenous History and Amazonian Biodiversity," in Harold K. Steen and Richard P. Tucker, eds., *Changing Tropical Forests* (Forest History Society, 1992), pp. 185–97. That same proceedings also holds a survey, Leslie E. Sponsel, "The Environmental History of Amazonia: Natural and Human Disturbances, and the Ecological Transition," pp. 233–52.

The recent developments in the Amazon constitute a small publishing industry. It is emblematic that two of the more popular both exploit fire in their titles: Alex Shoumatoff, *The World Is Burning* (Avon, 1990), and Andrew Revkin, *The Burning Season* (Houghton Mifflin, 1990). Unusual because it includes a sympathetic review of indigenous fire practices and beliefs ("Heritage of Fire") is Susanna Hecht and Alexander Cockburn, *The Fate of the Forest* (Verso, 1989). The journalistic literature on Amazonian fire is traced in Helena Lucarelli et al., "Queimadas e Incendios Florestais na Imprensa Brasileira: Uma Analise de Conteudo" (unpublished report, 1993).

Svedjebruk

The history and ecology of Scandinavian fire are profuse but not well represented in English. Good introductions can be found in Olle Zackrisson, "Forest Fire History: Ecological Significance and Dating Problems in the North Swedish Boreal Forest," in Stokes and Dieterich, *Proceedings of the Fire History Workshop,* pp. 120–25; Jussi Raumolin, ed., "Special Issue on Swidden Cultivation," *Suomen Anthropologi;* Roger Engelmark, et al., "Palaeo-Ecological Investigations in Coastal Västerbotten, N. Sweden," *Early Norrland* 9 (1976); and Kimmo Tolonen, "The Post-Glacial Fire Record," pp. 21–44, in Wein and MacLean, eds., *The Role of Fire in Northern Circumpolar Ecosystems.* An excellent popular introduction is available in Swedish, however: *Skog & Forskning* NR 4/91, special issue on fire ("Tänd eld på skogen!" literally, "tend fire in the forest"). Also very useful is A. G. Högbom, "Om skogseldar förr och nu och deras roll i skogarnas utvecklingshistoria," *Nörrlandskt handbibliotek* XIII (Almqvist & Wiksell, 1934).

Many descriptions of *svedjebruk* exist, and can be found in most landscape histories. English distillations are available in A. M. Soininen, "Burn-Beating as the Technical Basis of Colonisation in Finland in the 16th and 17th Centuries,"

Scandinavian Economic History Review 7 (2) (1959), pp. 150–66, and S. Montelius, "The Burning of Forest Land for the Cultivation of Crops," *Meddelanden från Uppsala Universitets Geografiska Institution,* series A, no. 87 (1953). Two particularly detailed summaries come from Finland: G. Grotenfelt, *Det primitiva jordbrukets metoder i Finland under den historiska tiden* (Methods of Primitive Agriculture in Finland During Historical Times) (Helsinki, 1899), and Ollie Heikinheimo, *Kaskeamisen vaikutus Suomen metsiin* (The Impact of Swidden Cultivation on Forests in Finland) (Helsinki, 1915). The former I studied with the help of Gunilla Oleskog; the latter was selectively translated, from the Finnish, by Hannele Mortensen. For Sweden specifically see also J. Arrhenius, *Handbook i Svenska Jordbruket,* 3 volumes (Uppsala, 1862). A number of published studies date back to the 1750s. A contemporary digest in popular format is L. Kardell, et al., "Svedjebruk förr och nu," rapport 20, Avdelningen för landskapsvård, SLU (1980).

Miscellaneous fire practices can be found scattered through Vilhelm Moberg, *A History of the Swedish People,* 2 volumes (Dorset Press, 1972), and for medieval times (especially in warfare) in Olaus Magnus, *Historia om de Nordiska Folken* (1555, reprinted 1980). See also Olof Nordström, et al., *Skogen och smålänningen,* Historiska föreningens i Kronobergs län skriftserie 6 (SmpTRYCK AB, 1989). The Finnish connection is documented in Richard Broberg, *Finsk invandring till mellersta Sverige,* Skrifter Utgivna av Föreningen för Värmlandslitteratur 7 (Uppsala, 1988), which includes an English summary. Linnaeus's travels serve as a convenient horizon for fire practices in preindustrial Sweden, particularly his expedition to Lappland and to Skåne. The references to fire are, in fact, almost every place one chooses to look. An introduction to Norwegian fire is available in Andreas Vevstad, ed., *Skogbrannvern og Skogforsikring i Norge* (Oslo, 1987).

Greek Fire

The ecology of fire in the Mediterranean has attracted an enormous literature. For an introduction see Conrad and Oeschel, *Proceedings of the Symposium on Dynamics and Management of Mediterranean-Type Ecosystems;* Mooney and Conrad, *Proceedings of the Symposium on the Environmental Consequences of Fire and Fuel Management in Mediterranean Ecosystems;* and Goldammer and Jenkins, eds., *Fire in Ecosystem Dynamics.* A useful summary is available in L. G. Liacos, "Present Studies and History of Burning in Greece," *Proceedings, Tall Timbers Fire Ecology Conference* 13, pp. 65–95. For an overview of forestry and pastoralism, with a number of fire references for Greece and Cyprus, see J. V. Thirgood, *Man and the Mediterranean Forest* (Academic Press, 1981).

Most of the references are self-explanatory. My main source of data was the conference (the proceedings of which has been recently published as *Seminar on Forest Fire Prevention, Land Use and People*) held in Athens in 1991, and its field trip. Of particular importance were papers by D. Kailidis, "Forest Fires in Greece," pp. 27–40; Alexander Dimitrakopoulos, "Fuel Management Planning for Accomplishing Fire Prevention and Forest Policy Objectives in the Mediterranean Region," pp. 185–97; and F. Castro Rego, "Fuel Management," pp. 209–21.

A history of fire warfare can be found in Stockholm International Peace Research Institute, *Incendiary Weapons* (MIT Press, 1975); George J. B. Fisher, *Incendiary Warfare* (McGraw-Hill, 1946); and J. P. Partington, *History of Greek Fire and Gunpowder* (W. Heffer and Sons, 1961).

La Nueva Reconquista

The references listed for Mediterranean fire ecology and history under Greece also apply here. The literature is large, and growing. Of special interest for Iberia is, of course, scholarship regarding the Mesta. The classic study is still Julian Klein, *The Mesta* (Harvard University Press, 1927). Supplementary works include Erich Bauer Manderscheid, *Los Montes de España en la Historia* (Ministerio de Agricultura, 1980); Catherine Delano Smith, *Western Mediterranean Europe* (Academic Press, 1979); Angus MacKay, *Spain in the Middle Ages: From Frontier to Empire, 1000–1500* (St. Martin's Press, 1977); Teodoro Marañon, "Agro-Sylvo-Pastoral Systems in the Iberian Peninsula: *Dehesas* and *Montados,*" *Rangelands* 10, no. 6 (1988): 255–58; and M. Ruiz and J. P. Ruiz, "Ecological History of Transhumance in Spain," *Biological Conservation* 37 (1986): 73–86; and for modern transport by rail, Antonio Abellan Garcia y Ana Olivera Poll, "La Trashumancia por Ferrocarril en España," *Estudios Geographicos* 40 (1979): 395–413. For a larger context of Mediterranean transhumance, see Fernand Braudel, *The Mediterranean and the Mediterranean World in the Age of Philip II,* vol. 1, trans. by Sian Reynolds (Harper & Row, 1972), pp. 85–102, and S. M. Rafiullah, *The Geography of Transhumance* (Aligarh Muslim University, 1966).

Excellent summaries of the contemporary scene are contained in Viegas, ed., *International Conference on Forest Fire Research. Proceedings* (Coimbra, 1990); *Ecologia,* Fuera de Serie Num. 1 (1990); ICONA, *Contra El Incendio Forestal. 30 Años de Lucha en España* (1985); *Montes. Revista de Ambito Forestal,* no. 22 Extraordinario (May 1989); and FAO, *Seminar on Forest Fire Prevention, Land Use and People* (Athens 1992), particularly the paper by Ricardo Velez, "Forest Fire Prevention: Policies and Legislation," pp. 251–63.

Red Skies of Irkutsk

The ecology and history of fire in Russia, or in the lands of the former Soviet Union, have not been studied systematically. Most research has had as its goal the improvement of fire control. Probably the best starting point is Wein and MacLean, eds., *The Role of Fire in Northern Circumpolar Ecosystems,* which has a couple of Russian contributions; E. S. Artsybachev, *Forest Fires and Their Control;* G. H. Korovin and N. A. Andreev, *Aerial Forest Protection* (in Russian); and E. N. Valendik, *The Struggle with Big Forest Fires* (also in Russian). Published ethnographic studies are noticeably poor, though whether for reasons of subject or ideology is not clear. The usual citations can be found in travel accounts, many of which are translated into English. See, for example, Chekhov's travels to Siberia and Sakhalin Island, and Fridtjof Nansen's Siberian journey, which includes photos of active fires.

Most of my essay derives from travels as a guest of Avialesookhrana, the aerial fire-protection service, and conversations with members of that organization and researchers affiliated with the Academy of Sciences, especially its Laboratory of Forest Fire Research in Krasnoyarsk. Other material comes by way of contacts with V. V. Furyaev, a fire ecologist at that lab.

Nataraja

The literature regarding fire, forestry, and other related matters in India is truly monumental. Useful points of departure are E. O. Shebbeare, "Fire Protection and Fire Control in India," *Third British Empire Forestry Conference* (Canberra, 1928), and Pyne, *Burning Bush,* pp. 257–60. The major source is undoubtedly the *Indian Forester* (1875–present). As a guide to the larger history, see Edward Stebbing, *The Forests of India,* 3 volumes; a fourth volume appeared under the editorship of Harry Champion and F. C. Osmastion in 1962. Recent overviews include Ajay S. Rawat, *History of Forestry in India* (New Delhi, 1991); Ajay S. Rawat, ed., *Indian Forestry: A Perspective* (Indus Publishing, 1993); and J. B. Lal, *India's Forest: Myth and Reality,* 2d ed. (Nataraj Publishers, 1992). Behind these stand the India Records Office in London; the Indian Institute Library, Bodleian Library, Oxford; and the Commonwealth Forestry Institute, also in Oxford.

For the ecological background see M. S. Mani, ed., *Ecology and Biogeography in India* (Dr. W. Junk Publishers, 1974), and G. S. Puri, et al., *Forest Ecology,* 2 volumes. My profile of endemism derives from M. P. Nayar, "Changing Patterns of the Indian Flora," *Bulletin of the Botanical Survey of India* 19 (1–4) (1977), pp.

145–55. Critical studies of caste, landscape, and protest are found in Madhav Gadgil, "Ecology of a Pastoral Caste: The Gavli Dhangar of Peninsular India," *Human Ecology* 10 (1982); Ramachandra Guha, *The Unquiet Woods: Ecological Change and Peasant Resistance in the Himalaya* (Oxford India Paperbacks, 1991); and more broadly still in Madhav Gadgil and Ramchandra Guha, *This Fissured Land: An Ecological History of India* (University of California Press, 1992). Descriptions of *jhum* are abundant in H. H. Bartlett, *Fire in Relation to Primitive Agriculture and Grazing in the Tropics. Annotated Bibliography,* 3 volumes; an intensive survey in the northeast is P. S. Ramakrishnam, *Shifting Agriculture and Sustainable Development,* Man and Biosphere Series, vol. 10 (Parthenon, 1992). For an exhaustive inquiry into the fire ceremony, consult Fritts Staal, *Agni,* 2 volumes (Asian Humanities Press, 1983).

A contemporary summary of fire—the source of most of my cited statistics—is Veena Joshi, "Biomass Burning in India," in Levine, ed., *Global Biomass Burning,* pp. 185–93. Efforts to install high-tech fire protection are described in R. Saigal, "Modern Forest Fire Control: The Indian Experience," *Unasylva* 162 (41) (1990), pp. 21–27. For an environmentalist critique, see Centre for Science and Environment, *The State of India's Environment 1984–85: The Second Citizens' Report* (New Delhi, 1986).

Other quotations, in order of appearance: Alfred Radcliffe-Brown, *The Andaman Islanders* (Free Fress, 1948), p. 258; G. F. Pearson, "Progress Report of Forest Administration in the Central Provinces, 1863–64" (Calcutta, 1865), p. 27; Clarence H. Hamilton, ed., *Buddhism* (Bobbs-Merrill, 1968), pp. 40–41; Ajay S. Rawat, "Indian Wild Life Through the Ages," p. 134, in Rawat, ed., *History of Forestry in India;* E. O. Shebbeare, "Fire Protection and Fire Control in India," *Third British Empire Forestry Conference* (Canberra, 1928), p. 1; Joseph Hooker, *Himalayan Journals,* vol. 2 (London, 1855), p. 3; F. B. Bradley-Birt, *Chota Nagpore* (London, 1910), p. 184; Benjamin Heyne, *Tracts, Historical and Statistical, in India* (London, 1814), p. 302; B. Ribbentrop, *Forestry in British India,* p. 150; S. Cox, in A.A.F. Minchin, "Working Plan for the Ghumsur Forests, Ganjam District" (Madras, 1921), pp. 38–40; D. E. Hutchins, *A Discussion of Australian Forestry* (Perth, 1916), p. 45; Rudyard Kipling, "In the Rukh," *The Jungle Books* (Oxford, 1987), pp. 326–27, 343; Shebbeare, "Fire Protection and Fire Control in India," p. 1; C. K. Hewetson, "Fires and Their Ecological Effects in Madhya Pradesh," *Indian Forester* 80 (4) (1964), p. 238; Pearson, quoted in Shebbeare, "Fire Protection and Fire Control in India," p. 1; Lt. Col. G. F. Pearson, "Report on the Administration of the Forest Department in the Several Provinces Under the Government of India, 1871–72" (Calcutta, 1872), p. 9; Dietrich Brandis, Memorandum No. 263, *Forest Conference of 1875* (Simla, 1875), p. 5; Pearson, *Forest Conference of 1871–72,* p. 9; B. H. Baden-Powell, "Report on the Administration of the Forest Department in the Several Provinces Under the Government of India 1872–73," vol. 1 (1874), p. 67; W. Burns, et al., "A Study of Some Indian Grasses and Grasslands," *Memoirs of the Department of Agriculture in India, Botanical Series* 14 (1928), pp. 1–57; An Aged Junior, "Some Remarks on Titles and Tigers," *Indian Forester* 16 (1–3) (1890), pp. 182–84; E. A. Greswell, "The Constructive Properties of Fire in Chil (Pinus longifolia) Forests," *Indian Forester* 52 (1926), pp. 502–5; M. D. Chaturvedi, "The Progress of Forestry in the United Provinces," *Indian Forester,* XI (1925), p. 365.

This essay was written in late 1992 and revised after generous commentary from Richard Eaton. In spring 1993 I discovered two newly published summaries of Indian forest history—Richard Haeuber, "Indian Forestry Policy in Two Eras: Continuity or Change?" *Environmental History Review* 17 (1) (Spring 1993), pp. 49–76, and Madhav Gadgil and Ramachandra Guha, *This Fissured Land: An Ecological History of India* (University of California Press, 1992). The latter contains a very useful distillation of how British forest policy evolved and how it altered traditional biota and village life, and is particularly useful in considering the status of tribal peoples. Pages 78–80 forced me to reevaluate my treatment of Agni, and helped better explain to me the allegory behind the burning of the Khundava forest. I revised my subsequent drafts accordingly. My primary data on fire practices, however, remained the *Indian Forester* and the official reports of the Indian Forest Service.

Control

AMERICAN FIRE

These two essays reexamine and build upon my analysis of Forest Service fire policy and research in *Fire in America.* I refer the reader to that book for background references and documentation for material up through the wilderness and intermix fire eras. For these, please consult the essays under those titles. The Keith Arnold quote on the Riverside lab comes from Harold K. Steen, ed., *View from the Top: Forest Service Research* (Forest Historical Society, 1994), p. 20.

A painless introduction to the Blue Mountains debacle is available in Robert W. Mutch, et al., "Forest Health in the Blue Mountains: A Management Strategy for Fire-Adapted Ecosystems," General Technical Report PNW-GTR-310 (U.S. Forest Service, 1993). For valuable background reading on fire-specific issues, see Frederick C. Hall, "Fire and Vegetation in the Blue Mountains: Implications for Land Managers," pp. 155–70, *Proceedings, Tall Timbers Fire Ecology Conference,* vol. 15 (Tall Timbers Research Station, 1974); Dean A. Shinn, "Historical

Perspectives on Range Burning in the Inland Pacific Northwest," *Journal of Range Management* 33 (6) (1980), pp. 415–23; and Boone J. Kauffman, "Ecological Relationships of Vegetation and Fire in Pacific Northwest Forests," pp. 39–52, in John D. Walstad, et al., eds., *Natural and Prescribed Fire in Pacific Northwest Forests* (Oregon State University Press, 1990).

The Southern California scene is covered in Pyne, *Fire in America.* But, of course, a great deal of research has occurred since that review was published. Virtually every national fire conference will include material from the region. Among larger studies see Ronald Lockman, *Guarding the Forests of Southern California* (Clark Publishing, 1981), and Richard A. Minnich, *The Biogeography of Fire in the San Bernardino Mountains of California: A Historical Study,* University of California Publications in Geography, vol. 28 (University of California, Department of Geography, 1988). An important study of presettlement burning is found in Jan Timbrook, et al., "Vegetation Burning by the Chumash," *Journal of California and Great Basin Anthropology* 2 (1982), pp. 163–86; to place these practices in the larger scope of aboriginal burning, see Thomas Blackburn and Kat Anderson, eds., *Before the Wilderness: Environmental Management by Native Californians* (Ballena Press, 1993). For pre-Anglo patterns, see Richard A. Minnich, "Fire Mosaics in Southern California and Northern Baja California," *Science* 219 (1983), pp. 1287–94. The quotation from William Mulholland comes from Richard A. Minnich, "Fire Behavior in Southern California Chaparral Before Fire Control: The Mount Wilson Burns at the Turn of the Century," *Annals of the Association of American Geographers* 77 (4) (1987), p. 613. On the most recent fires, see Daniel Gomes, et al., *Sifting Through the Ashes: Lessons Learned from the Painted Cave Fire* (Graduate Program in Public Historical Studies, University of California—Santa Barbara, 1993); Margaret Sullivan, *Firestorm! The Story of the 1991 East Bay Fire in Berkeley* (City of Berkeley, 1993); and Robert E. Martin, "The 1993 Southern California Fires," *International Forest Fire News* 10 (January 1994), pp. 21–22. The quotation from Joan Didion comes from *Slouching Towards Bethlehem* (Dell, 1968), p. 220.

WILDERNESS FIRE

The most concise way to profile changes in thinking is to compare the proceedings of three symposia held in Missoula, Montana, in 1974, 1983, and 1993, respectively: "Fire and Land Management Symposium," *Proceedings, Tall Timbers Fire Ecology Conference* 14 (1974); James E. Lotan, et al., tech. coords., *Proceed-*

ings—Symposium and Workshop on Wilderness Fire; and *Proceedings—Fire in Wilderness and Park Management* (in press).

The Yellowstone fires sparked a flare-up of picture books and media publications, few of which are likely to survive, almost none of which really investigated either the issues or the park's execution of its plan. Ideologically, the opposing sides are probably well represented by Alston Chase, *Playing God at Yellowstone* (Atlantic Monthly Press, 1986) on one side, and on the other Don Despain, et al., *Wildlife in Transition: Man and Nature on Yellowstone's Northern Range* (Roberts Rinehart, 1986), and Douglas B. Houston, *The Northern Yellowstone Elk* (Macmillan, 1982). That so much of the controversy has centered on elk (which was also the subject of the seminal Leopold Report for the National Park Service) probably helps explain why fires got to be such a problem. Other relevant documents include Ronald H. Wakimoto, "The Yellowstone Fires of 1988: Natural Process and Natural Policy," *Northwest Science* 64 (5) (1990), pp. 239–42, which summarizes the findings of the Fire Management Policy Review Time, and the GAO's report "Federal Fire Management: Limited Progress in Restarting the Prescribed Fire Program," GAO/RCED-91-42 (December 1990). Although written before the Yellowstone events, a careful appraisal of the evolution of thinking and programs is found in Bruce M. Kilgore, "The Role of Fire in Wilderness: A State-of-Knowledge Review," in Robert C. Lucas, compiler, *Proceedings—National Wilderness Research Conference: Issues, State-of-Knowledge, Future Directions,* Gen. Tech. Report INT-220 (U.S. Forest Service, 1986).

Among journals that devoted special issues to the fires, consider the following: *BioScience* 39 (10) (1989); *Journal of Forestry* 87 (12) (1989); *Western Wildlands* 152 (1989); and *Forum for Applied Research and Public Policy* 4 (2) (1989). The Tall Timbers Research Station resurrected its fire ecology conferences to host one devoted to "High Intensity Fires in Wildlands: Management Challenges and Options," vol. 17 (1989).

In the interests of full disclosure, I should state that I was the fire specialist sent by the Rocky Mountain Regional Office to help rewrite Yellowstone's fire plan in the summer of 1985. I had worked the two previous summers, also at the request of the Denver office, to help reconstruct Rocky Mountain National Park's fire plan after the Ouzel fire disaster. I was also a member of the Greater Yellowstone Postfire Ecological Assessment Committee. My knowledge of how Yellowstone actually conducted its fire program, what the concerns of the outside fire community regarding that program were, and what the park was obligated to do under its transitional fire plans all derive from these experiences. The 1985 memoranda recommending rechartering of the Yellowstone program and the creation of a Yellowstone fire center dedicated to the peculiar demands of natural fire programs were, understandably, never released.

INTERMIX FIRE

Most of the major fire conferences of the past five years, and not a few general forestry symposia, include abundant references to the "wildland/urban interface" and its fire problems. In addition to those mentioned in the general bibliography, consider William C. Fischer and Stephen F. Arno, compilers, *Protecting People and Homes from Wildfire in the Interior West: Proceedings of the Symposium and Workshop,* Gen. Tech. Report INT-251 (U.S. Forest Service, 1988), a fair summary of what the issues are. Similar studies exist for virtually all parts of the country. The defining document for the National Wildland/Urban Interface Fire Protection Initiative launched in 1986 was published as *Wildfire Strikes Home,* which became the title of a newsletter, later renamed *Wildfire News & Notes.* An excellent summary is available in James B. Davis, "The Wildland-Urban Interface: What It Is, Where It Is, and Its Fire Management Problems," *Fire Management Notes* 50 (2) (1989), pp. 22–33. For a chronicle of the Oakland tragedy, see Margaret Sullivan, *Firestorm! The Story of the 1991 East Bay Fire in Berkeley* (City of Berkeley, 1993). For the history of afforestation, see David J. Nowak, "Historical Vegetation Change in Oakland and Its Implications for Urban Forest Management," *Journal of Arboriculture* 19 (5) (1993), pp. 313–19.

Background reading for a fire history of the Southwest can be found in Pyne, *Fire in America,* pp. 514–29. Two recent publications highlight the state of contemporary thinking: Merrill R. Kaufmann, et al., tech. coords., *Old-Growth Forests in the Southwest and Rocky Mountain Regions, Proceedings of a Workshop,* Gen. Tech. Report RM-213 (U.S. Forest Service, 1992), and J. S. Krammes, tech. coord., *Effects of Fire Management of Southwestern Natural Resources,* Gen. Tech. Report RM-191 (U.S. Forest Service, 1990). The latter contains a rich bibliography of the technical literature. For the Dude Creek fire, see Eldon W. Ross, et al., "Accident Investigation Report: Dude Fire Incident. Multiple Firefighter Fatality" (Tonto National Forest, 1990).

The classic study remains Charles F. Cooper, "Changes in Vegetation, Structure, and Growth of Southwestern Pine Forests Since White Settlement," *Ecological Monographs* 30 (1960), pp. 129–64. Other (and newer) publications cited include: S. J. Holsinger, "The Boundary Line Between the Desert and the Forest," *Forestry and Irrigation* 8 (1902), pp. 21–27; Clarence Dutton, *Tertiary History of the Grand Cañon District,* Monograph 2, U.S. Geological Survey (1882); Lt. E. F. Beale, *Wagon Road from Fort Defiance to the Colorado River,* Sen. Exec. Doc. 124, 35 Congress 1 Session (1858); Aldo Leopold, "Grass, Brush, Timber, and Fire in Southern Arizona," *Journal of Forestry* 22 (1924), pp. 2–8; T. N. Johnson, "One Seed Juniper Invasion of Northern Arizona Grasslands," *Ecological Monographs* 32 (1962), pp. 187–207; Harold H. Biswell, et al., "Ponderosa

Fire Management: A Task Force Evaluation of Controlled Burning in Ponderosa Pine Forests of Central Arizona," *Misc. Publ. No. 2* (Tall Timbers Research Station, 1973); and Lumholtz in C. W. Hartmann, "Indians of Northwestern Mexico," *Congress International des Americanisters* 10 (1897), pp. 115–136. The quote from Gifford Pinchot comes from *Breaking New Ground* (University of Washington Press, 1972), p. 179. The Chuska Mountains comparison is contained in Melissa Savage and Thomas W. Swetnam, "Early 19th-century Fire Decline Following Sheep Pasturing in a Navajo Ponderosa Pine Forest," *Ecology* 71 (6) (1990), pp. 2374–78, and Melissa Savage, "Structural Dynamics of a Southwestern Pine Forest Under Chronic Human Influence," *Annals of the Association of American Geographers* 81 (2) (1991), pp. 271–89. The Leopold quote comes from "The Virgin Southwest," p. 179, in Susan L. Flader and J. Baird Callicott, eds., *The River of the Mother of God and Other Essays* (University of Wisconsin Press, 1991).

A recent addition to the controversy over the relative impacts of climate and humans is Conrad Bahre, *Legacy of Change* (University of Arizona Press, 1991). An interesting discussion of grazing as an ecological force in wildlife is David E. Brown, *The Grizzly in the Southwest* (University of Oklahoma Press, 1985). As a measure of the relative strength of lightning, consult Jack S. Barrows, "Lightning Fires in Southwestern Forests," unpublished report to Northern Forest Fire Laboratory (1978).

Mop-Up

As an attempt to summarize how I think about fire, the essay plays off hundreds of references for which there is little point in massing documentation. The specific quotations have their sources as follows: J. DeBrahm, *Report of the General Survey in the Southern District of North America*, ed. Louis De Vorsey, Jr. (University of South Carolina Press, 1971), p. 81; P. L. Barbour, ed., *The Complete Works of Captain John Smith* (University of North Carolina Press, 1987), vol. 2, p. 176; T. L. Mitchell, *Journal of an Expedition into the Interior of Tropical Australia* (London, 1848), pp. 412–13; and Alfred Radcliffe-Brown, *The Andaman Islanders* (Free Press, 1948), p. 258.

Index